T0263531

# Job Hazard Analysis

## A Guide for Voluntary Compliance and Beyond

# Job Hazard Analysis

## A Guide for Voluntary Compliance and Beyond

### From Hazard to Risk: Transforming the JHA from a Tool to a Process

James E. Roughton
Nathan Crutchfield

AMSTERDAM • BOSTON • HEIDELBERG • LONDON
NEW YORK • OXFORD • PARIS • SAN DIEGO
SAN FRANCISCO • SINGAPORE • SYDNEY • TOKYO

ELSEVIER          Butterworth-Heinemann is an imprint of Elsevier

Butterworth-Heinemann is an imprint of Elsevier
30 Corporate Drive, Suite 400, Burlington, MA 01803, USA
Linacre House, Jordan Hill, Oxford OX2 8DP, UK

⊗ Recognizing the importance of preserving what has been written, Elsevier
prints its books on acid-free paper whenever possible.

**Library of Congress Cataloging-in-Publication Data**
Roughton, James E.
    Job hazard analysis : a guide for voluntary compliance and beyond / James
Roughton and Nathan Crutchfield.
        p. cm.
    Includes bibliographical references and index.
    ISBN 978-0-7506-8346-3 (hardback : alk. paper)  1.  Industrial safety.
I.  Crutchfield, Nathan.   II.   Title.
    T55.R693 2007
    658.3'8–dc22

                                                              2007033646

**British Library Cataloguing-in-Publication Data**
A catalogue record for this book is available from the British Library.

ISBN: 978-0-7506-8346-3

For information on all Butterworth-Heinemann publications
visit our Web site at www.books.elsevier.com

Transferred to Digital Printing 2012

Working together to grow
libraries in developing countries

www.elsevier.com | www.bookaid.org | www.sabre.org

ELSEVIER    BOOK AID
            International    Sabre Foundation

To my wife, my friend and my lifelong partner of 37 years and counting, who has always been patient with me in my passion for safety. She has always given me the freedom to pursue my dreams of promoting safety.

Thanks to the many reviewers of this book: Neal Lettre, my Six Sigma Black Belt friend who reviewed the chapter on Six Sigma; Jeff Swire, safety professional who reviewed the chapter on behaviors; several of my students (Roy McConnell, Brian Wood, Karen Barkley) who reviewed the entire book and provided constructive feedback.

Thanks to Nathan Crutchfield, my co-author for added his experience in risk assessment to this book.

Thanks to other professionals such as Jim Mercurio, a good friend of mine and co-author of *Developing an Effective Safety Culture: A Leadership Approach*, who provided some insight into this project.

Last but not least, thanks to other safety professionals who I have made contact with over the last several years and to the many OSHA public domain websites, all of which provide a vast amount of resources used in the book.

*James E. Roughton*

My efforts for this book are dedicated to my wife, Bonnie, who brings a loving presence to my life, and to Brian, a wonderful son who will achieve much in life. I owe much to the many colleagues and clients that have assisted me over the years. Their friendship, insights, encouragement, criticism and professionalism have meant so much to me. I am grateful to have learned and benefited from them all.

Special thanks to James Roughton for allowing me to participate in this project, an important element in an overall safety process.

*Nathan Crutchfield*

# Contents

# Part 4 Additional Tools That Can Be Used to Develop a Successful JHA................... 379

# About the Authors

**James E. Roughton** has a Master of Science (MS) in Safety from Indiana University of Pennsylvania (IUP), is a Certified Safety Professional (CSP), a Canadian Register Safety Professional (CRSP), a Certified Hazardous Materials Manager (CHMM), a Certified Environmental Trainer (CET) and a Certified Six Sigma Black Belt. His experience includes 4 years in the military and 40 years experience in industry, with the past 30 years in the safety area developing and implementing safety management programs and management systems.

Mr. Roughton has worked for various corporations and has served in the following capacities: providing consulting services for medium to large manufacturing facilities; developing and implementing safety programs and management systems; providing consulting services for hazardous waste remedial investigation and site cleanup regarding developing and implementing site-specific safety plans; conducting site health and safety assessments, and providing internal support for multiple office locations.

In addition, he is an accomplished author in various areas of safety, environmental, quality, security, computers, etc. He is the author of six books, most notably *How to Develop an Effective Safety Culture: A Leadership Approach.*

He also provides mentoring to other professionals who want to get published and is a frequent co-author with those professionals. He is a frequent speaker at conferences and professional meetings.

He is the past president of the ASSE Georgia Chapter, an active member of the safety advisory board of the Department of Labor of Georgia, Lanier Tech Safety Board, past president for the Gwinnett Safety Professionals Association, and co-founder and past president of the Heart of Georgia Safety Society. He is also an adjunct instructor for Lanier Tech and Georgia Tech in Atlanta, Georgia.

In addition, he maintains his own websites, Accident-Related, www.gotsafety.net, 24-7 Safety, Responsibility not Luck and Safety Culture-Related www.emeetingplace.com, Safety Exchange of America. He also has received several management and professional awards for safety related activities, including Safety Professional of the Year (SPY). He can be reached at safeday@emeetingplace.com.

**Nathan Crutchfield** has a professional history that encompasses a full range of risk control program design, development, implementation and evaluation. He has provided expertise to a broad array of clients that include public entities, associations, and general industry. He has a Master's Degree in Business Administration from Georgia State University with a Bachelor of Civil Engineering Technology, Southern Technical Institute (now Southern Polytechnic State University, Marietta, Georgia).

He holds the designations of Chartered Property and Casualty Underwriter (CPCU), Associate in Risk Management (ARM), Associate in Research & Planning (ARP), and Certified Safety Professional (CSP).

Nathan is an independent risk control consultant with his own practice and was a vice president with a major risk management and brokerage organization for over 20 years. He has been involved with the Georgia Department of Labor annual Environmental Health and Safety Conference as a Planning Board member, was on the National Safety Council Board of Directors (1993–95) and has been a speaker at the various risk and safety conferences. He was awarded the NSC Distinguished Service to Safety Award in 2001.

He can be reached at Nathan@crutchfieldconsulting.com.

# Foreword

When you begin to write a book, you know that it will be a lot of work, with many nights of research, reading, getting approvals from various sources, then writing and editing the final efforts. With this in mind, you are always thinking about how to make the book better than other resources on the market and how to convey your concepts more clearly to the reader.

Part of this process for me involved finding another safety professional who shared my vision and could contribute to the quality of the book and help to convey the correct message to other safety professionals. This is how I (James) came up with my co-author, Nathan. He has a wealth of experience in the risk management field and I am lucky to find a co-author who complements my work on this book so well. We have had our agreements and disagreements, but in the end this makes a better programmed learning-type of textbook that can be used in all types of safety training, from the college class to many different industries.

Norm Abram, master carpenter from the TV program "This Old House," has many tools in his tool box that allow him to perform wonders when repairing old houses. Each tool he uses has a specific purpose and use. These tools allow Norm to build many different things, such as furniture, molding, siding for a project, etc. Think about it: if Norm used a rubber mallet instead of a normal hammer when driving nails, would it be as effective? There are some tools discussed in this book, such as Six Sigma, that are not usually considered for a safety "tool box." This set of tools has many new and useful features that will allow you to perform tasks for a wide variety of situations in the safety process.

In addition, many tools are available in the safety arena that are not fully utilized. Many are from various public domain web sites, such as the OSHA website, www.osha.gov; Oklahoma Department of Labor Safety Management website, http://www.ok.gov/~okdol/; Oregon OSHA Safety and Health Education website, http://www.cbs.state.or.us/osha/education.html; Washington Safety OSHA website, http://www.lni.wa.gov/wisha/rules/corerules/HTML/296-800-100.htm; Missouri Department of Labor website, http://www.dolir.mo.gov/ls/safetyconsultation/, to name a few. Many more websites were used in researching this book, but these are some of the best public domain resources. References are provided for each site where resource material was found to be of benefit.

In addition to the hazard recognition and JHA development concepts, this book provides in Chapter 13 a very brief overview of various Six Sigma tools that can be used in the continuous safety improvement process. Many different examples of specific tools such as diagrams, charts, analysis techniques, and methods provide step-by-step help to establish a process that can be continually improved.

We mentioned the TV show "This Old House." To take that concept one step further, if you walked into a hardware store and asked for a table saw, you might be asked what type of table saw you need. There are large saws that do many tasks and small saws that do specific small tasks, as well as many vendors, types, and prices. Take that example into the Six Sigma world and you will find a similar concept. Some Six Sigma tools are simple to use. As you get more deeply into a project, you will discover that the tools will become more detailed and complex. For example, a Pareto chart is easy, but using the XY Matrix and the FMEA requires much more effort, taking a lot more time and resources to complete. Many types of graphs can be used (line charts, histograms, etc) and they should be evaluated to determine which best presents the data. Further into Six Sigma are statistical tools that vary widely and can be quite useful in safety/hazard analysis.

The authors hope that this book can be used by management, supervision, safety professionals, educators, and students of safety management as a "road map" that provides an overview as well as new ideas for developing what we believe should be the focal point for a successful safety management process, the Job Hazard Analysis (JHA).

We hope that you enjoy this book and look forward to assisting you in your efforts to improve your professional safety skills.

Good luck on your journey to success!

James E. Roughton, MS, CSP, CRSP, CHMM, CET, 6σ Black Belt, safeday@emeetingplace.com, safeday@emeetingplace.com and www.Gotsafety.net

Nathan Crutchfield, CPCU, ARM, ARP, CSP Nathan@ crutchfieldconsulting.com

# Preface

"I've done it this way a thousand times, ten thousand times, a hundred thousand times without getting hurt." Sound familiar? Maybe, on the thousand and one, ten thousand and one, or hundred thousand and one time, someone does get hurt.

The truth is that we may not be doing our jobs in the safest possible way or even conducting our personal business in a safe manner all of the time. We tend to put ourselves at risk each day and so often do not know it because we have done something risky so many times it simply becomes the right way of doing things. If you were to review all of the accidents that still occur you would be amazed. This is the reason that I create the website, GotSafety.net to help highlight accident and methods to prevent them.

To help you understand the importance of developing a job hazard analysis, we have divided this book into four parts to help you understand the process.

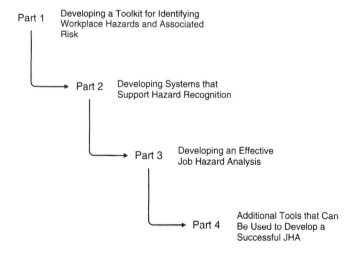

Part 1 — Developing a Toolkit for Identifying Workplace Hazards and Associated Risk

Part 2 — Developing Systems that Support Hazard Recognition

Part 3 — Developing an Effective Job Hazard Analysis

Part 4 — Additional Tools that Can Be Used to Develop a Successful JHA

# PART 1. DEVELOPING A TOOLKIT FOR IDENTIFYING WORKPLACE RISK AND HAZARDS

## Chapter 1. Preparing for the Risk and Hazard Assessment

Leadership and management skills are critical to maintaining and keeping a safety process viable in today's business environment. With the constant theme of organizational change, you will always face an array of internal obstacles, departmental political issues, and regulatory requirements that will appear and hinder your best efforts. Add in the behavioral quirks of human nature and the plot really thickens!

The need to have foundational skills that go beyond knowledge of compliance requirements is surprisingly found within the various compliance mandates. The Federal Occupational Safety and Health (OSHA) Act states that "Employers must furnish a place of employment free of recognized hazards that are causing or are likely to cause death or serious physical harm to employees," "OSHA ACT OF 1970, 29 CFR 1903.1." Further, the American National Standards Institute's Z10-2005, Occupational Health and Safety Management System Standard, 4.3 Objectives, states that: "The organization shall establish and implement a process to set documented objectives, quantified where practicable, based on issues that offer the greatest opportunity for Occupational Health and Safety Management System improvement and risk reduction."

Job hazard analysis (JHA) is an essential safety management tool. Used consistently and correctly, it will increase your ability to build an inventory or portfolio of hazards and risks associated with various jobs, job steps and the detailed tasks performed by your employees. Your professional "mental map" and skills will improve as you begin to use JHA to determine the interrelations between the job steps and tasks and the dynamics of the organization. As your portfolio of JHAs increase, you will improve your safety "tool box" and the skill sets that increase your effectiveness in implementing your programs in the face of continual organizational change.

The JHA provides the basic methodology and structure needed to recognize hazards and the elements of personal choices that are associated with each job. Introducing a JHA process will greatly enhance an organization's evaluation of hazards and their associated risks and should be an essential, fundamental part of any safety process.

Chapter 1 will focus on identifying existing and potential hazards that may be associated with your workplace.

# Chapter 2. Workplace Risk and Hazard Reviews

The JHA, as the centerpiece of your safety program, provides the blueprint to design the workplace review. A JHA enhances your ability to anticipate and understand how all job elements combine and allows you to develop effective control programs and procedures. It is important in your process to ensure that the work environment is actively analyzed and monitored. To build your process, a workplace review should be conducted by designated knowledgeable employees who physically review the operations and activities. By asking specific questions concerning their observations, they develop an insight on conditions within their work areas that may cause harm or damage.

Safety reviews should do more than identify visible hazards. They should provide useful data for the purpose of effective analysis and evaluation of the safety management system. The analysis should also attempt to understand how our own personal behavior affects potential harm. Our personal workplace behaviors can be driven by the importance management places on correcting identified hazards or controls to risks.

Chapter 2 will focus on workplace analysis and how a variety of workplace review methods can identify existing hazards or potential hazards, which are the conditions and operations where changes could create hazards.

# Chapter 3. Developing Systems to Manage Hazards

An effective JHA management system provides for the continual analysis of the workplace and anticipates changes needed to modify or develop policies and procedures to control new, existing, or reoccurring hazards. The JHA provides the structured format that determines the variety of job steps required to complete a task and the conditions needed for its safe completion.

For better or worse, the safety professional usually inherits an ongoing workplace as it is, with or without management support, operational hazards, and an array of employee and management behaviors that have developed over a long period of time. Many levels of risk exist in the workplace and stem from things such as chemicals, materials, equipment, tools, and environment and, of course, the long term behaviors of employees and management within the organization.

Hazard and risk measures provide the information related to of define specific hazard training. Hazards that employees are exposed to should be systematically identified and evaluated.

Chapter 3 will discuss the systems used in a hazard analysis of the work environment.

# PART 2. DEVELOPING SYSTEMS THAT SUPPORT HAZARD RECOGNITION

## Chapter 4. Understanding the Human Role in the Safety Process

After the risk and hazards have been identified and assessed, behaviors of the individuals in the operation must be reviewed. A great control program is worthless if the individuals will not or cannot follow its criteria. The controls defined for the safety process must be followed to be successful and this requires building in knowledge of what influences behavior.

Chapter 4 is designed to assist those not familiar with the human role in the safety process and provide some background information on how the process works and its value in providing for continuous improvement of the JHA process.

## Chapter 5. Effective Use of Employee Participation

The success of any business depends on the total involvement of every employee in the operation. Without the involvement of the employees, the potential for developing a full understanding of the job and how it is currently completed is limited. This chapter looks at the reasons behind employee participation and suggests methods and activities that can help to increase the potential for the successful implementation of JHA process.

Chapter 5 will outline the objective of employee participation and how it is used to encourage everyone to help in the structuring and effective functioning of the safety process and with the decisions that directly affect their personal safety.

## Chapter 6. Defining Associated Risk

Risk management principles have been used for many years and in many high hazard industries and operations. However, as many programs have been typically designed around regulatory compliance or losses, the risk management concepts are still new to many employees and still are rarely used in many organizations to assess events that could cause an injury. Even experienced safety professionals still go on their "gut" instinct—"I think, I feel"—or prior knowledge to develop safety programs. A shift must take place in our thought process. We must understand the need to collect risk-related data, analyze the data, and make decisions based on risk assumptions. Risk principles are used to prioritize and clarify the importance and objectives of hazard control.

Risk can be defined as a measure of the probability and severity of adverse effects. We will provide several simple, logical formats to provide understanding on how to use effective risk management principles. These formats will outline how hazards are associated with specific job steps and related tasks.

Chapter 6 will discuss risk and the measurement of probability and severity of adverse effects.

## Chapter 7. Assessing Safety and Health Training Needs

The JHA process requires the transfer of knowledge about specific job risks in a way that can be easily and readily understood by all levels of the organization. Having knowledge of risk, safety management, and the JHA process is not enough. You must be able to clearly communicate the importance of the JHA tools, methods, and concepts. To do this, you must have an understanding of training and learning theory.

Chapter 7 provides a basic overview of the knowledge and skills needed to succeed as a safety trainer. The basic information relevant to planning, preparing, presenting, and evaluating the classroom is provided.

## PART 3. DEVELOPING AN EFFECTIVE JOB HAZARD ANALYSIS

## Chapter 8. Planning for the Job Hazard Analysis

Today, a wide array of safety material is available from many safety vendors who specialize in developing compliance-related safety programs, presentations, supervisor handbooks, general safety tips, safety slogans, etc. Compliance-related safety material has become the basis for many organizations' safety programs. These generic programs can be easily purchased, allowing you to put your company name on the program and quickly produce it. We refer to this type of program as "plug-n-play."

Chapter 8 will discuss planning an effective JHA. It will discuss how to plan for the JHA and why it is important to keep it current when changes require it to be modified to ensure that it continues to be an effective procedure.

## Chapter 9. Breaking the Job Down into Individual Components

You may have developed a management system built around traditional hazard recognition programs and communicated specific hazards to employees. You must be aware of the perceptions that management and employees may have regarding a hazardous task. When planning a systematic process to analyze each task in your workplace, you will want to ensure that the individual evaluations are handled consistently, thoroughly, and thoughtfully. By establishing a structured procedure, the results of your analysis program will provide consistent and reliable information.

Chapter 9 will discuss fundamental issues that need to be considered when developing a JHA program.

## Chapter 10. Putting Together the Puzzle Pieces

To add consistency to JHA development, the information detailed in this book is a combination of information collected from many public domain resources and also from personal and professional experiences. You will see that every safety professional has a different way of doing a JHA, but we believe that you will also find that if you follow the elements outlined in this book, the end result will provide a more comprehensive JHA.

Chapter 10 will detail the complete JHA and how to use it in your environment.

## Chapter 11. Safe Operating Procedure

Why develop a safe operating procedure (SOP)? Isn't the JHA the end result of the process? The JHA is the methodology used to pull together all aspects and elements of the job. However, the information the SOP provides must be put into the format of your organization's SOP.

Chapter 11 will discuss a simple method of outlining an SOP that will close the gap between it and the JHA.

# PART 4. ADDITIONAL TOOLS THAT CAN BE USED TO DEVELOP A SUCCESSFUL JHA

## Chapter 12. Overview of a Safety Management Process

Risk of all types of exists in business operations. In many organizations, the goal is long-term success and profitability. In many cases, employee safety is not normally considered as part of the overall business plan.

Chapter 12 will discuss management systems and how they can be used to help you get started on the right track to developing a successful safety management system.

## Chapter 13. Six Sigma as a Management System: A Tool for Effectively Managing a JHA Process

Chapter 13 will provide a simple overview of the Six Sigma process and provide a snapshot of the tools that can be used in JHA development.

# FINAL WORDS: CAN YOU DEVELOP A CULTURE THAT WILL SUSTAIN ITSELF?

The big question comes down to this: "Can You Develop a Culture That Will Sustain Itself?" With a structured process that goes to the core of how jobs and tasks are done, we believe you will increase the probability of a self-sustaining process. We have provided you with basic information as well as detailed information to help you establish and maintain a successful safety program as part of an overall process and provided methods to integrate your programs into your management system.

Good luck in your new safety process!

# Introduction

"We shape our buildings, thereafter they shape us."

*—THE DAILY GURU*

"Life is filled with opportunities for practicing the inexorable, unhurried rhythm of mastery, which focuses on process rather than product, yet which, paradoxically, often ends up creating more and better products in a shorter time than does the hurried, excessively goal-oriented rhythm that has become the standard in our society."

—-*Mastery* by George Leonard, published by the Penguin Group, 1992

This is what the book is about: using a structured, methodical approach to job risks and hazards in lieu of just doing programs. Figure I-1 provides an overview of various JHA formats used in the safety arena today.

## WHY DO YOU NEED THIS BOOK?

"Life is like a box of chocolates. You never know what you're gonna get."

—Forrest Gump

Picture this: your facility contracts with a consultant to conduct a review of your safety program. This consultant spends days in your facility interviewing many employees and reviewing records. When the review is complete, a draft report is written. The consultant meets with the management team and discusses the findings, such as: "Here are the things that you are doing right and here are opportunities for improvements." After the closing meeting, the consultant leaves the facility. In a few weeks management receives a final written report listing all of the opportunities for improvements that were discussed in the closing meeting. What happens then?

We have seen many of these reports that are voluminous, provided with the hope that the management team will know exactly how to implement improvement. Analyzing these types of reports can be a challenge even to the seasoned professional and, from our experience, often the management team

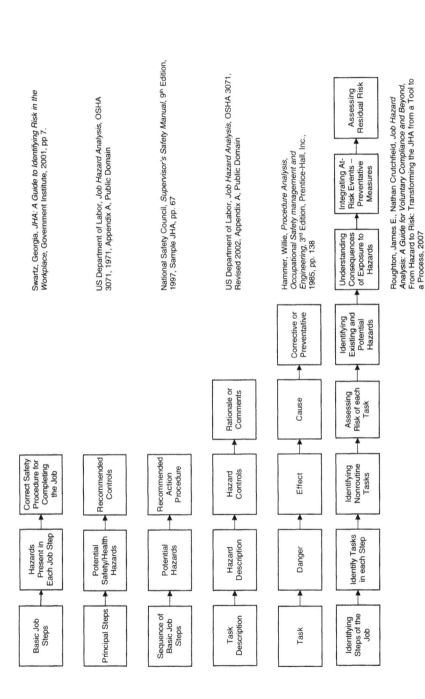

**Figure I-1** Overview of Various JHA Formats

says, "There is so much material here," and then asks the question: "Where do we start?" We believe that the place to begin is with a comprehensive structured JHA process that combines an understanding of the operation, risk assessment, behavioral aspect of people, and safety.

## THE VALUE OF THE JHA

The intent of this book is to provide a process that can be integrated into the overall management system and bring more value to an organization. The JHA is an essential tool in the overall management system that will help to prevent injuries and in turn establish more effective job procedures that provide value to the organization.

In this book we will discuss many types of programs that support the overall process. We want everyone to keep in mind that the JHA program that we will be discussing in the book is part of an overall process that supports the management system.

## WHAT IS A JOB HAZARD ANALYSIS?

The JHA is used to assess the existing and potential hazards of a job, understand the consequences of risk, and act as an aid in helping identify, eliminate, or control hazards. The JHA is a tool that is used to focus on a specific job, define the steps required to do that job, and ultimately define each task required to perform each step. The JHA focuses on the relationship between:

- The employee.
- The job as a whole unit.
- The steps that make up the job.
- The tasks that are defined in each step.
- The tools, materials and equipment being used.
- Existing and potential hazards.
- The consequences of exposure to those hazards.
- Potential at-risk events associated with each task.
- Existing policies and procedures.
- The nature of the physical environment that the job is completed within.

By assessing all of the components of a task, a comprehensive overview is developed that allows one to focus on the essential areas where changes may be needed.

By using a comprehensive JHA process, you can provide the foundation for a more effective safety culture. Our approach uses both quality management and contemporary safety concepts. This JHA process bridges the traditional systems of risk management and at-risk events. [2]

We have found from our experience that, for many organizations, the JHA is vastly under-utilized. We believe the JHA is an essential component of the safety management system.

## WHAT IS IN THIS BOOK?

Effective management of a workplace is a decisive factor in reducing the extent and severity of work-related injuries and related costs. An effective safety process consists of core fundamental elements necessary to build safety directly into each job task.

Our approach to JHA development will build on the traditional JHA format and show how to integrate a risk assessment and behavioral component into the process by also addressing at-risk events (behaviors) that put us at risk to specific hazards.

This book is divided into four sections to provide you with a logical flow from hazard recognition to JHA development, and help to develop a successful safety culture and programs that support the management process. Your safety process must be aligned with the goals and objectives of your organization.

Part 1 will set the stage for hazard recognition and will provide the basic foundation necessary to identify and correct hazards in the workplace. In Chapters 1, 2, and 3 we will discuss how to identify hazards in the workplace and methods on how to control identified hazards.

Part 2 provides proven methods and techniques to help strengthen the JHA process. In Chapter 4 we will provide an overview of how at-risk events affect employees and how these at-risk events interact in every aspect of our lives every day. Chapter 5 provides a detailed discussion on the importance of employee participation, face-to-face contact, one of the most powerful tools in a safety tool kit. Chapter 6 will discuss how to identify risk associated with each job task and will introduce several methods for identifying and assessing risk. In Chapter 7, we will discuss methods for effective training techniques.

Part 3 details the JHA development and implementation in detail. In Chapter 8, we will provide a detailed overview of the JHA process and how to plan for the JHA. Chapter 9 will discuss how to break the job down into its individual components, the step and the task. Chapter 10 take all of the information presented and will provide a detailed overview of developing an actual JHA. Chapter 11 provides an overview of Standard Operating Procedures.

Part 4 provides an overview of management systems to set the stage for an organization to move from the compliance to a strong safety culture. In Chapter 12 we will present management systems that are simple to implement with proper management commitment. Chapter 13 will provide an overview of Six Sigma and provide a snapshot of the tools that the safety professional can use and how these tools fit into the JHA development.

## CONNECTING THE DOTS BETWEEN THE SAFETY PROGRAM AND A SAFETY PROCESS

The concepts in this book can be utilized by anyone interested in safety, including those that are just getting started in safety, seasoned safety professionals, supervision, and those who are looking for a cutting-edge methodology to address hazards in the workplace.

This book is a valuable resource for students of safety, facility management, project managers, industrial engineers, supervisors, employers, or other individuals who have some level of responsibility for safety and risk control. The bottom line: this book can be used to enhance your safety management system, thereby promoting a safer work environment.

This book will provide you with a portfolio of tools and methods to help you to assess the risk and hazards of your organization. By considering the JHA as a process, not a program, that combines physical elements, behavioral issues, and risk assessment, your ability to identify, reduce and eliminate hazards can be enhanced. You will gain knowledge of new techniques that will enhance your ability and that of your employees to recognize hazards more effectively and to control identified hazards. In addition, the material presented in this book can be used to encourage the use of specific organized data to analyze job steps and related tasks by learning how to interrelate with the job elements more creatively.

Figure I-2 gives an overview of this approach. As we stated in the beginning, we believe that the new method that we are introducing in this book allows the user to capture existing and potential hazards and identify the consequences of exposure, thereby integrating at-risk events and preventive measures in a more proactive manner.

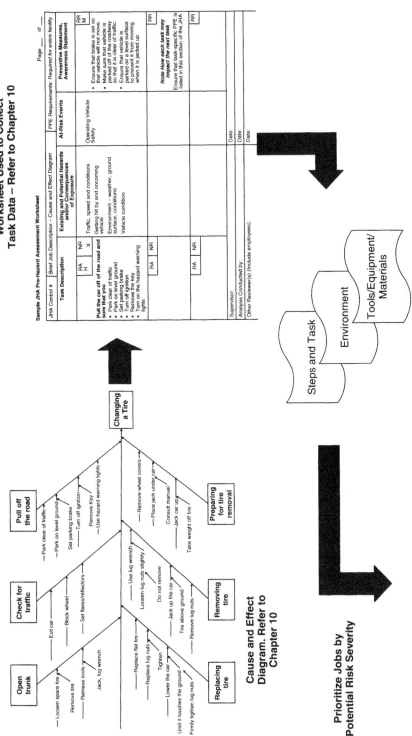

| Brief Description: Changing a Tire | Department: Highway | Date April 2007 | Last Revision: |
|---|---|---|---|
| Performed By: J. Roughton/Nathan Crutchfield | Employee: Joe Tire | Supervisor: Jim Production | |
| Personal Protective Equipment: None required | Note: Recommended PPE will be listed under each specific task based PPE Assessment | | |

*Note: NR = Non-Routine Task – Place a check mark for these types of task. RA = Risk Analysis, RR = Residual Risk (After JHA Development)*

| Job Steps and Task-Specific Description | NR | RA | Existing and Potential Hazards and/or Consequences of Exposure | Potential At-Risk Events and Preventive Measures | RR |
|---|---|---|---|---|---|
| **Pull car off the road**<br>• Park clear of traffic<br>• Park on level ground<br>• Set parking brake<br>• Turn off ignition<br>• Remove the key<br>• Turn on the hazard warning lights | | H | Traffic – speed and conditions<br>Getting hit by an oncoming vehicle<br>Environment – weather, ground surface, conditions<br>Vehicle condition | **Operating Vehicle Safety**<br>• Ensure that brake is set so that vehicle will not move.<br>• Make sure that vehicle is parked off of the roadway so that it is clear of traffic.<br>• Ensure that vehicle is parked on a level surface to prevent it from moving when it is jacked up | M |
| **Check for Traffic**<br>• Exit the car<br>• Block the wheel<br>• Set flares/reflectors on the road | X | H | Traffic speed<br>Getting hit by an oncoming vehicle | **Exit/Entry from Car**<br>• Pay attention to on-coming traffic when existing vehicle | M |
| **Open Trunk**<br>• Loosen spare tire<br>• Remove tire<br>• Retrieve jack and lug wrench | | M | Strain/sprain from removing tire and tools from trunk | **Eyes on Task**<br>• Keep eyes on task when removing tire<br>**Tools and Equipment**<br>• Keep back as straight possible when removing tire an tools from trunk<br>**Lifting and Lowering**<br>• Use proper lifting techniques when lifting tire from trunk | L |
| **Preparing for Tire Removal**<br>• Remove wheel covers<br>• Place jack under car<br>• Jack car up<br>• Take weight off tire | | M | Strain/sprain from removing wheel covers<br>Struck against vehicle when placing jack under car<br>Struck by vehicle when placing jack up car<br>Terrain/ground surface and condition, shift off of jack | **Tools and Equipment**<br>Select the correct tool for removing wheel covers<br>**Eyes on Task**<br>• Keep eyes on task when removing wheel covers<br>• Keep eyes on task when placing jack under car<br>• Keep eyes on task and highway when jacking up car | L |

|  | Catastrophic | Critical | Marginal | Negligible |
|---|---|---|---|---|
| Frequent | HIGH | HIGH | SERIOUS | SERIOUS |
| Probable | HIGH | HIGH | SERIOUS | SERIOUS |
| Occasional | HIGH | SERIOUS | MEDIUM | LOW |
| Remote | SERIOUS | MEDIUM | MEDIUM | LOW |
| Improbable | MEDIUM | LOW | LOW | LOW |

**Defines Risk Assessment, Refer to Chapter 6**

**Determine Modification or Controls**

**Hierarchy of Controls, Refer to Table 9-2**
- ☐ Avoid or reduce the hazard strength
- ☐ Engineering controls
- ☐ Administrative controls
- ☐ PPE

**SOP, Refer to Chapter 11**

**Completed JHA, Refer to Chapter 10**

Standard or Safe Operating Procedures

**Figure I-2** An Overview of the Entire JHA Process, Soups to Nuts

# REFERENCES

1. ASSE - Technical Publications (Management), Professional Safety Review, October 2002, pp. 17, Karl A. Jacobsen, PE., CSP, Boston, MA.
2. Roughton, James, James Mercurio, *Developing an Effective Safety Culture: A Leadership Approach*, Butterworth-Heinemann, 2002.

Note: The material in this book is a collection of many public domain documents and information found throughout the internet and/or on various state and federal OSHA websites. These documents can be found on various websites (for example www.OSHA.gov under publications, Oregon OSHA website, Job Hazard Analysis, http://www.cbs.state.or.us/ external/osha/pdf/workshops/103w.pdf), public domain and other related sources. Many of the chapters in this book are based on the U.S. Department of Labor, Occupational Safety and Health Administration OSHA 3071 Job Hazard Analysis, 1971, 1998, 1999, 2002 revised documents.

# Acronyms

| | |
|---|---|
| ABSS | Activity-Based Safety System |
| AHA | Activity Hazard Analysis |
| AIHA | American Institute of Hygiene Associate |
| ANOVA | Analysis of Variances |
| ANSI | American National Standards Institute |
| BBS | Behavior-Based Safety |
| BMP | Best Management Practice |
| C&E | Cause and Effect Matrix |
| CPI | Continuous Process Improvement |
| CRM | Continuous Risk Management |
| DAMIC | Define, Analyze, Measure, Improve, and Control |
| DPMO | Defects per Million Opportunities |
| DOE | Design of Experiment in Six Sigma |
| DOE | Department of Energy, Government |
| DOT | Department of Transportation |
| EEO | Equal Opportunity Objectives |
| EPA | Environment Protection Administration |
| FAA | Federal Aviation Administration |
| FMEA | Failure Mode and Effect Analysis |
| FTA | Fault Tree Analysis |
| HAZOP | Hazard and Operability Studies |
| IH | Industrial Hygienist |
| JHA | Job Hazard Analysis |
| JSA | Job Safety Analysis |
| KSA's | Knowledge, Skills, and Attitudes/Abilities |
| MSDS | Material Safety Data Sheet |
| OJT | On-the-Job Training |
| ORM | Operational Risk Management |
| OSHA | Occupational Safety and Health Administration |
| OIR | OSHA Incident Rate |
| OSH | Act Occupational Safety and Health Act |
| OSPP | OSHA's Strategic Partnership Program |

| | |
|---|---|
| PCA | Process Capability Analysis |
| NR | Non-routine |
| POA | Plan of Action |
| PDCA | Plan-Do-Check-Act |
| PDC | Professional Development Course |
| PHA | Process Hazard Analysis |
| PPE | Personal Protective Equipment |
| PSM | Process Safety Management |
| RA | Risk Assessment |
| R&R | Gauge Repeatability & Reproducibility |
| ROI | Return on Investment |
| RR | Residual Risk |
| SPC | Statistical Process Control |
| SOP | Standard Operating Procedure |
| TCIR | Total Case Incident Rates |
| THA | Task-specific Hazards Analysis |
| TQM | Total Quality Management |
| VOE | Voice of the Employee |
| VPP | Voluntary Protection Programs |

*"We are what we repeatedly do. **Excellence**, then, is not an act but a habit."*

—Aristotle

# Part 1

## Developing a Toolkit for Identifying Workplace Hazards and Associated Risk

# 1

# Preparing for the Hazard Analysis and Risk Assessment

In this chapter, you will begin the JHA process by preparing for the risk and hazard assessment. At the end of the chapter you will be able to:

- Explain why the JHA should be the centerpiece of your safety process
- Assess the objectives of hazard recognition
- Outline how the JHA can provide the core criteria for the assessment
- Develop a systematic approach to the assessment process.

"Git-R-done!"

—Larry, the Cable Guy

The Federal Occupational Safety and Health (OSHA) Act states: "Employers must furnish a place of employment free of recognized hazards that are causing or are likely to cause death or serious physical harm to employees" (OSHA Act of 1970, 29 CFR 1903.1). Further, the American National Standards Institute's Z10-2005, Occupational Health and Safety Management System Standard, 4.3 Objectives states: "The organization shall establish and implement a process to set documented objectives, quantified where practicable, based on issues that offer the greatest opportunity for Occupational Health and Safety Management System improvement and risk reduction."

We believe that safety management must become part of how organizations improve by acting as a change agent. Safety professionals can effect changes by developing three fundamental skills that will allow them to be able to assess and understand an organization's dynamics. The first is having the skills to develop and maintain a personal "mental map" of goals and objectives necessary to control hazards and related risks. Second is the ability to have a clear and concise understanding of the work environment. The third skill is developing and maintaining a "tool box" of techniques, methods, and concepts that include not only a compliance approach to safety but also include problem solving, analysis, and communications. With these skill sets, one's profession becomes a work of art as well as a professional craft.

## 1.1 THE CENTERPIECE OF A SAFETY PROCESS

Before we begin our discussion on hazard analysis and risk assessment, we need to set the stage for what is to be accomplished in this book by asking several questions:

- Have you considered all of the existing and potential hazards and/or the consequences of exposures that are associated with your organization and/or your specific industry?
- Do you have injuries where you "scratch your head" and think, "How could this injury have happened again?"
- Have you uncovered hazards or other issues that you did not know previously existed when you investigated an injury?
- Have you investigated an injury and found other existing and potential hazards that were not associated with this particular injury?
- Have your employees reported hazards that you did not think were important, for which the warning signs were not all that clear, and later . . . . . . . . . . WHAM, you are blindsided by a loss-producing event?
- Have you ever investigated an injury to find out that the employees knew that the hazard existed and that they thought it was just part of doing their job, and had always just accepted the risk?
- Have you ever investigated an injury and heard, "But we've always done it this way," or "I have been doing it this way for "x" years, so don't tell me I have to change now"?

Do these questions sound familiar? Probably "YES." These are questions that you have or will have to address at some point in time. An effective JHA process will help you to address these issues and provide you with a better mechanism to understand workplace hazards that exist in your operation.

The JHA is an essential safety management tool that, if used consistently, will build an inventory or portfolio of hazards and associated risk, provide an understanding of how things really get done in your workplace, and reduce the potential for undesired events as controls are implemented.

The JHA process revises and improves preventive measures as an environment or workplace changes. This is a never-ending battle as new hazards and risks are constantly created and opportunities for improvement will always exist if your hazard identification and control system is working properly [3].

Therefore, we believe that the JHA program must become the centerpiece of an overall safety process.

## 1.2 HAZARD RECOGNITION AND CONTROL SYSTEMS

The objective of hazard recognition is to identify perceived, existing, and potential hazards and/or the consequences of exposure to hazards.

One of the most important elements of any hazard recognition system is to help management and employees to have some knowledge of operational hazards and associated risks. This knowledge is essential to ensure that hazards are controlled, reduced, or eliminated as they are identified. The systematic assessing and analyzing of workplace hazards uses a process strategy that includes the following analysis:

- Conducting a risk assessment of the workplace.
- Prioritizing the risk assessment findings.
- Developing controls to resolve risk-related issues.
- Recommending and implementing controls.
- Monitoring the results of the controls implemented.

## 1.2.1 Conducting a Risk Assessment of the Workplace

An assessment of the workplace is accomplished by conducting structured and routine physical reviews. The physical hazards survey process identifies the presence or absence of any specific-related hazards and begins by asking simple questions:

- What is currently happening? What are we currently doing?
- Have there been any changes in tools, equipment, materials, or the environment?

- Are there violations of policies, procedures, protocols, rules, and guidelines?
- Are we doing what we should or think we should be doing?

## 1.2.2 Prioritizing the Risk Assessment Findings

An analysis is made of assessment findings to determine the nature and impact of hazardous conditions, unsafe practices and at-risk events involved in the operation. Methods include risk analysis of the existing and potential hazards in facilities, equipment, materials, and processes. A process is the "method for doing something, generally involving a number of steps or operations" (*New World Dictionary*, Second Edition). The goal is a process strategy that, when systematically applied, will help to reduce the employee's exposure to hazards. Using data from the assessment, we develop process maps and flow charts that detail and make tangible the sequence of actions and components necessary to complete a job task [3]. Refer to Figure 1-1 for an overview of hazard identification. Refer to Figure 1-2 for an overview of a typical process strategy.

## 1.2.3 Developing Solutions to Resolve Risk-Related Issues

Once specific hazards and the risks have been identified, the JHA is used to determine how the job step and task elements are interrelated. The JHA establishes the sequence of controls necessary for effectively reducing hazards and risks.

## 1.2.4 Recommending and Implementing Controls

As controls are identified, it is important to define and offer effective recommendations and obtain management buy-in. Action plans are developed to control, reduce or eliminate the hazard. Figure 1-3 provides a hazard analysis flow diagram for developing these controls.

## 1.2.5 Monitoring the Results

Hazards may not have been detected during the initial hazard assessment and other hazards may have slipped out of control due to "normal" operations.

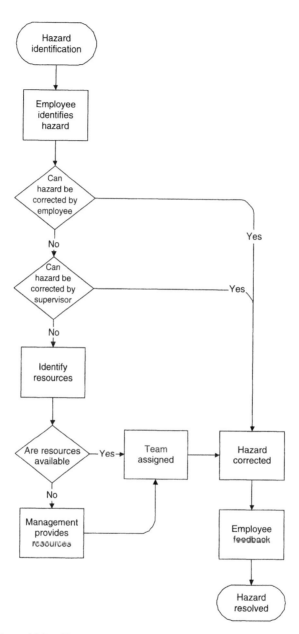

**Figure 1-1** Hazard Identification Flow Diagram

Adapted from Hazard Identification Flowchart, Oklahoma Department of Labor, Safety and Health Management: Safety Pays, 2000, http://www.state.ok.us/~okdol/ Chapter 7, pp. 43, public domain
Adapted and Modified, Roughton, James, James Mercurio, *Developing an Effective Safety Culture: A Leadership Approach*, Butterworth-Heinemann, Figure 10-2, pp. 186, 2002

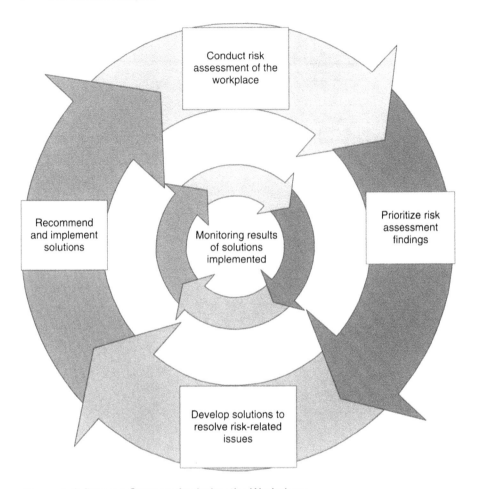

**Figure 1-2** Process Strategy Analyzing the Workplace

Such slippage is common and is why continuous monitoring and employee feedback is important. Monitoring is usually the weakest part of the safety processes and requires a structure that maintains focus to ensure that the safety process remains in place [3].

After hazards have been identified and controls are developed, the emphasis shifts to methods that can be used to help ensure that all controls stay in place and that other hazards do not develop. This can be accomplished by continually scanning the work environment, using employee participation (we will discuss this further in Chapter 5), the process, work practices, and revising controls to help reduce workplace hazards. Refer to Table 1-1 for methods for preventing hazards from developing.

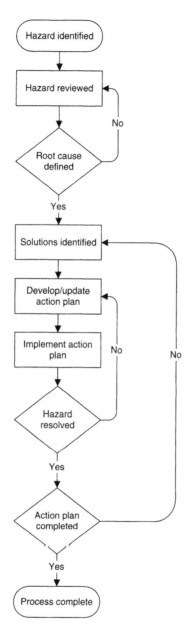

**Figure 1-3** An Overview of Hazard Analysis

Adapted from Developing a Solution to Solve the Problem, DOE Guideline Root Cause Analysis Guidance Document, DOE-NE-STD-1004-92, February 1992, U.S. Department of Energy Office of Nuclear Energy, Office of Nuclear Safety Policy and Standards, Appendix E, Figure E-1, pp. E-1, Six Steps Involved in Change Analysis, adapted and modified public domain version

Adapted from Roughton, James, James Mercurio, *Developing an Effective Safety Culture: A Leadership Approach*, Butterworth-Heinemann, Figure 12-1, pp. 229, 2002

**Table 1-1**
**Methods for Preventing Hazards from Developing**

- Perform routine preventive and regular maintenance. Warning signs: Delaying maintenance or repairs deferred.
- Implement hazard correction procedures and controls that are effectively used. Warning signs: Items noted from inspections are not corrected and/or remain open for a long period of time.
- Ensure that all employees understand why and how to use and maintain specific Personal Protective Equipment (PPE). Warning signs: Observation shows PPE is not being worn, cleaned, or maintained properly.
- Ensure that all employees can demonstrate the safe methods of identifying work rules, procedures, and protocols. Warning signs: Observation and loss analysis data find administrative controls bypassed or weakly enforced. High employee turnover, absenteeism and ergonomic issues, etc.
- Implement medical surveillance programs that are tailored to fit your operation. Warning sign: No medical surveillance program exists or is waived to meet hiring needs. Employees are sent to treating physician without any company representative.

Adapted from OSHA website http://www.osha-slc.gov/SLTC/safetyhealth_ecat/comp3.htm, public domain
Roughton, James, J. Mercurio, *Developing an Effective Safety Culture: A Leadership Approach*, Butterworth-Heinemann, 2002, Chapter 10, pp. 176, public domain

## 1.3 DEVELOPING A SYSTEM TO IDENTIFY AND REPORT HAZARDS

Hazards are essentially anything that increases the potential for injury or harm. Therefore, developing a reliable system for reporting hazards is an essential element of an effective safety process. Such a system can be characterized by the following elements:

- Develop a policy that is consistent with other protocols to encourage employees to report their concerns about quality, harassment, working conditions, or other operational issues.
- Create activities to involve employees in the safety process.
- Protect employees from harassment, removing the fear of reporting.
- Ensure that there is a timely and appropriate response to employees who report hazards.
- Ensure that timely and appropriate corrective action plans are developed where valid concerns exist.

- Develop a system to track and follow-up on all corrective action plans to ensure completion [5].
- Follow up on action plans to ensure proper implementation.

To ensure that these elements are implemented, your portfolio of methods and techniques should include the following tools:

- Quality Control Methodology: Cause and effect analysis, data analysis, process mapping, and flow charts.
- Basic knowledge of behavioral theory of individuals, groups, and organizations.
- Risk management and hazard assessment: risk matrix concepts, developing and using the hierarchy of controls.

These techniques and methods must be designed to ensure that honest feedback is provided about what is going on, no matter how unpleasant or how much work is involved. With current information technology, the Internet and other forms of communications, the defense of claiming "not knowing" about a hazard or concerns has been removed. Getting issues into the open allows for in-depth review and action planning.

## 1.3.1 Company Safety Policy

A company safety policy must be written that is simple and easy to interpret. It must be incorporated into your overall safety system and communicated to all employees. If the current safety policy does not outline how employees are to report hazards, it should be modified to document appropriate procedures. Once completed, the policy can be communicated through all types of media: posting on both electronic and hardcopy, bulletin boards and/or websites, distributing to all employees, and, most importantly, discussed in daily, weekly, and monthly meetings, and/or other plant-wide meetings. You will know that you have communicated your safety policy when every employee is able to tell you what the policy states.

A positive policy for reporting safety issues affirms your intentions to protect employees from harassment or reprisal. Refer to Appendix 1-A for a model policy statement and sample guidance in writing a complete statement [4, 5].

## 1.3.2 Involving Employees in the JHA Process

Employees have a unique and valuable perspective on hazard recognition, the control of physical conditions and safe work procedures. Use the specific job knowledge and skill that employees have gained from their involvement with a process, equipment, task, tools and materials. This will be discussed in more detail in Chapter 5, "The Effective Use of Employee Participation."

## 1.3.3 Protecting Employees from Harassment

One very important consideration is to separate the hazard-reporting policy from any disciplinary system. In addition, do not place policy statements dealing with reporting hazards adjacent to an employee discipline policy in any media, in sequence in the employee handbook, or when discussing these subjects in the same meeting. Physical proximity can give employees a perception that reporting a hazard can get themselves or another employee in trouble [5]. Fear of reprisal changes the dynamics of the process from a positive effort to a negative position. As the old saying goes, "You can do 1000 things right, but the one thing that you do wrong will not be forgotten."

It is important that employees clearly understand that reporting hazards will not result in discipline, harassment, or reprisal from management, supervision or their co-workers. The hazard-reporting policy should make this clear. Other steps that you can take to ensure that this issue is properly handled can include:

- Avoid implementation of performance indicators or measurements that rate supervision negatively.
- Approach all discussions and written descriptions of employee hazard reports as a group problem-solving effort to keep the worksite safe. Emphasize the positive benefits of getting the problem in the open.
- Emphasize the responsibility each employee has for co-workers' safety as well as for their own.
- Enforce the company policy clearly, emphatically, and consistently if you discover or suspect any case of harassment for reporting hazards [5]. Refer to Table 1-2 for some key points to remember about reporting hazards. Some variations in the reporting system may work better than others.

## 1.3.4 Identifying Workplace Hazards

As hazards are identified, a detailed analysis is conducted to examine each hazard in detail, understand the nature of its various risks, and develop an action plan to resolve the identified conditions. This analysis requires that each operational item or component be examined to determine how it relates to or influences the entire operation or process [4].

### Table 1-2
### Important Points to Remember about Reporting Hazards

The following are some important points to remember:

- Develop a policy that encourages employees to report hazards, no matter the nature of the hazard.
- Communicate the policy. Make sure that it is known and understood by all employees.
- Make hazard reporting a value built into your management system and not just a "priority."
- Respond to all reported hazards in a timely manner.
- Track all hazard-prevention activities to completion.
- Use the information collected to revise JHAs to include updating the hazard inventory and to improve the hazard recognition program.

Adapted from Missouri Department of Labor, Catching the Hazard that Escapes Control, http://www.dolir.state.mo.us/ls/oshaconsultation/ccp/chapter_9.html, public domain

All hazards should be assessed with feedback provided to employees. If management determines that no hazard exists, the findings and assessment should be immediately explained to the employees with full details as to why no hazard was present. You must come to an agreement with the employee that the hazard identified is or is not a safety issue that they should or should not be concerned about [2]. It is better to have employees erring on the side of safety with some non-hazards reported than to overlook even one real hazard because an employee believed that management would not respond [5].

### 1.3.4.1 Employee Reporting Systems

Employee hazard reporting methods that can be used include.

- Verbal reports: directly talking to supervision.
- Suggestion programs: suggestion box or e-mail systems.
- Hazard reporting card systems: directly turned into management.
- "Hazard wanted" program: designating blocks of time for seeking out operational hazards.

A combination of these activities is suggested to ensure a comprehensive system for reporting concerns that will include as many employees as possible [5]. Appendix 1-B provides sample forms for employee reporting hazards, how to tracking hazard corrections, and follow-up.

## Verbal Reports

Employees should be able to verbally report hazards directly to their supervisors. When supervisors are properly oriented and accept their responsibility for safety, informal verbal reporting can naturally occur as part of the normal flow of daily communications. When an employee's expressed concerns appear valid, the supervisor has the responsibility to correct the hazard, develop a corrective action plan, and, if necessary, request for assistance to correct hazards. As with employees, supervisors must not fear to report concerns or problems [5].

Verbal reporting used alone does not provide for corrective action tracking, nor does it enable you to look for trends and patterns of hazards that are reported. In addition, verbal reporting provides the employee little protection from a supervisor who may not be sufficiently concerned about safety [2].

In environments where "everybody knows everything and everyone," the verbal reporting system may be all that is needed. However, you should consider adopting a simple written system where the supervisor makes a short report of each hazard reported and the corrective action taken [5].

## Suggestion Programs

A frequently used method to encourage employees to report hazards is to implement an employee safety suggestion program. Not only does this provide a method for reporting unsafe conditions, but it encourages employees to think about innovative ways of improving safety. If a suggestion program is used, management must ensure that collection points are checked frequently, suggestions are reviewed on a timely basis, with feedback to employees provided in a timely manner. This will ensure that identified hazards get corrected promptly. Without a positive emphasis, suggestion programs can be just dust collectors [5].

## Hazard Card Program

Facilities may develop or purchase a system that includes a structured generic format for training employees in basic hazard recognition. These systems use preprinted cards for employees to document unsafe conditions and at-risk events. These cards are typically turned into supervision for review and tracking of related valid hazards [5].

Some organizations give awards for the highest number of cards with valid concerns while others set quotas for the number of hazards noted. Care must be taken to prevent a false use of the system and have it become a negative approach dismissed by employees. If properly supervised and used consistently, these systems can be successful. Their success is dependent on the "culture" of the organization [5].

### "Hazard Wanted" Program

This method makes a game out of finding specific hazards during a set time frame. Employees are encouraged to look for hazards, take a photo of the hazard, and report the hazard to their supervisor for review. Pictures and reports are posted and used in safety meetings. The best hazard that is identified and corrected can be rated and an award can be presented to the employee similar to the hazard card program.

Note that if pictures of hazards are taken, you must make certain that the hazard is controlled or eliminated. This is to demonstrate that something was fixed with back-up documentation as well as to avoid any liability issues.

### A Word of Caution with Regard to Hazard Reporting

The safety department can be saddled with becoming the gate keeper and the entire process viewed as something separate from "real" operational issues. Management must take the responsibility and work with the safety department providing overall guidance. Management must allow for employee time to look for hazards.

## 1.4 MAINTENANCE WORK ORDERS

Any program designed to report hazards will not work if employees cannot see action or results in a reasonable amount of time. Work order systems that assign special maintenance codes to safety-related issues can provide the tracking of corrective actions. These codes require the maintenance department to assign a higher priority to validate safety-related work orders. As specific safety work orders are generated, the list can be posted on the safety bulletin boards so that all employees can review the status of the work orders.

When complete resolution of a hazard requires ordering parts or materials and a lengthy delay, the employees must be provided a periodic status report. This ensures that the employees' concerns have not been overlooked or forgotten. Alternative or modified actions should be taken when work orders cannot be immediately accomplished. It is your responsibility to provide interim effective protection for your employees and you must take whatever steps are feasible to temporarily eliminate or control the hazardous situation [5].

The work order system, as with any other aspect of operations, must be reviewed for its reliability. It should clearly provide a way of setting priorities for high hazard and high risk areas.

## 1.5 FORMS USED TO REPORT HAZARDS

The best hazard reports are those customized for the specific workplace and operation. Appendix 1-B provides several sample reporting forms: Sample Forms for Employee Reporting of Hazards, Tracking Hazard Correction, and Follow-Up Documentation.

## 1.6 ACTION PLANNING

For every hazard identified, an action plan must be developed and tracked until the hazard is eliminated or controlled. Some hazards are quickly corrected and may not present a safety issue for employees. Corrections that are complicated with budget/time-consuming criteria require a tracking system. Developing a corrective action plan for hazards will help to determine if the management system was effective by assessing if the same or similar hazards reappear. As different corrective activities are reviewed, the cause of management system failures or hazards can be further analyzed. Tracking hazard correction enables management to keep employees informed of the action plan status [1].

An action plan is important, because it:

- Keeps the management team informed of the status of long-term corrective actions.
- Provides a record of actions taken and if the hazard reoccurs again.
- Assists in documenting the history of actions for future use.
- Provides timely and accurate information that can be supplied to employees who report hazards [1].
- Establishes alternative controls to be used until the problem is resolved.

When documenting information about your action plan, it is important to list all interim preventative measures and to include the anticipated date of completion. The weakness of a control plan is the possibility that incorrect or incomplete information is documented or that the corrective action activities will not be recorded properly.

Forms used for employee to report hazards should have a section for describing an action plan [1]. Refer to Appendix 1-C for an example of several action planning forms.

## 1.7 TRACKING HAZARDS

The success of any hazard identification and corrective action program is based on the quality of the follow-up and action planning. If the corrective actions cannot be accomplished immediately, measures must be put in place to ensure that all corrective actions have been completed and signed off by management.

---

An essential part of any safety system is the identification and correction of hazards that occur despite prevention and control procedures.

Documentation is important so that management and employees have a record of the corrective action.

Use a form that documents the original discovery of a hazard to track its correction. One method is to record hazard correction information on the inspection report next to the hazard description.

Employee reports of hazards should provide space for notations about hazard correction.

---

Adapted from the OSHA Website http://www.osha-slc.gov/SLTC/safetyhealth_ecat/comp3.htm#SafeWorkPractices, public domain

Roughton, James, J. Mercurio, *Developing an Effective Safety Culture: A Leadership Approach*, Butterworth-Heinemann, 2002, Chapter 11, pp. 194 [6]

## 1.7.1 Tracking by Committee

In large organizations, a central management safety committee or a joint employee-management committee will devote a part of their meeting to review inspection reports, employee hazard reports, and/or injury reports. The committee will document meeting minutes outlining uncorrected hazards [1]. The benefit of this type of system is the additional higher level of scrutiny applied to action planning and correction tracking. This system can be cumbersome, however, especially when information must be transferred from a report to the committee. The possibility can occur of information being lost in transit, information being incomplete, or incorrect information being conveyed [1].

## 1.7.2 Follow-up Reviews

Follow-up reviews identify hazards that may have developed after the initial processes or procedures have been implemented. The frequency of any follow-up surveys depends on the scope and complexity of the organization [5].

# 1.8 CODES OF SAFE WORK PRACTICES

To tie all of this together, an organization must develop a code of safety work practices that will be general in nature and inclusive of many types of activities. Refer to Appendix 1-D for a set of codes of safety work practices.

# 1.9 SUMMARY

This chapter has highlighted the important points of hazard recognition and control and has examined methods and techniques for improving knowledge and learning more about risks, hazards, their avoidance, correction, and effective control [1].

We have discussed how and when the relationship between the employee and management must be open, candid, and interactive, how safety concerns are identified and discussed, and how solutions are mutually agreed upon with management's buy-in. This type of relationship between management and the employees will help to foster a safer work environment and helps to create a change in the culture where discipline is rarely needed. If discipline is required, employees are more likely to perceive it as a positive as opposed to a negative and to perceive that the offender is deserving of the reprimand.

A hazard analysis of the work environment requires a variety of concepts and methods to identify existing and potential hazards and/or consequences of exposure. Effective management systems continually analyze the workplace to anticipate and develop policies, procedures and protocols to help minimize the introduction and reoccurrence of hazards.

When identifying workplace hazards, the human element becomes a part of the assessment. Whenever one employee is replaced by another employee, the difference in skill and experience can increase the risk to both the new employees and their co-worker. Managers need to be sensitive to changes and be willing to provide training and orientation, physical and administrative adjustments, and other accommodations.

In the next chapter we will continue our discussion on how the JHA, as the centerpiece of your safety process, provides the blueprint to better designs of the workplace, and how a JHA process enhances your ability to anticipate and understand how all job elements combine and allows you to develop effective programs and procedures.

## CHAPTER REVIEW QUESTIONS

1. What are the most critical elements in maintaining and keeping your process visible?
2. What are the three essential skills needed in order for a safety professional to be successful?
3. What is a process?
4. What is the objective of hazard recognition?
5. What is the best method for assessing a workplace?
6. Why is it important to monitor the results of the JHA process?
7. What are the methods for preventing hazards from developing?
8. What are some of the reliable methods for developing a system to identify and report hazards?
9. What is a company safety policy and why is it necessary?
10. Why is it important to have employee involved in the JHA process?

## REFERENCES

1. U.S. Department of Labor, Office of Cooperative Programs, Occupational Safety and Administration (OSHA), Managing Employee Safety and Health, November 1994, public domain
2. OSHA Web Site, http://www.osha-slc.gov/SLTC/etools/safetyhealth/index.html, public domain
3. Oregon Website, Hazard Identification and Control, OR-OSHA 104, http://www.cbs.state.or.us/osha/educate/training/pages/courses.html, public domain
4. Missouri Department of Labor, Establishing Complete Hazard Inventories, Chapter 7, http://www.dolir.mo.gov/ls/safetyconsultation/ccp/, public domain
5. Missouri Department of Labor, *Catching the Hazard That Escapes Control*, Chapter 9, http://www.dolir.mo.gov/ls/safetyconsultation/ccp/, public domain
6. Roughton, James, James Mercurio, *Developing an Effective Safety Culture: A Leadership Approach*, Butterworth-Heinemann, 2002

# Appendix A

## A.1 SAMPLE GUIDANCE IN WRITING A POLICY STATEMENT

### A.1.1 Introduction

Generally, a written safety policy statement will run 6 to 12 sentences in length. It includes some or all of the five elements:

- An introductory statement
- A statement of the purpose or philosophy of the policy
- A summary of management responsibilities
- A summary of employee responsibilities
- A closing statement.

#### A.1.1.1 Introductory Statement

The written policy statement generally starts with a clear, simple expression of your concern for an attitude about employee safety. Examples of introductions to policy statements include:

- This company considers no phase of its operation or administration more important than the safety of our employees. We will provide and maintain safe working conditions, and establish and insist on safe work methods and practices at all times.
- Injury prevention is a primary job of management, and management is responsible for establishing safe working conditions.
- This company has always believed that our employees are our most important assets. We will always place the highest value on safe operations and on the safety of our employees.

- The company will, at all times and at every level of management, attempt to provide and maintain a safe working environment for all employees. All safety protection programs are aimed at preventing injuries and exposures to harmful atmospheric contaminants.
- All members of management and all employees must make safety protection a part of their daily concerns.

## A.1.2 Purpose/Philosophy

An effective safety process will have a stated purpose or philosophy. This is included in a written policy statement so that both you and your employees are reminded of the purpose and value of the program. You may wish to incorporate the purpose/philosophy into your policy. Examples of purpose and philosophy include:

- We have established our safety program to eliminate work-related injuries and damage. We expect it to improve operations and reduce personal and financial losses.
- Safety protection shall be an integral part of all operations, including planning, procurement, development, production, administration, sales, and transportation. Injuries have no place in our company.
- We want to make our safety protection efforts so successful that we make elimination of injuries a way of life.
- We aim to resolve safety problems through prevention.
- We will involve both management and employees in planning, developing, and implementing safety protection.

## A.1.3 Management Responsibilities

A safety action plan will describe in detail who is to develop the program and make it work, as well as who is assigned specific responsibilities, duties, and authority. The policy statement may include a summary of the following responsibilities:

- Each level of management must reflect an interest in company safety and must set a good example by complying with company rules for safety protection. Management interest must be vocal, visible, and continuous from top management to departmental supervisors.
- The company management is responsible for developing an effective safety program.

- Plant superintendents are responsible for maintaining safe and healthful working conditions and practices in areas under their jurisdiction.
- Department heads and supervisors are responsible for preventing injuries.
- Supervisors are responsible for preventing exposures to hazard in their specific work areas. Supervisors will be held accountable for the safety of all employees.
- The safety director has the authority and responsibility to provide guidance to supervisors and to help management prevent injuries.
- Management representatives who have been assigned safety responsibilities will be held accountable for meeting those responsibilities.

## A.1.4 Employee Responsibilities

Many companies acknowledge the vital role of their employees in the operation of a successful safety process by developing specific employee roles and contributions in the policy statement. Employees have the unique responsibility to assist management with all injury prevention efforts through active participation in all risk control activities, especially providing direct feedback regarding the effectiveness of these efforts.

The following are examples of employee responsibilities:

- All employees are expected to follow safe practices, obey rules and regulations, and work in a way that maintains the high safety standards developed and sanctioned by the company.
- All employees are expected to give full support to risk control activities.
- Every employee must observe established safety regulations, procedures, and protocols.
- All employees are expected to take an active interest in the safety process, participate in program activities, and abide by the rules and regulations of this company.
- All employees must recognize their responsibility to prevent injuries and must take necessary actions. Their performance in this regard will be measured as part of their overall job performance.

## A.2 CLOSING STATEMENT

The closing statement is often a reaffirmation of your commitment to provide a safe workplace. It appeals for the cooperation of all company employees in support of the safety program.

- I urge all employees to make this safety process an integral part of your daily job tasks and activities.
- By accepting our individual responsibilities to operate safely, we will all contribute to the well-being of one another and consequently the company.
- We must be successful in our efforts for the elimination of injuries and for safety management to become a way of life.

Source: Adapted from Roughton, James, J Mercurio, *Developing an Effective Safety Culture: A Leadership Approach*, Butterworth-Heinemann, 2002, Appendix A, pp. 389–392.

## A.2.1 Sample Model Policy Statements

The following are some samples of Model Policy Statements. Some may have been presented above. These statements are general in nature and inclusive of many types of business activities. These statements are intended as a model only and should be adapted to describe your own site-specific work environment.

### Sample #1

"The Occupational Safety and Health Act (OSHA) of 1970 clearly requires that the company shall provide a workplace with safe and healthful working conditions. The safety of our employees continues to be the first consideration in the operation of this business."

### Sample #2

"Safety in our business must be a part of every operation. Without question it must be every employee's responsibility at all levels of the organization."

### Sample #3

"It is the intent of our organization to comply with all rules and applicable laws. To do this we must constantly be aware of conditions in all work areas that can produce injuries. No employee is required to work at a job that is not safe. Your cooperation in detecting hazards is part of your job. You must inform your supervisor immediately of any situation beyond your ability or authority to correct."

### Sample #4

"The personal safety of each employee of this company is of primary importance. The prevention of occupationally induced injuries is of such consequence that it will be given precedence over operating productivity. To the greatest degree possible, management will provide the resources required for personal safety in keeping with the highest standards."

## Sample #5

"We will maintain a safety process conforming to the best practices we can achieve. To be successful, our program will embody the proper attitudes toward injury prevention on the part of supervision and employees. It requires cooperation in all safety matters, not only between supervisor and employee, but between each employee and their co-workers. Only through such a cooperative effort can a safety program in the best interest of all be established and preserved."

## Sample #6

"Our objective is a safety process that will reduce the number of injuries to an absolute minimum, not merely in keeping with, but surpassing, the best experience of operations similar to ours. Our goal is zero injuries."

"Our safety process will include the following elements:

- Providing mechanical and physical safeguards to the maximum extent possible.
- Conducting a program of safety inspections to find and eliminate unsafe working conditions or at-risk events, to control safety hazards, and to comply fully with all safety standards for every job.
- Training all employees in good safety practices.
- Providing necessary PPE and instructions for its proper use and care.
- Developing and enforcing safety rules and requiring that employees cooperate with all rules as a condition of employment.
- Investigating, promptly and thoroughly, every injury to find out what caused it and to correct the problem so that it will not happen again.
- Setting up a system of recognition and awards for outstanding safety service or performance."

## Sample #7

"We recognize that the responsibilities for safety must be shared by everyone as such:

- As a corporation we will accept the responsibility for management and leadership of the safety program, for its effectiveness and improvement, and for providing the safeguards required to ensure safe conditions.

- Supervisors are responsible for developing the proper attitudes toward safety in themselves and in those they supervise, and for ensuring that all operations are performed with the utmost regard for the safety of all personnel involved, including them.
- Employees are responsible for wholehearted, genuine cooperation with all aspects of the safety program including compliance with all rules and regulations and for continuously practicing safety while performing their duties."

Adapted from OSHA Handbook for Small Businesses, Safety Management Series, U.S. Department of Labor, Occupational Safety Administration, OSHA 2209, 1996 (Revised), Appendix C, pp. 51, public domain.

## A.3 SUMMARY

Policy statements can vary in length and content. The briefest are typically basic statements of policy only. Longer statements may include company philosophy. Others will address the safety responsibilities of management and other employees.

Policy statements can cover in detail items such as specific assignment of safety duties, description of specific duties, delegation of authority, safety rules and procedures, and encouragement of employee involvement. While some companies may wish to include these additional items in the policy statement, we believe it usually is best to leave these details for later discussion.

These examples may give you an idea for a policy statement that can be written in your style, expressing your attitudes and value to the safety process.

Adapted from Roughton, James, J. Mercurio, *Developing an Effective Safety Culture: A Leadership Approach*, Butterworth-Heinemann, 2002, Appendix A, pp. 392–393.

# Appendix B

## B.1 SAMPLE FORMS FOR EMPLOYEE REPORTING OF HAZARDS

### B.1.1 Tracking Hazard Corrections

#### B.1.1.1 Follow-Up Documentation

Adapted from Roughton, James, J. Mercurio, *Developing an Effective Safety Culture: A Leadership Approach*, Butterworth-Heinemann, 2002, Appendix C, pp. 397–402.

**Example #1**
**Form for Employees to Report Hazards**

**Part 1**

| Hazard or safety problem |
| --- |
|  |
|  |
|  |

| Department where hazard or problem is observed | |
| --- | --- |
| Date: | Time: |
| Suggested action: | |
|  | |
|  | |
|  | |
| Employee's Signature (Optional) | |

Employee: Completes Part 1 and Gives to Supervisor

**Part 2**

| |
|---|
| Action taken: |
| |
| |
| Department: |
| Date: |
| Supervisors Signature: |

Supervisor: Completes and Gives to Manager

**Part 3**

| |
|---|
| Date: |
| Review/Comments: |
| |
| |
| Manager's signature: |

## Example #2
## Reporting Safety or Health Problem

| |
|---|
| Date |

| |
|---|
| Description of problem (include exact location, if applicable) |
| |
| |

| | |
|---|---|
| Note any previous attempts to notify management of this problem and provide name of person notified: | |
| | |
| Date | |
| Optional: Submitted by: | Name |

| Safety Department Findings: |
|---|
| Actions Taken |
| Safety Committee Review Comments: |

| All actions completed by: | Date |
|---|---|

**Example #3**
**Employee Report of Hazard**

This form is provided to assist employees in reporting hazards.

I believe that a condition or practice at the following location is a hazard or safety problem.

| Is there an immediate threat of serious physical harm? | Yes | No |
|---|---|---|
| Provide information that will help locate the hazard, such as building or area of building or the supervisor's name. | | |

Describe briefly the hazard you believe exists and the approximate number of employees exposed to it.

If this hazard has been called to anyone's attention, as far as you know, please provide the name of the person or committee member notified and the approximate date.

Signature (optional)

Type or print name (optional)

Date

Management evaluation of reported hazard

Final action taken

| All actions completed by Initials | Date |

**Sample**
**Follow-Up Review**

This form can be used as part of sample forms 1, 2, or 3 or separately

| | |
|---|---|
| Hazard | |
| Possible injury or damage | |
| Exposure | Frequency of exposure |
| Duration | |
| Interim protection provided | |
| Corrective action taken | |
| Follow-up review made on | Date |
| Any additional action taken? | |
| Signature of Manager or Supervisor | Date |
| Is corrective action still in place? | Yes | No |
| Three month follow-up check made on | Date |

# Appendix C

## C.1 ACTION PLANNING: THREE SAMPLE VERSIONS ARE INCLUDED

Adapted from Roughton, James, J. Mercurio, *Developing an Effective Safety Culture: A Leadership Approach*, Butterworth-Heinemann, 2002, Appendix B, pp. 395–396.

### Sample Action Plan #1

| Activity | Responsibility | Target Date | Status |
|---|---|---|---|
| *Management Leadership* | | | |
| | | | |
| | | | |
| | | | |
| *Employee Participation* | | | |
| | | | |
| | | | |
| | | | |
| *Job Hazard Analysis to Be Completed* | | | |
| | | | |
| | | | |
| | | | |
| *Hazard Reduction and Control Measures* | | | |
| | | | |
| | | | |
| | | | |
| *Training* | | | |
| | | | |
| | | | |
| | | | |

## Sample #2
### Tracking Hazard Corrective Actions Form

Instructions: Under the column headed "System", note how the hazard was found. Enter Insp. for inspection, ERH and name of employee reporting hazard, or INC. for incident investigation.

Under the column Hazard Description (column 2), use as many lines needed to describe the hazard. In the third column, provide the name of the person who has been assigned responsibility for corrective actions (column 3). In column 4 list any interim corrective actions to correct the hazard and the date performed. In the last column, enter the completed corrective action and the date that final correction was made.*

| 1 | 2 | 3 | 4 | 5 |
|---|---|---|---|---|
| System | Hazard Description | Responsibility: Assigned | Interim Completed Action | Action With Date |
| | | | | |
| | | | | |
| | | | | |
| | | | | |
| | | | | |
| | | | | |
| | | | | |
| | | | | |

**Sample Action Plan #3**

| Activity | Conduct monthly employee safety meetings | Establish procedures for management and employee participation in inspections and injury investigations | Provide hazard recognition & injury investigation training to management and employees |
|---|---|---|---|
| Time Commitment | Begin by June | Inspections and investigations begin by Sept. 30 | Complete by December 31 |
| Responsible Employees | Plant Manager | Plant/Safety Manager | Safety Manager |
| Resources Needed | 1-hour safety meeting each month for all employees. audiovisual equipment | Time spent by employees on inspections and investigations | Time for developing training materials, possibly outside professional/trainer |
| Results Expected | Employee input on safety matters; volunteers for participation programs | Inspections and investigations being performed | Inspections that ID all hazards; investigations that uncover root causes; fewer injuries |
| Possible Roadblocks | Employee mistrust at first; limited knowledge about all potential hazards | Lack of employee interest | Cost of reduced production; unwillingness to spend funds for improvements |
| Status/Evaluation | Monthly reports to Plant Manager; assessment annually | Quarterly reports to CEO, look for patterns or trends | Keep training records; retrain periodically |

* Place activities to be completed on Gantt chart, calendar, and/or individual performance objectives.

# Appendix D

## D.1 CODES OF SAFE WORK PRACTICES

A site-specific code of safe work practices should be developed for every facility to ensure that everyone understands all safety requirements established for the facility.

This section will outline a list of suggested code of safe work practices. It is general in nature and inclusive of many types of activities. It is intended only as a model which you can redraft to describe your own site-specific work environment.

## D.1.1 General Policy

- All employees shall follow safe work rules, render every possible aid to safe operations, and report all unsafe conditions or practices to the supervisor.
- Supervisors shall insist that employees observe and obey every rule and regulation necessary for the safe conduct of the work, and shall take such action necessary to obtain compliance.
- All employees shall be given frequent injury prevention instructions. Instructions, practice drills and articles concerning workplace safety shall be given at least once every _____ working days.
- Anyone known to be under the influence of alcohol and/or illegal drugs shall not be allowed on the job. Persons with symptoms of alcohol and/or drug abuse are encouraged to discuss personal or work-related problems with the supervisor/employer. (See your specific employee assistance program.)
- No one shall knowingly be permitted or required to work while his or her ability or alertness is impaired by fatigue, illness, medications, or other causes that might expose the individual or others to injury.

- Employees should be alert to ensure that all guards and other protective devices are in proper places and adjusted, and shall report deficiencies. Approved personal protective equipment (PPE) shall be worn in specified work areas.
- Horseplay, scuffling, and other such unsafe practices which tend to endanger the safety or well-being of employees are prohibited.
- Work shall be well-planned and supervised to prevent injuries when working with equipment and handling heavy materials. All injuries shall be reported promptly to the supervisor so that arrangements can be made for medical and/or first-aid treatment. First-aid materials are located in _____, emergency, fire, ambulance, rescue squad, and doctor's telephone numbers are located on _____, and fire extinguishers are located at _____.

Adapted from OSHA Handbook for Small Businesses, Safety Management Series, U.S. Department of Labor, Occupational Safety Administration, OSHA 2209, 1996 (Revised), Appendix C, pp. 52–53, public domain.

# 2

## Workplace Hazard Analysis and Review of Associated Risk

The Job Hazard Analysis process begins with the development of a full understanding of the job and the specific data concerning the steps leading to the task, tools, equipment, environment and operational conditions. This chapter provides insights on sources and methods useful in identifying issues and areas needing an in-depth analysis. At the end of the chapter you will be able to:

- Identify the use of audits and inspections
- Use the tools to gather information and analyze for the workplace review
- Define the roles played by outside specialists, supervisors, employees
- Determine who should be involved in the review
- Identify sources for risk and hazard data.

"Hope is not a strategy."

—Rick Page

It is important in your safety process to ensure that the work environment is actively analyzed and monitored. As discussed in Chapter 1, the identification of hazards begins the initial development of areas for further analysis, as the JHA is the centerpiece of a safety process. The JHA provides the blueprint for designing workplace reviews. The JHA process enhances the ability for an organization to anticipate and understand how job elements (steps and task) combine and interact with each other. To build this process, a workplace review should be conducted by designated employees who physically review the operations and activities. By asking specific questions concerning their observations, employees develop an insight into conditions at that point in time and place [2]. Employee participation will be discussed in detail in Chapter 5. Traditionally, several methods are currently used, as discussed in the following section.

## 2.1 ANALYSIS OF THE WORKPLACE

Hazard reviews provide useful data to further enhance analysis and evaluation of the safety management system [4]. Refer to Figure 2-1 for an overview of a comprehensive baseline review. This will be helpful as we discuss analysis of the workplace.

The starting point in analysis and evaluation development is to consider the workplace as a system of cause and effect with inherent risk built into the work environment. Each of its parts has specific characteristics and hazards

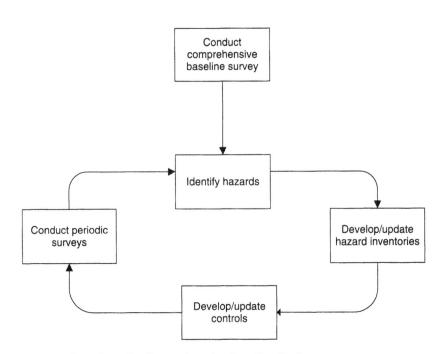

**Figure 2-1** Overview of a Comprehensive Baseline Review

Adapted from OSHA Web site, http://osha.gov/SLTC/safetyhealth_ecat/comp2.htm#, public domain

and risks. Step one is to select the methods or tools that can be used to identify and analyze the work environment. These tools can include:

- Reviewing current job descriptions and protocols that identify what employees are expected to accomplish or tasks they are assigned. This includes interviewing experienced employees who have knowledge of the operation.
- Performing specific task analysis, taking into consideration hazards and associated risks.
- Utilizing workplace surveys based on existing standards, procedures, regulations or guidelines that have been customized and tested for consistency.
- Utilizing change analysis of planned, new, modified or relocated facilities, processes, materials, and equipment to update procedures and/or guidelines.
- Utilizing "walk-around" inspections conducted by knowledgeable employees.
- Reviewing documentation of the history of similar industries and past workplace experience [4].

---

Site analysis has important elements that are critical for an effective overview:

- Analysis should cover all areas of the workplace.
- Analysis should be conducted at regular intervals. The frequency will depend on the size of the workplace and the potential hazards and severity of risks.

All reviewers should be trained to recognize hazards and associated risks and encouraged to bring fresh ideas to the analysis.

Information from the analysis should be used to expand the site's hazard inventory and used to modify and/or update JHAs and/or improve the hazard recognition prevention and control.

---

Adapted from the U.S. Department of Labor, Office of Cooperative Programs, Occupational Safety and Administration (OSHA), Managing Employees' Safety and Health, November 1994, public domain

The next step is to use process flow charts and cause and effect diagrams that show the interrelations of the elements under review. We will discuss these tools later in the book.

## 2.2 INSPECTIONS AND AUDITS

Think of inspections and audits as similar to "tactics" and "strategy." Inspections are a "tactic" program within a process, a specific task designed to accomplish a specific goal. An audit, like "strategy," covers the overall plan and reviews the entire process for function and effectiveness: i.e., are the specific tasks being done and are they effective? Inspections involve identifying hazards, while audits involve determining program areas that are effective or ineffective.

Inspections are the most frequently recommended activity used to assess hazards and have been the centerpiece of regulatory and organizational programs. The inspection is usually a general "walk-around" to identify conditions that do not comply with defined safety procedures and requirements. They can range from very informal as in office areas or light industry, to highly complex as found in extremely high-risk operations, space flight, aircraft, and nuclear power plants, etc.

Inspections are also typically the worst-implemented part of any safety process, as they are time-consuming and take individuals away from "productive" tasks. The question is: Do they provide value to the process? As many organizations have a low perception of risk and have not experienced severe loss, bypassing or taking shortcuts in inspection results in only partially effective reviews.

The audit is a process evaluation method that attempts to determine program status by assigning a numerical rating to program elements. Both the audit and inspection are designed to assess conformance to pre-established standards or regulations. These two methods are primarily designed to find something wrong, and if not presented correctly, can be counterproductive to safety if employees believe the outcomes are only considered as negatives. Remember that we are dealing with human behaviors!

## 2.3 THE CHECKLIST

Checklists are a good starting point and are usually based on the experience and knowledge of specific hazards developed from years of experience. Most of this is from negative experience, where systems failed or unidentified hazards resulted in injuries or other loss-producing events. Your hazards and risks are probably not exactly the same as those envisioned by a generic format. Generic inspection checklists are used to save time and are therefore not usually customized for the specific environment. In many cases, employees and supervisors may not feel that these checklists are of benefit or apply to their work environment. Therefore, generic checklists must be modified to fit your specific work environment. Refer to Appendix E for a sample safety checklist.

## 2.3.1 Consultants and Outside Specialists

To supplement internal resources, you may find that from time to time a variety of outside services may be needed. These services may be independent consultants, engineers, equipment or other specialists. For example, insurance companies provide (typically in your insurance contracts) consulting services, such as general surveys, workers' compensation guidance, property and other specialty loss control reviews. In addition, a host of other services may be available under current insurance programs. These consultants range in skills and the project or action plan must be clearly communicated to ensure they understand what you are trying to accomplish. You and the consultant must have a clear concise game plan to achieve the objectives and agree on the parameters of the project. Their services must be managed and all recommendations clearly defined.

In addition, the insurance brokerages or agents may have access to private consultants that can provide specialized services to help determine workplace hazards. They work on behalf of your organization and are not subject to the insurance underwriter.

Industry experts and specialists with extensive knowledge of specific disciplines may be needed to review complex engineering, safety and health related issues. These consultants may include such disciplines as industrial hygiene, ergonomics, and environmental sciences. This outside professional assistance can bring depth of expertise not found within the organization, as well as bring in a fresh set of eyes. A combination of outside specialists and internal expertise has greater impact for identifying and analyzing targeted areas [4].

Third party team reviews should be conducted on a scheduled basis. These reviews should follow a consistent approach, using customized audit components designed with enough detail for the specific operations under review.

For an IH survey, all occupational health risks and hazards must be inventoried. The IH process should fully review chemicals and hazardous materials, the nature and volume of their use, and their potential health impacts.

The IH review should establish a baseline of all identified potential health hazards from materials in use. IH testing must be routinely conducted for specific air, dust, fume, vapor, noise levels, etc. to establish a baseline for the facility [2].

Ergonomists provide methods and assistance in redesigning of tasks to better fit the limits and constraints of the human body. They take into account body mechanics, movement, fatigue, and strengths needed to complete a task. For example, if you have a high injury rate due to cuts and lacerations from using cutting devices, you may need to have a knife expert assess your cutting task.

## 2.3.2 Employee Interviews

One of the most important and still often overlooked as part of the workplace review is interviewing the employee. These interviews can be as simple as an informal conversation to gain current perceptions, as information based on the employee's knowledge can sometimes be more important than the written survey.

The interview is essential when completing a change analysis, as the employees are the ones doing the actual physical implementation of any change.

General guidelines that can be used when conducting employee interviews include:

- Put employees at ease: Try to keep the interview informal. Try to develop a rapport with the employees, if you don't already have it. Know in advance a little background information, name, job, hobbies, etc. This will make a smooth transition into the actual interview. Be friendly, understanding, and open minded. Make sure that the employee knows that you care about their input. Explain the purpose of the interview and your role in the process. Explain exactly why you are conducting the interview (i.e., JHA development, injury review, change analysis, etc.). This discussion should reduce any reluctance to participate.
- Stay focused on the purpose of the interview: The intent of an interview is to open and maintain a line of communication with employees. Your focus is on hazard reduction. As the interview develops, information not pertinent to your inquiry may develop. You must be prepared to direct this information to other resources for the appropriate response. However, do not "blow off" the employee, as they may have an issue that is affecting safe behavior. Evaluate the information to see how it can impact the work environment and possibly degrade safety efforts. Make an effort to help. The key: If you can direct or assist the employee with their problem or issue, *do it*! You will be seen as a problem solver and a "go to" person and that will be of benefit to you.
- Be calm and do not rush the interview: If you are agitated, or in a hurry to get the interview done, you will be sending the wrong message, signaling a lack of sincerity. Make sure you have scheduled your time and have flexibility with your schedule. If you are having issues or problems that will impact your ability to listen, reschedule the interview. If you have

a cell phone or Blackberry, make sure that it is turned off so that you can give the employee your full attention.

- Use active listening: Let the employees talk! Put your ego aside. It is an interview, not your opportunity to discuss what you think. Many important facts may not be identified if you interrupt the employees. Repeat the information given to you. Talk with the employees until you understand what they are trying to convey to you. In observing interviews, we have found safety directors sometimes unconsciously steer the employee's responses into their way of thinking, do an "information dump" to impress the employees, and, in some cases, don't pick up on what the employees actually said.

- Try to avoid "yes and no" questions. Instead, ask open-ended questions. One effective question is: "Tell me about the procedures for......" "Help me understand how it....." Ask the employees to tell you about potential hazards that they think or know exist. Do not ask them if they know of any hazards, as they could easily just say "no." Do not ask leading questions. Do not put the employees on the defensive. If there are hazards present, do not question the employees in a manner that might accuse or blame them of wrongdoing. The employee is not on trial.

- Ask follow-up questions: Follow-up questions will help to clarify particular areas of concern or allow the interview to get specifics. This may be as simple as a response "Why do you think that is happening?" "Why do we have to do it that way?" or "What would you do different?"

- At the end of the interview, ask the employee if they have any questions for you. Answer these questions as best as you can, or tell them that you will find the answer and get back to them as soon as possible.

- Take notes: Notes should be taken very carefully, and as casually as possible. Offer to let the employees read the notes if they desire.

- Provide feedback: Thank the employees for their time and insight. Conclude the interview with a statement of appreciation for their contribution. Advise the employees of the outcome of the interview as soon as possible.

- Be available: Ask the employee to contact you if they think of anything else that may be of interest or concern [4].

In our experience, interviews almost always result in surprises. In one case, a safety director was following up on a forklift accident. An employee asked him if the "blue stuff in the boiler room can hurt you." When the safety director asked him why he was asking, the employee said that he and his co-workers had found the chemical to be a good laxative and he just wanted to know if it was OK to drink. Needless to say, the safety director was appalled and began an immediate investigation. Stay open and alert for surprises!

One important note: Conduct the interview in the work area or facility. Do not call the employee into your office or a conference room, as this may be intimidating if they are the only other person in the room. Being familiar with the location or the employee's job does not allow you to assume that discussions can easily identify and describe any issues. Be in the work area so the employee can show you exactly where the concerns may be and introduce you to other possible issues. The objective is communication and clarity about the workplace conditions.

## 2.4 TYPES OF INSPECTIONS

There are many types of inspections that can be performed. These include:

General Walk-Around Inspections
Verification Reviews
Focus Reviews
Self-Assessment
Document Review

## 2.4.1 General Walk-Around Inspections

"Walk-around" inspections are used to conduct periodic and daily safety inspections of the workplace. Walk-around inspections identify obvious blatant hazards and visible at-risk events that can be related to unsafe work practices or conditions.

Who should conduct a "walk-around" inspection? Management and supervision are the key individuals who should conduct routine safety walk-around inspections. Other key individuals could be members of safety committees. These individuals are instrumental in establishing accountability by assigning responsibility for conducting daily inspections in specific areas [1]. Walk-around inspections allow the employees to feel, observe and take full stock of an area and its conditions, and observe how things are being accomplished. Their input from such inspections provides a good method to review the safety culture and can help identify areas for improvement.

Information from inspections should be used to improve the hazard prevention and control program [2]. Refer to Table 2-1 for important elements of site reviews.

One important consideration for use of a checklist is that it provides a consistent approach to conducting a review and begins the process of documentation for action planning. No matter how simple the operation is, without a checklist you are depending on memory and the mood of the individual doing the review.

**Table 2-1**
**Important Elements of Site Reviews**

The following are important elements of regular site reviews and show how critical follow-up reviews are to the process. It is important that reviewers are trained to recognize hazards and to bring fresh ideas to the process.

- Reviews should cover all areas of the workplace, targeting high-hazard operations.
- Reviews should be conducted at regular intervals. The frequency will depend on the size of the workplace and the nature of the risks and hazards.
- Use JHAs to review the steps and tasks that are being completed and ensure that the correct tools, equipment, materials, etc. are in use.

Information from reviews should be used to expand the hazard inventory, modify/update JHAs and/or improve the hazard prevention and control program.

OSHA Website, http://www.osha-slc.gov/SLTC/safetyhealth_ecat/mod4_factsheets.htm#, public domain

## 2.4.2 Verification Reviews

Verification is an essential action that must be included in your inspection process. Verification review is performed to ensure that issues identified in a previous review have been appropriately completed.

Unfortunately, verification reviews are needed to prevent "pencil whipping" of the inspection process. The phrase "to pencil whip" means to fill out the report without actually doing the inspection just to get management or the safety department out of your hair. If the perception of risk is low and a serious loss or injury has not occurred, even though the risk may be high, employees and management alike have been known to "pencil whip" a report or inspection in the belief that a loss event cannot happen since it has never happened directly to them.

In addition, these reviews ensure that the management systems are in place and that the corrective actions have been implemented and maintained. However, the scope of this type of review is limited to previously identified issues.

Refer to Figure 2-2 for a flow diagram of a risk toolkit. In addition, review Figure 1-1 in Chapter 1, for an overview of hazard identification.

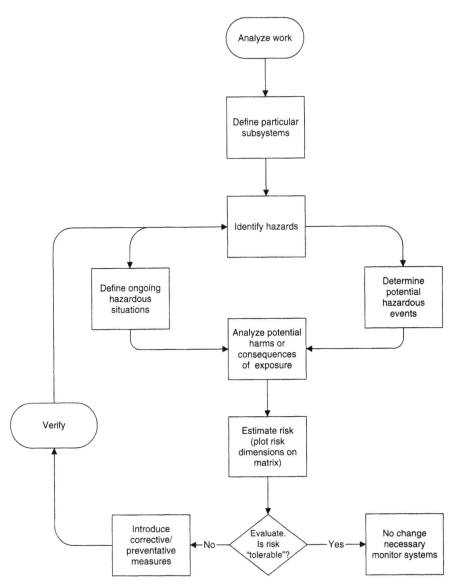

**Figure 2-2** Flow Diagram of the Risk Toolkit

Adapted from Cox Sue, Tom Cox, *Safety Systems and People*, Butterworth-Heinemann, First Published, 1996, Figure 2.6, Flowchart of the Risk Toolkit, pp. 41

## 2.4.3 Focus Reviews

Focus reviews are conducted in conjunction with a comprehensive or verification review, a result of a significant risk or hazard finding, or an enforcement action by a regulatory agency. The purpose of a focus review is to concentrate on a particular process or audit component that needs greater assistance in implementing a sustainable hazard control. The JHA is used to target job steps and tasks where the hazards of specific actions or elements pose the greatest risk.

## 2.4.4 Self-Assessment

Self-assessments review the activities of management and employees or the operation of equipment and process to ensure continued compliance with applicable audit components. At a minimum, the self-assessment should cover administrative management, baseline safety and health assessment, as well as specialized operations and industry best practices that may impact the facility.

## 2.4.5 Document Review

Documents and records are inspected to verify that what you said you are going to document has been documented, completed, properly stored and secured. The document review is not a hazard assessment, only the determination that policy statements, safety policies and/or procedures, training, recordkeeping, JHAs, etc. are being used as intended. One problem we have found is the failure to develop an organized filing system, either hardcopy or electronic, that allows desired documents to be properly stored for use and retrieval. If documents are not completely filled out, are missing, or obviously are not meeting your program guidelines and standards, there is a strong probability that these shortcomings reflect similar shortcomings and issues within general operations, such as in quality, security, or general administrative areas.

## 2.4.6 Written Inspection Reports

Written inspection reports are necessary to record hazards identified, responsibility assigned for correction, and tracking of corrective actions to completion. A well-designed document will help to ensure that:

- Responsible individuals are assigned to make sure that the hazard is corrected in a timely manner.
- Methods of tracking corrective actions to completion are implemented.
- Problems in the hazard control system are identified when the same type of hazard continues to occur after an action plan is completed and verified.
- Problems are identified in the accountability system.
- Hazards where no prevention or controls have been planned are identified.

The administrative system must ensure that documented records are used and someone knowledgeable in the safety process reads and analyzes them and begins the action plan. Documenting hazards or risks and then not prioritizing them for severity and potential frequency leads to charges of negligence, regulatory fines and lawsuit should a severe loss event happen [4, 5].

Reviews of any type can only be effective if a level of trust exists between all the participants, management, supervision, and employees. They must not be viewed as an adversarial experience or an attempt at fault finding. Rather, they should be viewed as a means for mutual improvement; understanding how the management system is working is critical to any valid review [6].

## 2.5 WHO SHOULD REVIEW THE WORKPLACE?

Remember the old saying, "You can't see the forest for the trees"? Those who work in an area may not see hazards as they gradually develop and associated risk may become an expected way of life whenever supervision and employees become complacent. It is always good to have cross-functional reviews conducted, having supervisors and employees from one area review another area. This may also reduce the potential for "pencil whipping," as discussed previously.

An array of employees with specific skills and perspectives should be involved in reviewing the workplace. The following sections will provide a brief overview of some of those responsibilities [4, 5].

### 2.5.1 Supervisors

Supervisors are responsible for reviewing their work areas at the beginning of each shift or day of operation. A daily shift review is essential when multiple shifts or departments use the same work area and equipment or when after-hours maintenance and cleaning are routinely conducted. Supervisors must be given responsibility and held accountable through their formal performance or merit reviews for the completion and documentation of the safety process [6].

All supervisors must have hazard recognition training and training in how to control identified hazards. When supervision is responsible for area inspections, they must have specific training in how to inspect. A frequent major issue is in knowing how to match the inspection criteria with the observations during an inspection; supervision may or may not have the skills needed to document and manage the process. Formal course work may not be necessary, but the training should be provided by someone who is qualified with experience and the skills to provide the necessary information [4, 5].

In addition, supervision must be trained in developing and monitoring the corrective action system. Ongoing monitoring of work environment allows the supervisor the opportunity to correct safety concerns before serious issues develop [1].

## 2.5.2 Employees

Employee participation uses the knowledge that your employees have gained from their continued involvement and experience with equipment, materials, and the process. As discussed, if outside consultants or individuals are conducting a review, you should encourage your employees to communicate openly with whoever is doing the review [1, 4, 5].

An inspection review process is one of the best ways to solicit employees' knowledge of the safety process. By allowing employees to participate in a facility walk-through, the awareness, concern for employees and methods to resolve problems can be conveyed. Employees can be involved in safety committees to conduct routine inspections. Implementing these methods:

- Expands the number of eyes looking at the facility to improve hazard recognition and identification.
- Increases employee awareness to hazardous situations as a well as the appropriate control methods [1, 4, 5].

The key is that all employees should have training in the hazards that they may be exposed to when performing their work activities. As with supervisors, when employees are responsible for reviews or inspections, they should have specific training in how to inspect. On-the-job training with an experienced and trained supervisor can assure that adequate operational safety education with understanding of consequences of exposure is maintained [1, 4, 5].

## 2.5.3 Safety Professionals

The safety professional can act as a mentor and provide guidance on methods to conduct inspections. A safety professional's guidance can expand and improve the lessons learned based on their knowledge from prior reviews, change analysis, and routine hazard analysis [1, 4, 5].

# 2.6 PREVENTIVE MAINTENANCE PROGRAMS

Many managers may not associate preventive maintenance with the safety process. However, good preventive maintenance plays a major role in ensuring that hazard controls continue to function effectively. Preventive maintenance helps keep new hazards from occurring due to equipment malfunction or product defect [1]. Refer to Table 2-2 for an overview of preventive maintenance [5].

Preventive maintenance records can assist you in targeting areas of potential high risk. Records concerning equipment or facilities that have recurring problems, delayed repair, bypassed systems, etc. are an excellent source for identifying jobs to analyze and inspect. By paralleling your inspections with scheduled preventive maintenance activity, knowledge of maintenance problems that impact safety can be developed [1, 5]. Review maintenance records periodically to understand how much reactive maintenance (repair or replacement of defective parts after failure (breakdown maintenance)) has been done. Poor maintenance and poor housekeeping are direct indicators that things are amiss.

A preventive maintenance program should include a workplace survey to identify all equipment or processes that may require routine maintenance.

### Table 2-2
### Preventive Maintenance

| |
|---|
| Preventive maintenance programs play an essential and major role in ensuring that the hazard control system continues to function properly. |
| It helps keep new hazards from occurring due to equipment malfunction. |
| Reliable scheduling and documentation of maintenance activity is necessary. Scheduling depends on knowledge of what operational elements need maintenance and how often. |
| The point of preventive maintenance is to be proactive and take action before breakdown and provide repairs or replacement before a problem develops. |
| Documentation is not just a good idea, but is a necessity as quality, best practice and regulatory standards may require that preventive maintenance be effectively completed. |

Adapted from the OSHA Website http://www.osha-slc.gov/SLTC/safetyhealth_ecat/comp3.htm#SafeWork Practices, public domain, Roughton, James, J. Mercurio, *Developing an Effective Safety Culture: A Leadership Approach*, Butterworth-Heinemann, 2002, Chapter 11, pp. 195, public domain

To be effective, maintenance should be performed on a routine basis and as recommended by the process designer and equipment manufacturer [2, 5].

The JHA process can be used in the preventive maintenance program. Since every piece of equipment or part of a system needs maintenance (mechanical, electrical, oiling, cleaning, testing, replacement of worn parts, evaluation of parts, etc.), the steps and task used in maintenance should be part of the JHA process. You will need a complete list of all items that are to be maintained. If this list does not exist, work with the maintenance manager to conduct a survey and develop the list. The survey should be repeated periodically to update the list of maintenance items. Whenever new equipment is installed in the workplace, the list and associated JHAs should be revised and updated accordingly [1, 5].

After the record review, make new estimates of average time requirements, and work with the maintenance staff to adjust your maintenance timetable accordingly [1, 5].

## 2.7 OTHER THINGS TO CONSIDER DURING A SITE INSPECTION

After hazards are identified and controls are put in place, other analysis tools can help ensure that controls stay in place and new hazards do not appear. These tools are used to collect information and analyze it to see if the management system is working:

- Incident investigations.
- Injury and illness trend analysis [2].

### 2.7.1 Incident Investigations

Incident investigations are used to identify and uncover hazards that were missed or created when a process or operation has slipped out of control. Many benefits can be derived when the investigation process is used as a positive problem-solving tool and focuses on finding the root cause, not placing blame. You should also attempt to gather information on "near-misses" as, given any slight change in time or position, injury or damage could have been the result from the same set of circumstances.

As you continue through this book, you will see how the JHA process can aid in the review of an incident by providing a means to beak the job apart and determine how the interactions of the steps and tasks may have created the conditions that led to the loss-producing event. Using the traditional six "primary" questions remains the foundation for investigations, as with problem solving: Who, What, When, Where, Why, and How. When used in conjunction with a JHA, a wealth of information about the situation, behavior of the employees, and/or other conditions is collected [2].

## 2.7.2 Trend Analysis

The analysis of trends over time provides for the determination of the effectiveness of your safety process. Are losses decreasing? Are they most likely to reoccur and at what possible intervals? Are they fluctuating as might be expected? Knowledge of basic statistics and quality control is essential to help quantify safety data. Without data that is properly analyzed, much of your efforts will be based on opinion and guesswork. The following areas can be used to analyze the management system:

- Injury and illness records analysis will help to determine patterns with common causes. A review of the injury and illness report forms is the most common source for injury and damage pattern analysis. Sources of data include regulatory forms such as the OSHA 300 logs, insurance claims reports, first aid logs, and internal damage reports. Quality control and maintenance work order reports on defects and problems are a further source for risk and hazard data.
- You must have enough data to see a pattern emerge. A smaller facility may require a review of 3–5 years of records. Larger sites may find useful trends in yearly, quarterly, monthly, or even weekly data. Drilling down by sorting the database by categories and using charts, along with basic statistical tools to identify the variations and trends, provides for the business case study you need to convince management of the need for change.
- When analyzing injury and illness records, look for patterns of similar injuries, illnesses, and damage. If the job elements are similar, these will generally indicate a similar lack of hazard specific controls. Look for similarities where injuries occurred, parts of the body injured, types of work being performed, time of day, types of equipment, etc.

Analysis of Other Records:

- Repeat hazards uncovered by workplace reviews, like repeat injuries, mean that management and physical controls are not working properly. Patterns in hazard identification records can show up over a shorter period of time than incidents. Upgrading a control may involve something as basic as improving communication or accountability [2].

## 2.8 SUMMARY

In this chapter we defined the term "inspection" as looking closely at something to understand how and if it meets specific defined standards or requirements. We introduced the term "regular site inspection" as a general inspection of every part of the workplace to locate hazards that need correction. To be effective, you must consider going beyond regulatory requirements, as these are generally only minimum standards.

Hazards found during the workplace analysis should be analyzed to determine failures in the safety system. System failures should be corrected to ensure that similar hazards do not reoccur [2].

A methodical inspection will follow a checklist based on the hazard inventory of your facility that has been identified through various methods. This should include preventive actions and controls measures designed to reduce or eliminate employee exposure to specific hazards. Regular site inspections and JHA reviews should be designed to evaluate each control to make sure that hazards are contained [5].

Preventive maintenance can be a complicated matrix of timing and activity. Keeping track of completed maintenance tasks can be as simple as adding a date and initials to a posted work schedule. Use of a computer to keep track of completed maintenance activity is now relatively easy and routine [2].

Records must be maintained for all programs, inspections, corrections, and preventative maintenance that involve safety issues. Documentation provides a means of tracking and checking program effectiveness when combined with a facility review. It can also serve as a way of identifying and recognizing employees whose efforts to keep a safe workplace have been instrumental in preventing costly repairs and injuries [3].

In the next chapter we continue the JHA process by outlining how an effective JHA management system allows an organization to conduct continual analysis of the workplace. The system anticipates changes that may be needed to modify or develop safety policies and procedures regarding new, existing or reoccurring hazards. We will demonstrate how the JHA provides the format that allows an organization to determine the variety of job elements required to complete a task and the conditions needed for its safe completion. We will discuss in more detail how the JHA is a toolkit of techniques and how to use various workplace hazard review methods.

## CHAPTER REVIEW QUESTIONS

1. Why is the JHA considered to be a centerpiece of your safety program?
2. What are some of the tools that can be used to analyze the workplace?
3. What are the four main parts of a worksite analysis?
4. What are the differences between inspections and audits?
5. Why is it sometimes necessary to use consultant and outside specialist surveys?
6. What is change analysis?
7. When is change analysis useful?
8. Why is it important to learn techniques for interviewing employees?
9. What is a general walk-around inspection?
10. What are important elements concerning site reviews?

## REFERENCES

1. U.S. Department of Labor, Office of Cooperative Programs, Occupational Safety and Administration (OSHA), *Managing Employees' Safety and Health*, November 1994, public domain
2. OSHA Web Site, http://www.osha-slc.gov/SLTC/etools/safetyhealth/index.html, public domain
3. Oregon Website, Hazard Identification and Control, OR-OSHA 104, http://www.cbs.state.or.us/osha/educate/training/pages/courses.html, public domain
4. Missouri Department of Labor, *Establishing Complete Hazard Inventories*, Chapter 7, http://www.dolir.mo.gov/ls/safetyconsultation/ccp/, public domain
5. Missouri Department of Labor, *Catching the Hazard That Escapes Control*, Chapter 9, http://www.dolir.mo.gov/ls/safetyconsultation/ccp/, public domain
6. Roughton, James, James Mercurio, *Developing an Effective Safety Culture: A Leadership Approach*, Butterworth-Heinemann, 2002

# Appendix E

## E.1 SELF-INSPECTION CHECKLIST

This checklist is by no means all-inclusive. Add or delete items that apply to your facility; however, carefully consider each item as you come to it and then make your decision. You also will need to refer to OSHA standards for complete and specific standards that may apply to your work condition. (**NOTE**: This checklist is typical for general industry but not intended for construction or maritime.)

**Sample Checklist**

| EMPLOYER POSTING | Yes | No |
|---|---|---|
| Is the required OSHA workplace poster displayed in a prominent location where all employees are likely to see it? | | |
| Are emergency telephone numbers posted where they can be readily found in case of emergency? | | |
| Where employees may be exposed to any toxic substances or harmful physical agents, has appropriate information concerning employee access to medical and exposure records and Material Safety Data Sheets been posted or otherwise made readily available to affected employees? | | |
| Are signs concerning exiting from buildings, room capacities, floor loading, biohazards, exposures to x-ray, microwave, or other harmful radiation or substances posted where appropriate? | | |
| Is the Annual Summary of the OSHA 300 Log using the form OSHA 300-A posted from February 1 until April 30 of the year in which it was posted. | | |

| RECORDKEEPING | Yes | No |
|---|---|---|
| Is all occupational injury or illness, except minor injuries requiring only first aid, being recorded as required on the OSHA log? | | |
| Are employee medical records and records of employee exposure to hazardous substances or harmful physical agents up-to-date and in compliance with current OSHA standards? | | |
| Are employee training records kept and accessible for review by employees, when required by OSHA standards? | | |
| Have arrangements been made to maintain required records for the legal period of time for each specific type record? (Some records must be maintained for at least 40 years.) | | |
| Are operating permits and records up-to-date for such items as elevators, air pressure tanks, and liquefied petroleum gas tanks? | | |

| SAFETY AND HEALTH PROGRAM | Yes | No |
|---|---|---|
| Do you have an active safety and health program in operation that deals with general safety and health program elements as well as the management of hazards specific to your worksite? | | |
| Is one person clearly responsible for the overall activities of the safety and health program? Do you have a safety committee or group made up of management and labor representatives that meets regularly and reports in writing on its activities? | | |
| Do you have a working procedure for handling in-house employee complaints regarding safety and health? | | |
| Are you keeping your employees advised of the successful effort and accomplishments you and/or your safety committee have made in assuring they will have a workplace that is safe and healthful? | | |
| Have you considered incentives for employees or workgroups who have excelled in safety activities to help reduce workplace injury/illnesses? | | |

| MEDICAL SERVICES AND FIRST AID | Yes | No |
|---|---|---|
| Is there a hospital, clinic, or infirmary for medical care in proximity to your workplace? | | |
| If medical and first aid facilities are not in proximity to your workplace, is at least one employee on each shift currently qualified to render first aid? | | |
| Have all employees who are expected to respond to medical emergencies as part of their work* (1) received first-aid training; (2) had hepatitis B vaccination made available to them; (3) had appropriate training on procedures to protect them from bloodborne pathogens, including universal precautions; and (4) have available and understand how to use appropriate personal protective equipment to protect against exposure to bloodborne diseases? | | |
| Where employees have had an exposure involving bloodborne pathogens, did you provide an immediate post-exposure medical evaluation and follow-up? | | |
| Are medical personnel readily available for advice and consultation on matters of employees' health? | | |
| Are emergency phone numbers posted? | | |
| Are first-aid kits easily accessible to each work area, with necessary supplies available, periodically inspected and replenished as needed? | | |
| Have first-aid kit supplies been approved by a physician, indicating that they are adequate for a particular area or operation? | | |
| Are means provided for quick drenching or flushing of the eyes and body in areas where corrosive liquids or materials are handled? | | |

| FIRE PROTECTION | Yes | No |
|---|---|---|
| Is your local fire department well acquainted with your facilities, their location and specific hazards? | | |
| If you have a fire alarm system, is it certified as required? | | |

| | | |
|---|---|---|
| If you have a fire alarm system, is it tested at least annually? | | |
| If you have interior standpipes and valves, are they inspected regularly? | | |
| If you have outside private fire hydrants, are they flushed at least once a year and on a routine preventive maintenance schedule? | | |
| Are fire doors and shutters in good operating condition? | | |
| Are fire doors and shutters unobstructed and protected against obstructions, including their counterweights? | | |
| Are fire door and shutter fusible links in place? | | |
| Are automatic sprinkler system water control valves, air and water pressure checked weekly/periodically as required? | | |
| Is the maintenance of automatic sprinkler systems assigned to responsible persons or to a sprinkler contractor? | | |
| Are sprinkler heads protected by metal guards, when exposed to physical damage? | | |
| Is proper clearance maintained below sprinkler heads? | | |
| Are portable fire extinguishers provided in adequate number and type? | | |
| Are fire extinguishers mounted in readily accessible locations? | | |
| Are fire extinguishers recharged regularly and noted on the inspection tag? | | |
| Are employees periodically instructed in the use of extinguishers and fire protection procedures? | | |

| PERSONAL PROTECTIVE EQUIPMENT AND CLOTHING | Yes | No |
|---|---|---|
| Are employers assessing the workplace to determine if hazards that require the use of personal protective equipment (e.g., head, eye, face, hand, or foot protection) are present or are likely to be present? | | |

| | | |
|---|---|---|
| *Pursuant to an OSHA memorandum of July 1, 1992, employees who render first aid only as a collateral duty do not have to be offered pre-exposure hepatitis B vaccine only if the employer puts the following requirements into his/her exposure control plan and implements them: (1) the employer must record all first-aid incidents involving the presence of blood or other potentially infectious materials before the end of the work shift during which the first-aid incident occurred; (2) the employer must comply with post-exposure evaluation, prophylaxis, and follow-up requirements of the standard with respect to "exposure incidents," as defined by the standard; (3) the employer must train designated first-aid providers about the reporting procedure; and (4) the employer must offer to initiate the hepatitis B vaccination series within 24 hours to all unvaccinated first-aid providers who have rendered assistance in any situation involving the presence of blood or other potentially infectious materials. | | |
| If hazards or the likelihood of hazards are found, are employers selecting and having affected employees use properly fitted personal protective equipment suitable for protection from these hazards? | | |
| Has the employer been trained on PPE procedures, i.e., what PPE is necessary for a job task, when they need it, and how to properly adjust it? | | |
| Are protective goggles or face shields provided and worn where there is any danger of flying particles or corrosive materials? | | |
| Are approved safety glasses required to be worn at all times in areas where there is a risk of eye injuries such as punctures, abrasions, contusions or burns? | | |
| Are employees who need corrective lenses (glasses or contacts) in working environments having harmful exposures, required to wear *only* approved safety glasses, protective goggles, or use other medically approved precautionary procedures? | | |
| Are protective gloves, aprons, shields, or other means provided and required where employees could be cut or where there is reasonably anticipated exposure to corrosive liquids, chemicals, blood, or other potentially infectious materials? See 29 CFR 1910.1030(b) for the definition of "other potentially infectious materials." | | |

| | | |
|---|---|---|
| Are hard hats provided and worn where danger of falling objects exists? | | |
| Are hard hats inspected periodically for damage to the shell and suspension system? | | |
| Is appropriate foot protection required where there is the risk of foot injuries from hot, corrosive, poisonous substances, falling objects, crushing or penetrating actions? | | |
| Are approved respirators provided for regular or emergency use where needed? | | |
| Is all protective equipment maintained in a sanitary condition and ready for use? | | |
| Do you have eye wash facilities and a quick Drench Shower within the work area where employees are exposed to injurious corrosive materials? | | |
| Where special equipment is needed for electrical workers, is it available? | | |
| Where food or beverages are consumed on the premises, are they consumed in areas where there is no exposure to toxic material, blood, or other potentially infectious materials? | | |
| Is protection against the effects of occupational noise exposure provided when sound levels exceed those of the OSHA noise standard? | | |
| Are adequate work procedures, protective clothing and equipment provided and used when cleaning up spilled toxic or otherwise hazardous materials or liquids? | | |
| Are there appropriate procedures in place for disposing of or decontaminating personal protective equipment contaminated with, or reasonably anticipated to be contaminated with, blood or other potentially infectious materials? | | |

| **GENERAL WORK ENVIRONMENT** | Yes | No |
|---|---|---|
| Are all worksites clean, sanitary, and orderly? | | |
| Are work surfaces kept dry or appropriate means taken to assure the surfaces are slip-resistant? | | |

| | | |
|---|---|---|
| Are all spilled hazardous materials or liquids, including blood and other potentially infectious materials, cleaned up immediately and according to proper procedures? | | |
| Is combustible scrap, debris and waste stored safely and removed from the worksite promptly? Is all regulated waste, as defined in the OSHA blood borne pathogens standard (29 CFR 1910.1030), discarded according to federal, state, and local regulations? | | |
| Are accumulations of combustible dust routinely removed from elevated surfaces including the overhead structure of buildings, etc.? | | |
| Is combustible dust cleaned up with a vacuum system to prevent the dust going into suspension? | | |
| Is metallic or conductive dust prevented from entering or accumulating on or around electrical enclosures or equipment? | | |
| Are covered metal waste cans used for oily and paint soaked waste? | | |
| Are all oil and gas fired devices equipped with flame failure controls that will prevent flow of fuel if pilots or main burners are not working? | | |
| Are paint spray booths, dip tanks, etc., cleaned regularly? | | |
| Are the minimum number of toilets and washing facilities provided? | | |
| Are all toilets and washing facilities clean and sanitary? | | |
| Are all work areas adequately illuminated? | | |
| Are pits and floor openings covered or otherwise guarded? | | |
| Have all confined spaces been evaluated for compliance with 29 CFR 1910.146? | | |

| **WALKWAYS** | Yes | No |
|---|---|---|
| Are aisles and passageways kept clear? | | |
| Are aisles and walkways marked as appropriate? | | |
| Are wet surfaces covered with non-slip materials? | | |

| | | |
|---|---|---|
| Are holes in the floor, sidewalk or other walking surface repaired properly, covered or otherwise made safe? | | |
| Is there safe clearance for walking in aisles where motorized or mechanical handling equipment is operating? | | |
| Are materials or equipment stored in such a way that sharp projectives will not interfere with the walkway? | | |
| Are spilled materials cleaned up immediately? | | |
| Are changes of direction or elevations readily identifiable? | | |
| Are aisles or walkways that pass near moving or operating machinery, welding operations or similar operations arranged so employees will not be subjected to potential hazards? | | |
| Is adequate headroom provided for the entire length of any aisle or walkway? | | |
| Are standard guardrails provided wherever aisle or walkway surfaces are elevated more than 30 inches (76.20 centimeters) above any adjacent floor or the ground? | | |
| Are bridges provided over conveyors and similar hazards? | | |

| **FLOOR AND WALL OPENINGS** | Yes | No |
|---|---|---|
| Are floor openings guarded by a cover, a guardrail, or equivalent on all sides (except at entrance to stairways or ladders)? | | |
| Are toeboards installed around the edges of permanent floor opening (where persons may pass below the opening)? | | |
| Are skylight screens of such construction and mounting that they will withstand a load of at least 200 pounds (90 kilograms)? | | |
| Is the glass in the windows, doors, glass walls, etc., which are subject to human impact, of sufficient thickness and type for the condition of use? | | |
| Are grates or similar type covers over floor openings such as floor drains of such design that foot traffic or rolling equipment will not be affected by the grate spacing? | | |

| | | |
|---|---|---|
| Are unused portions of service pits and pits not actually in use either covered or protected by guardrails or equivalent? | | |
| Are manhole covers, trench covers and similar covers, plus their supports designed to carry a truck rear axle load of at least 20,000 pounds (9000 kilograms) when located in roadways and subject to vehicle traffic? | | |
| Are floor or wall openings in fire resistive construction provided with doors or covers compatible with the fire rating of the structure and provided with a self-closing feature when appropriate? | | |

| STAIRS AND STAIRWAYS | Yes | No |
|---|---|---|
| Are standard stair rails or handrails on all stairways having four or more risers? | | |
| Are all stairways at least 22 inches (55.88 centimeters) wide? | | |
| Do stairs have landing platforms not less than 30 inches (76.20 centimeters) in the direction of travel and extend 22 inches (55.88 centimeters) in width at every 12 feet (3.6576 meters) or less of vertical rise? | | |
| Do stairs angle no more than 50 and no less than 30 degrees? | | |
| Are stairs of hollow-pan type treads and landings filled to the top edge of the pan with solid material? | | |
| Are step risers on stairs uniform from top to bottom? | | |
| Are steps on stairs and stairways designed or provided with a surface that renders them slip resistant? | | |
| Are stairway handrails located between 30 (76.20 centimeters) and 34 inches (86.36 centimeters) above the leading edge of stair treads? | | |
| Do stairway handrails have at least 3 inches (7.62 centimeters) of clearance between the handrails and the wall or surface they are mounted on? | | |

| | | |
|---|---|---|
| Where doors or gates open directly on a stairway, is there a platform provided so the swing of the door does not reduce the width of the platform to less than 21 inches (53.34 centimeters)? | | |
| Are stairway handrails capable of withstanding a load of 200 pounds (90 kilograms), applied within 2 inches (5.08 centimeters) of the top edge, in any downward or outward direction? | | |
| Where stairs or stairways exit directly into any area where vehicles may be operated, are adequate barriers and warnings provided to prevent employees stepping into the path of traffic? | | |
| Do stairway landings have a dimension measured in the direction of travel, at least equal to the width of the stairway? | | |
| Is the vertical distance between stairway landings limited to 12 feet (3.6576 centimeters) or less? | | |

| ELEVATED SURFACES | Yes | No |
|---|---|---|
| Are signs posted, when appropriate, showing the elevated surface load capacity? | | |
| Are surfaces elevated more than 30 inches (76.20 centimeters) above the floor or ground provided with standard guardrails? | | |
| Are all elevated surfaces (beneath which people or machinery could be exposed to falling objects) provided with standard 4-inch (10.16 centimeters) toeboards? | | |
| Is a permanent means of access and egress provided to elevated storage and work surfaces? | | |
| Is required headroom provided where necessary? | | |
| Is material on elevated surfaces piled, stacked or racked in a manner to prevent it from tipping, falling, collapsing, rolling or spreading? | | |
| Are dock boards or bridge plates used when transferring materials between docks and trucks or rail cars? | | |

| EXITING OR EGRESS | Yes | No |
|---|---|---|
| Are all exits marked with an exit sign and illuminated by a reliable light source? | | |
| Are the directions to exits, when not immediately apparent, marked with visible signs? | | |
| Are doors, passageways or stairways, that are neither exits nor access to exits, and which could be mistaken for exits, appropriately marked "NOT AN EXIT," "TO BASEMENT," "STOREROOM," etc.? | | |
| Are exit signs provided with the word "EXIT" in lettering at least 5 inches (12.70 centimeters) high and the stroke of the lettering at least 1/2-inch (1.2700 centimeters) wide? | | |
| Are exit doors side-hinged? | | |
| Are all exits kept free of obstructions? | | |
| Are at least two means of egress provided from elevated platforms, pits or rooms where the absence of a second exit would increase the risk of injury from hot, poisonous, corrosive, suffocating, flammable, or explosive substances? | | |
| Are there sufficient exits to permit prompt escape in case of emergency? | | |
| Are special precautions taken to protect employees during construction and repair operations? | | |
| Is the number of exits from each floor of a building and the number of exits from the building itself, appropriate for the building occupancy load? | | |
| Are exit stairways that are required to be separated from other parts of a building enclosed by at least 2-hour fire-resistive construction in buildings more than four stories in height, and not less than 1-hour fire-resistive constructive elsewhere? | | |
| Where ramps are used as part of required exiting from a building is the ramp slope limited to 1 foot (0.3048 meters) vertical and 12 feet (3.6576 meters) horizontal? | | |
| Where exiting will be through frame less glass doors, glass exit doors, or storm doors are the doors fully tempered and meet the safety requirements for human impact? | | |

| EXIT DOORS | Yes | No |
|---|---|---|
| Are doors that are required to serve as exits designed and constructed so that the way of exit travel is obvious and direct? | | |
| Are windows that could be mistaken for exit doors made inaccessible by means of barriers or railings? | | |
| Are exit doors openable from the direction of exit travel without the use of a key or any special knowledge or effort when the building is occupied? | | |
| Is a revolving, sliding or overhead door prohibited from serving as a required exit door? | | |
| Where panic hardware is installed on a required exit door, will it allow the door to open by applying a force of 15 pounds (6.75 kilograms) or less in the direction of the exit traffic? | | |
| Are doors on cold storage rooms provided with an inside release mechanism that will release the latch and open the door even if it's padlocked or otherwise locked on the outside? | | |
| Where exit doors open directly onto any street, alley or other area where vehicles may be operated, are adequate barriers and warnings provided to prevent employees from stepping into the path of traffic? | | |
| Are doors that swing in both directions and are located between rooms where there is frequent traffic, provided with viewing panels in each door? | | |

| PORTABLE LADDERS | Yes | No |
|---|---|---|
| Are all ladders maintained in good condition, joints between steps and side rails tight, all hardware and fittings securely attached and moveable parts operating freely without binding or undue play? | | |
| Are non-slip safety feet provided on each ladder? | | |

| | | |
|---|---|---|
| Are non-slip safety feet provided on each metal or rung ladder? | | |
| Are ladder rungs and steps free of grease and oil? | | |
| Is it prohibited to place a ladder in front of doors opening toward the ladder except when the door is blocked open, locked or guarded? | | |
| Is it prohibited to place ladders on boxes, barrels, or other unstable bases to obtain additional height? | | |
| Are employees instructed to face the ladder when ascending or descending? | | |
| Are employees prohibited from using ladders that are broken, missing steps, rungs, or cleats, broken side rails or other faulty equipment? | | |
| Are employees instructed not to use the top step of ordinary stepladders as a step? | | |
| When portable rung ladders are used to gain access to elevated platforms, roofs, etc., does the ladder always extend at least 3 feet (0.9144 meters) above the elevated surface? | | |
| Is it required that when portable rung or cleat type ladders are used, the base is so placed that slipping will not occur, or it is lashed or otherwise held in place? | | |
| Are portable metal ladders legibly marked with signs reading "CAUTION – Do Not Use Around Electrical Equipment" or equivalent wording? | | |
| Are employees prohibited from using ladders as guys, braces, skids, gin poles, or for other than their intended purposes? | | |
| Are employees instructed to only adjust extension ladders while standing at a base (not while standing on the ladder or from a position above the ladder)? | | |
| Are metal ladders inspected for damage? | | |
| Are the rungs of ladders uniformly spaced at 12 inches, (30.48 centimeters) center to center? | | |

| HAND TOOLS AND EQUIPMENT | Yes | No |
|---|---|---|
| Are all tools and equipment (both company and employee owned) used by employees at their workplace in good condition? | | |
| Are hand tools such as chisels and punches, which develop mushroomed heads during use, reconditioned or replaced as necessary? | | |
| Are broken or fractured handles on hammers, axes and similar equipment replaced promptly? | | |
| Are worn or bent wrenches replaced regularly? | | |
| Are appropriate handles used on files and similar tools? | | |
| Are employees made aware of the hazards caused by faulty or improperly used hand tools? | | |
| Are appropriate safety glasses, face shields, etc. used while using hand tools or equipment which might produce flying materials or be subject to breakage? | | |
| Are jacks checked periodically to ensure they are in good operating condition? | | |
| Are tool handles wedged tightly in the head of all tools? | | |
| Are tool cutting edges kept sharp so the tool will move smoothly without binding or skipping? | | |
| Are tools stored in dry, secure location where they won't be tampered with? | | |
| Is eye and face protection used when driving hardened or tempered spuds or nails? | | |

| PORTABLE (POWER OPERATED) TOOLS AND EQUIPMENT | Yes | No |
|---|---|---|
| Are grinders, saws and similar equipment provided with appropriate safety guards? | | |
| Are power tools used with the correct shield, guard, or attachment, recommended by the manufacturer? | | |

| | | |
|---|---|---|
| Are portable circular saws equipped with guards above and below the base shoe? | | |
| Are circular saw guards checked to assure they are not wedged up, thus leaving the lower portion of the blade unguarded? | | |
| Are rotating or moving parts of equipment guarded to prevent physical contact? | | |
| Are all cord-connected, electrically operated tools and equipment effectively grounded or of the approved double insulated type? | | |
| Are effective guards in place over belts, pulleys, chains, sprockets, on equipment such as concrete mixers, and air compressors? | | |
| Are portable fans provided with full guards or screens having openings 1/2 inch (1.2700 centimeters) or less? | | |
| Is hoisting equipment available and used for lifting heavy objects, and are hoist ratings and characteristics appropriate for the task? | | |
| Are ground-fault circuit interrupters provided on all temporary electrical 15 and 20-ampere circuits, used during periods of construction? | | |
| Are pneumatic and hydraulic hoses on power operated tools checked regularly for deterioration or damage? | | |

| ABRASIVE WHEEL EQUIPMENT GRINDERS | Yes | No |
|---|---|---|
| Is the work rest used and kept adjusted to within 1/8 inch (0.3175 centimeters) of the wheel? | | |
| Is the adjustable tongue on the topside of the grinder used and kept adjusted to within 1/4 inch (0.6350 centimeters) of the wheel? | | |
| Do side guards cover the spindle, nut, and flange and 75 percent of the wheel diameter? | | |
| Are bench and pedestal grinders permanently mounted? | | |

| | | |
|---|---|---|
| Are goggles or face shields always worn when grinding? | | |
| Is the maximum RPM rating of each abrasive wheel compatible with the RPM rating of the grinder motor? | | |
| Are fixed or permanently mounted grinders connected to their electrical supply system with metallic conduit or other permanent wiring method? | | |
| Does each grinder have an individual on and off control switch? | | |
| Is each electrically operated grinder effectively grounded? | | |
| Before new abrasive wheels are mounted, are they visually inspected and ring tested? | | |
| Are dust collectors and powered exhausts provided on grinders used in operations that produce large amounts of dust? | | |
| Are splashguards mounted on grinders that use coolant to prevent the coolant-reaching employees? | | |
| Is cleanliness maintained around grinders? | | |

| **POWDER-ACTUATED TOOLS** | Yes | No |
|---|---|---|
| Are employees who operate powder-actuated tools trained in their use and carry a valid operator's card? | | |
| Is each powder-actuated tool stored in its own locked container when not being used? | | |
| Is a sign at least 7 inches (17.78 centimeters) by 10 inches (25.40 centimeters) with bold face type reading "POWDER-ACTUATED TOOL IN USE" conspicuously posted when the tool is being used | | |
| Are powder-actuated tools left unloaded until they are actually ready to be used? | | |
| Are powder-actuated tools inspected for obstructions or defects each day before use? | | |
| Do powder-actuated tool operators have and use appropriate personal protective equipment such as hard hats, safety goggles, safety shoes and ear protectors? | | |

| MACHINE GUARDING | Yes | No |
|---|---|---|
| Is there a training program to instruct employees on safe methods of machine operation? | | |
| Is there adequate supervision to ensure that employees are following safe machine operating procedures? | | |
| Is there a regular program of safety inspection of machinery and equipment? | | |
| Is all machinery and equipment kept clean and properly maintained? | | |
| Is sufficient clearance provided around and between machines to allow for safe operations, set up and servicing, material handling and waste removal? | | |
| Is equipment and machinery securely placed and anchored, when necessary to prevent tipping or other movement that could result in personal injury? | | |
| Is there a power shut-off switch within reach of the operator's position at each machine? | | |
| Can electric power to each machine be locked out for maintenance, repair, or security? | | |
| Are the noncurrent-carrying metal parts of electrically operated machines bonded and grounded? Are foot-operated switches guarded or arranged to prevent accidental actuation by personnel or falling objects? | | |
| Are manually operated valves and switches controlling the operation of equipment and machines clearly identified and readily accessible? | | |
| Are all emergency stop buttons colored red? | | |
| Are all pulleys and belts that are within 7 feet (2.1336 meters) of the floor or working level properly guarded? | | |
| Are all moving chains and gears properly guarded? | | |
| Are splashguards mounted on machines that use coolant to prevent the coolant from reaching employees? | | |

| | | |
|---|---|---|
| Are methods provided to protect the operator and other employees in the machine area from hazards created at the point of operation, ingoing nip points, rotating parts, flying chips, and sparks? | | |
| Are machinery guards secure and so arranged that they do not offer a hazard in their use? | | |
| If special handtools are used for placing and removing material, do they protect the operator's hands? | | |
| Are revolving drums, barrels, and containers required to be guarded by an enclosure that is interlocked with the drive mechanism, so that revolution cannot occur unless the guard enclosures is in place, so guarded? | | |
| Do arbors and mandrels have firm and secure bearings and are they free from play? | | |
| Are provisions made to prevent machines from automatically starting when power is restored after a power failure or shutdown? | | |
| Are machines constructed so as to be free from excessive vibration when the largest size tool is mounted and run at full speed? | | |
| If machinery is cleaned with compressed air, is air pressure controlled and personal protective equipment or other safeguards utilized to protect operators and other workers from eye and body injury? | | |
| Are fan blades protected with a guard having openings no larger than 1/2 inch (1.2700 centimeters), when operating within 7 feet (2.1336 meters) of the floor? | | |
| Are saws used for ripping, equipped with anti-kick back devices and spreaders? | | |
| Are radial arm saws so arranged that the cutting head will gently return to the back of the table when released? | | |

| LOCKOUT/TAGOUT PROCEDURES | Yes | No |
|---|---|---|
| Is all machinery or equipment capable of movement, required to be de-energized or disengaged and locked-out during cleaning, servicing, adjusting or setting up operations, whenever required? | | |
| Where the power disconnecting means for equipment does not also disconnect the electrical control circuit: | | |
| Are the appropriate electrical enclosures identified? | | |
| Is means provided to assure the control circuit can also be disconnected and locked-out? | | |
| Is the locking-out of control circuits in lieu of locking-out main power disconnects prohibited? | | |
| Are all equipment control valve handles provided with a means for locking-out? | | |
| Does the lockout procedure require that stored energy (mechanical, hydraulic, air, etc.) be released or blocked before equipment is locked-out for repairs? | | |
| Are appropriate employees provided with individually keyed personal safety locks? | | |
| Are employees required to keep personal control of their key(s) while they have safety locks in use? | | |
| Is it required that only the employee exposed to the hazard place or remove the safety lock? | | |
| Is it required that employees check the safety of the lock-out by attempting a startup after making sure no one is exposed? | | |
| Are employees instructed to always push the control circuit stop button immediately after checking the safety of the lockout? | | |
| Is there a means provided to identify any or all employees who are working on locked-out equipment by their locks or accompanying tags? | | |

| | Yes | No |
|---|---|---|
| Are a sufficient number of accident preventive signs or tags and safety padlocks provided for any reasonably foreseeable repair emergency? | | |
| When machine operations, configuration or size requires the operator to leave his or her control station to install tools or perform other operations, and that part of the machine could move if accidentally activated, is such element required to be separately locked or blocked out? | | |
| In the event that equipment or lines cannot be shut down, locked-out and tagged, is a safe job procedure established and rigidly followed? | | |

| **WELDING, CUTTING AND BRAZING** | Yes | No |
|---|---|---|
| Are only authorized and trained personnel permitted to use welding, cutting or brazing equipment? | | |
| Does each operator have a copy of the appropriate operating instructions and are they directed to follow them? | | |
| Are compressed gas cylinders regularly examined for obvious signs of defects, deep rusting, or leakage? | | |
| Is care used in handling and storing cylinders, safety valves, and relief valves to prevent damage? | | |
| Are precautions taken to prevent the mixture of air or oxygen with flammable gases, except at a burner or in a standard torch? | | |
| Are only approved apparatus (torches, regulators, pressure reducing valves, acetylene generators, and manifolds) used? | | |
| Are cylinders kept away from sources of heat? | | |
| Are the cylinders kept away from elevators, stairs, or gangways? | | |
| Is it prohibited to use cylinders as rollers or supports? | | |
| Are empty cylinders appropriately marked and their valves closed? | | |

| | | |
|---|---|---|
| Are signs reading: DANGER—NO SMOKING, MATCHES, OR OPENLIGHTS, or the equivalent, posted? | | |
| Are cylinders, cylinder valves, couplings, regulators, hoses, and apparatus kept free of oily or greasy substances? | | |
| Is care taken not to drop or strike cylinders? | | |
| Unless secured on special trucks, are regulators removed and valve-protection caps put in place before moving cylinders? | | |
| Do cylinders without fixed hand wheels have keys, handles, or non-adjustable wrenches on stem valves when in service? | | |
| Are liquefied gases stored and shipped valve-end up with valve covers in place? | | |
| Are provisions made to never crack a fuel gas cylinder valve near sources of ignition? | | |
| Before a regulator is removed, is the valve closed and gas released from the regulator? | | |
| Is red used to identify the acetylene (and other fuel gas) hose, green for oxygen hose, and black for inert gas and air hose? | | |
| Are pressure-reducing regulators used only for the gas and pressures for which they are intended? | | |
| Is open circuit (No Load) voltage of arc welding and cutting machines as low as possible and not in excess of the recommended limits? | | |
| Under wet conditions, are automatic controls for reducing no load voltage used? | | |
| Is grounding of the machine frame and safety ground connections of portable machines checked periodically? | | |
| Are electrodes removed from the holders when not in use? | | |
| Is it required that electric power to the welder be shut off when no one is in attendance? | | |
| Is suitable fire extinguishing equipment available for immediate use? | | |
| Is the welder forbidden to coil or loop welding electrode cable around his body? | | |

| | | |
|---|---|---|
| Are wet machines thoroughly dried and tested before being used? | | |
| Are work and electrode lead cables frequently inspected for wear and damage, and replaced when needed? | | |
| Do means for connecting cable lengths have adequate insulation? | | |
| When the object to be welded cannot be moved and fire hazards cannot be removed, are shields used to confine heat, sparks, and slag? | | |
| Are firewatchers assigned when welding or cutting is performed in locations where a serious fire might develop? | | |
| Are combustible floors kept wet, covered by damp sand, or protected by fire-resistant shields? | | |
| When floors are wet down, are personnel protected from possible electrical shock? | | |
| When welding is done on metal walls, are precautions taken to protect combustibles on the other side? | | |
| Before hot work is begun, are used drums, barrels, tanks, and other containers so thoroughly cleaned that no substances remain that could explode, ignite, or produce toxic vapors? | | |
| Is it required that eye protection helmets, hand shields and goggles meet appropriate standards? | | |
| Are employees exposed to the hazards created by welding, cutting, or brazing operations protected with personal protective equipment and clothing? | | |
| Is a check made for adequate ventilation in and where welding or cutting is performed? | | |
| When working in confined places, are environmental monitoring tests taken and means provided for quick removal of welders in case of an emergency? | | |

| COMPRESSORS AND COMPRESSED AIR | Yes | No |
|---|---|---|
| Are compressors equipped with pressure relief valves, and pressure gauges? | | |
| Are compressor air intakes installed and equipped so as to ensure that only clean uncontaminated air enters the compressor? | | |
| Are air filters installed on the compressor intake? | | |
| Are compressors operated and lubricated in accordance with the manufacturer's recommendations? | | |
| Are safety devices on compressed air systems checked frequently? | | |
| Before any repair work is done on the pressure system of a compressor, is the pressure bled off and the system locked-out? | | |
| Are signs posted to warn of the automatic starting feature of the compressors? | | |
| Is the belt drive system totally enclosed to provide protection for the front, back, top, and sides? | | |
| Is it strictly prohibited to direct compressed air towards a person? | | |
| Are employees prohibited from using highly compressed air for cleaning purposes? | | |
| If compressed air is used for cleaning off clothing, is the pressure reduced to less than 10 psi? | | |
| When using compressed air for cleaning, do employees wear protective chip guarding and personal protective equipment? | | |
| Are safety chains or other suitable locking devices used at couplings of high-pressure hose lines where a connection failure would create a hazard? | | |
| Before compressed air is used to empty containers of liquid, is the safe working pressure of the container checked? | | |

| | | |
|---|---|---|
| When compressed air is used with abrasive blast cleaning equipment, is the operating valve a type that must be held open manually? | | |
| When compressed air is used to inflate auto ties, is a clip-on chuck and an inline regulator preset to 40 psi required? | | |
| Is it prohibited to use compressed air to clean up or move combustible dust if such action could cause the dust to be suspended in the air and cause a fire or explosion hazard? | | |

| **COMPRESSORS AIR RECEIVERS** | Yes | No |
|---|---|---|
| Is every receiver equipped with a pressure gauge and with one or more automatic, spring-loaded safety valves? | | |
| Is the total relieving capacity of the safety valve capable of preventing pressure in the receiver from exceeding the maximum allowable working pressure of the receiver by more than 10 percent? | | |
| Is every air receiver provided with a drainpipe and valve at the lowest point for the removal of accumulated oil and water? | | |
| Are compressed air receivers periodically drained of moisture and oil? | | |
| Are all safety valves tested frequently and at regular intervals to determine whether they are in good operating condition? | | |
| Is there a current operating permit used by the Division of Occupational Safety and Health? | | |
| Is the inlet of air receivers and piping systems kept free of accumulated oil and carbonaceous materials? | | |

| **COMPRESSED GAS CYLINDERS** | Yes | No |
|---|---|---|
| Are cylinders with a water weight capacity over 30 pounds (13.5 kilograms), equipped with means for connecting a valve protector device, or with a collar or recess to protect the valve? | | |
| Are cylinders legibly marked to clearly identify the gas contained? | | |

| | | |
|---|---|---|
| Are compressed gas cylinders stored in areas which are protected from external heat sources such as flame impingement, intense radiant heat, electric arcs, or high temperature lines? | | |
| Are cylinders located or stored in areas where they will not be damaged by passing or falling objects or subject to tampering by unauthorized persons? | | |
| Are cylinders stored or transported in a manner to prevent them from creating a hazard by tipping, falling or rolling? | | |
| Are cylinders containing liquefied fuel gas, stored or transported in a position so that the safety relief device is always in direct contact with the vapor space in the cylinder? | | |
| Are valve protectors always placed on cylinders when the cylinders are not in use or connected for use? | | |
| Are all valves closed off before a cylinder is moved, when the cylinder is empty, and at the completion of each job? | | |
| Are low-pressure fuel-gas cylinders checked periodically for corrosion, general distortion, cracks, or any other defect that might indicate a weakness or render it unfit for service? | | |
| Does the periodic check of low-pressure fuel-gas cylinders include a close inspection of the cylinder's bottom? | | |

| **HOIST AND AUXILLARY EQUIPMENT** | Yes | No |
|---|---|---|
| Is each overhead electric hoist equipped with a limit device to stop the hook travel at its highest and lowest point of safe travel? | | |
| Will each hoist automatically stop and hold any load up to 125 percent of its rated load if its actuating force is removed? | | |
| Is the rated load of each hoist legibly marked and visible to the operator? | | |
| Are stops provided at the safe limits of travel for trolley hoist? | | |
| Are the controls of hoist plainly marked to indicate the direction of travel or motion? | | |

| | | |
|---|---|---|
| Is each cage-controlled hoist equipped with an effective warning device? | | |
| Are close-fitting guards or other suitable devices installed on hoist to assure hoist ropes will be maintained in the sheave groves? | | |
| Are all hoist chains or ropes of sufficient length to handle the full range of movement of the application while still maintaining two full wraps on the drum at all times? | | |
| Are nip points or contact points between hoist ropes and sheaves which are permanently located within 7 feet (2.1336 meters) of the floor, ground or working platform, guarded? | | |
| Is it prohibited to use chains or rope slings that are kinked or twisted? | | |
| Is it prohibited to use the hoist rope or chain wrapped around the load as a substitute, for a sling? | | |
| Is the operator instructed to avoid carrying loads over people? | | |

| **INDUSTRIAL TRUCKS—FORKLIFTS** | Yes | No |
|---|---|---|
| Are only employees who have been trained in the proper use of hoists allowed to operate them? Are only trained personnel allowed to operate industrial trucks? | | |
| Is substantial overhead protective equipment provided on high lift rider equipment? Are the required lift truck operating rules posted and enforced? | | |
| Is directional lighting provided on each industrial truck that operates in an area with less than 2 foot-candles per square foot of general lighting? | | |
| Does each industrial truck have a warning horn, whistle, gong, or other device which can be clearly heard above the normal noise in the areas where operated? | | |
| Are the brakes on each industrial truck capable of bringing the vehicle to a complete and safe stop when fully loaded? | | |
| Will the industrial trucks' parking brake effectively prevent the vehicle from moving when unattended? | | |

| | | |
|---|---|---|
| Are industrial trucks operating in areas where flammable gases or vapors, or combustible dust or ignitable fibers may be present in the atmosphere, approved for such locations? | | |
| Are motorized hand and hand/rider trucks so designed that the brakes are applied, and power to the drive motor shuts off when the operator releases his or her grip on the device that controls the travel? | | |
| Are industrial trucks with internal combustion engine, operated in buildings or enclosed areas, carefully checked to ensure such operations do not cause harmful concentration of dangerous gases or fumes? | | |
| Are powered industrial trucks being safely operated? | | |

| **SPRAYING OPERATIONS** | Yes | No |
|---|---|---|
| Is adequate ventilation assured before spray operations are started? | | |
| Is mechanical ventilation provided when spraying operations are done in enclosed areas? | | |
| When mechanical ventilation is provided during spraying operations, is it so arranged that it will not circulate the contaminated air? | | |
| Is the spray area free of hot surfaces? | | |
| Is the spray area at least 20 feet (6.096 meters) from flames, sparks, operating electrical motors and other ignition sources? | | |
| Are portable lamps used to illuminate spray areas suitable for use in a hazardous location? | | |
| Is approved respiratory equipment provided and used when appropriate during spraying operations? | | |
| Do solvents used for cleaning have a flash point to 100 °F or more? | | |
| Are fire control sprinkler heads kept clean? | | |
| Are "NO SMOKING" signs posted in spray areas, paint rooms, paint booths, and paint storage areas? | | |

| | | |
|---|---|---|
| Is the spray area kept clean of combustible residue? | | |
| Are spray booths constructed of metal, masonry, or other substantial noncombustible material? | | |
| Are spray booth floors and baffles noncombustible and easily cleaned? | | |
| Is infrared drying apparatus kept out of the spray area during spraying operations? | | |
| Is the spray booth completely ventilated before using the drying apparatus? | | |
| Is the electric drying apparatus properly grounded? | | |
| Are lighting fixtures for spray booths located outside of the booth and the interior lighted through sealed clear panels? | | |
| Are the electric motors for exhaust fans placed outside booths or ducts? | | |
| Are belts and pulleys inside the booth fully enclosed? | | |
| Do ducts have access doors to allow cleaning? | | |
| Do all drying spaces have adequate ventilation? | | |

| **ENTERING CONFINED SPACES** | Yes | No |
|---|---|---|
| Are confined spaces thoroughly emptied of any corrosive or hazardous substances, such as acids or caustics, before entry? | | |
| Are all lines to a confined space, containing inert, toxic, flammable, or corrosive materials valved off and blanked or disconnected and separated before entry? | | |
| Are all impellers, agitators, or other moving parts and equipment inside confined spaces locked-out if they present a hazard? | | |
| Is either natural or mechanical ventilation provided prior to confined space entry? | | |
| Are appropriate atmospheric tests performed to check for oxygen deficiency, toxic substances and explosive concentrations in the confined space before entry? | | |

| | | |
|---|---|---|
| Is adequate illumination provided for the work to be performed in the confined space? | | |
| Is the atmosphere inside the confined space frequently tested or continuously monitored during conduct of work? | | |
| Is there an assigned safety standby employee outside of the confined space, when required, whose sole responsibility is to watch the work in progress, sound an alarm if necessary, and render assistance? | | |
| Is the standby employee appropriately trained and equipped to handle an emergency? | | |
| Is the standby employee or other employees prohibited from entering the confined space without lifelines and respiratory equipment if there is any question as to the cause of an emergency? | | |
| Is approved respiratory equipment required if the atmosphere inside the confined space cannot be made acceptable? | | |
| Is all portable electrical equipment used inside confined spaces either grounded and insulated, or equipped with ground fault protection? | | |
| Before gas welding or burning is started in a confined space, are hoses checked for leaks, compressed gas bottles forbidden inside of the confined space, torches lightly only outside of the confined area and the confined area tested for an explosive atmosphere each time before a lighted torch is to be taken into the confined space? | | |
| If employees will be using oxygen-consuming equipment—such as salamanders, torches, and furnaces—in a confined space, is sufficient air provided to assure combustion without reducing the oxygen concentration of the atmosphere below 19.5 percent by volume? | | |
| Whenever combustion-type equipment is used in a confined space, are provisions made to ensure the exhaust gases are vented outside of the enclosure? | | |
| Is each confined space checked for decaying vegetation or animal matter that may produce methane? | | |

| | | |
|---|---|---|
| Is the confined space checked for possible industrial waste that could contain toxic properties? | | |
| If the confined space is below the ground and near areas where motor vehicles will be operating, is it possible for vehicle exhaust or carbon monoxide to enter the space? | | |

| ENVIRONMENTAL CONTROLS | Yes | No |
|---|---|---|
| Are all work areas properly illuminated? | | |
| Are employees instructed in proper first aid and other emergency procedures? | | |
| Are hazardous substances, blood, and other potentially infectious materials identified, which may cause harm by inhalation, ingestion, or skin absorption or contact? | | |
| Are employees aware of the hazards involved with the various chemicals they may be exposed to in their work environment, such as ammonia, chlorine, epoxies, caustics, etc.? | | |
| Is employee exposure to chemicals in the workplace kept within acceptable levels? | | |
| Can a less harmful method or process be used? | | |
| Is the work area's ventilation system appropriate for the work being performed? | | |
| Are spray painting operations done in spray rooms or booths equipped with an appropriate exhaust system? | | |
| Is employee exposure to welding fumes controlled by ventilation, use of respirators, exposure time, or other means? | | |
| Are welders and other workers nearby provided with flash shields during welding operations? | | |
| If forklifts and other vehicles are used in buildings or other enclosed areas, are the carbon monoxide levels kept below maximum acceptable concentration? | | |
| Has there been a determination that noise levels in the facilities are within acceptable levels? | | |

| | | |
|---|---|---|
| Are steps being taken to use engineering controls to reduce excessive noise levels? | | |
| Are proper precautions being taken when handling asbestos and other fibrous materials? | | |
| Are caution labels and signs used to warn of hazardous substances (e.g., asbestos) and biohazards (e.g., bloodborne pathogens)? | | |
| Are wet methods used, when practicable, to prevent the emission of airborne asbestos fibers, silica dust and similar hazardous materials? | | |
| Are engineering controls examined and maintained or replaced on a scheduled basis? | | |
| Is vacuuming with appropriate equipment used whenever possible rather than blowing or sweeping dust? | | |
| Are grinders, saws, and other machines that produce respirable dusts vented to an industrial collector or central exhaust system? | | |
| Are all local exhaust ventilation systems designed and operating properly such as airflow and volume necessary for the application, ducts not plugged or belts slipping? | | |
| Is personal protective equipment provided, used and maintained wherever required? | | |
| Are there written standard operating procedures for the selection and use of respirators where needed? | | |
| Are restrooms and washrooms kept clean and sanitary? | | |
| Is all water provided for drinking, washing, and cooking potable? | | |
| Are all outlets for water not suitable for drinking clearly identified? | | |
| Are employees' physical capacities assessed before being assigned to jobs requiring heavy work? | | |
| Are employees instructed in the proper manner of lifting heavy objects? | | |

| | | |
|---|---|---|
| Where heat is a problem, have all fixed work areas been provided with spot cooling or air conditioning? | | |
| Are employees screened before assignment to areas of high heat to determine if their health condition might make them more susceptible to having an adverse reaction? | | |
| Are employees working on streets and roadways where they are exposed to the hazards of traffic, required to wear bright colored (traffic orange) warning vests? | | |
| Are exhaust stacks and air intakes so located that contaminated air will not be recirculated within a building or other enclosed area? | | |
| Is equipment producing ultraviolet radiation properly shielded? | | |
| Are universal precautions observed where occupational exposure to blood or other potentially infectious materials can occur and in all instances where differentiation of types of body fluids or potentially infectious materials is difficult or impossible? | | |

| **FLAMMABLE AND COMBUSTIBLE MATERIALS** | Yes | No |
|---|---|---|
| Are combustible scrap, debris, and waste materials (oily rags, etc.) stored in covered metal receptacles and removed from the worksite promptly? | | |
| Is proper storage practiced to minimize the risk of fire including spontaneous combustion? | | |
| Are approved containers and tanks used for the storage and handling of flammable and combustible liquids? | | |
| Are all connections on drums and combustible liquid piping, vapor and liquid tight? | | |
| Are all flammable liquids kept in closed containers when not in use (e.g., parts cleaning tanks, pans, etc.)? | | |
| Are bulk drums of flammable liquids grounded and bonded to containers during dispensing? | | |
| Do storage rooms for flammable and combustible liquids have explosion-proof lights? | | |

| | | |
|---|---|---|
| Do storage rooms for flammable and combustible liquids have mechanical or gravity ventilation? | | |
| Is liquefied petroleum gas stored, handled, and used in accordance with safe practices and standards? | | |
| Are "NO SMOKING" signs posted on liquified petroleum gas tanks? | | |
| Are liquified petroleum storage tanks guarded to prevent damage from vehicles? | | |
| Are all solvent wastes, and flammable liquids kept in fire-resistant, covered containers until they are removed from the worksite? | | |
| Is vacuuming used whenever possible rather than blowing or sweeping combustible dust? | | |
| Are firm separators placed between containers of combustibles or flammables, when stacked one upon another, to assure their support and stability? | | |
| Are fuel gas cylinders and oxygen cylinders separated by distance, and fire-resistant barriers, while in storage? | | |
| Are fire extinguishers selected and provided for the types of materials in areas where they are to be used?<br><br>• Class A Ordinary combustible material fires.<br>• Class B Flammable liquid, gas or grease fires.<br>• Class C Energized-electrical equipment fires. | | |
| Are appropriate fire extinguishers mounted within 75 feet (2286 meters) of outside areas containing flammable liquids, and within 10 feet (3.048 meters) of any inside storage area for such materials? | | |
| Are extinguishers free from obstructions or blockage? | | |
| Are all extinguishers serviced, maintained and tagged at intervals not to exceed 1 year? | | |
| Are all extinguishers fully charged and in their designated places? | | |

| | | |
|---|---|---|
| Where sprinkler systems are permanently installed, are the nozzle heads so directed or arranged that water will not be sprayed into operating electrical switchboards and equipment? | | |
| Are "NO SMOKING" signs posted where appropriate in areas where flammable or combustible materials are used or stored? | | |
| Are safety cans used for dispensing flammable or combustible liquids at a point of use? | | |
| Are all spills of flammable or combustible liquids cleaned up promptly? | | |
| Are storage tanks adequately vented to prevent the development of excessive vacuum or pressure as a result of filling, emptying, or atmosphere temperature changes? | | |
| Are storage tanks equipped with emergency venting that will relieve excessive internal pressure caused by fire exposure? | | |
| Are "NO SMOKING" rules enforced in areas involving storage and use of hazardous materials? | | |

| **HAZARDOUS CHEMICAL EXPOSURE** | Yes | No |
|---|---|---|
| Are employees trained in the safe handling practices of hazardous chemicals such as acids, caustics, etc.? | | |
| Are employees aware of the potential hazards involving various chemicals stored or used in the workplace such as acids, bases, caustics, epoxies, and phenols? | | |
| Is employee exposure to chemicals kept within acceptable levels? | | |
| Are eye wash fountains and safety showers provided in areas where corrosive chemicals are handled? | | |
| Are all containers, such as vats, and storage tanks labeled as to their contents, e.g., "CAUSTICS"? | | |
| Are all employees required to use personal protective clothing and equipment when handling chemicals (gloves, eye protection, and respirators)? | | |

| | | |
|---|---|---|
| Are flammable or toxic chemicals kept in closed containers when not in use? | | |
| Are chemical piping systems clearly marked as to their content? | | |
| Where corrosive liquids are frequently handled in open containers or drawn from storage vessels or pipelines, are adequate means readily available for neutralizing or disposing of spills or overflows and performed properly and safely? | | |
| Have standard operating procedures been established, and are they being followed when cleaning up chemical spills? | | |
| Where needed for emergency use, are respirators stored in a convenient, clean, and sanitary location? | | |
| Are respirators intended for emergency use adequate for the various uses for which they may be needed? | | |
| Are employees prohibited from eating in areas where hazardous chemicals are present? | | |
| Is personal protective equipment provided, used and maintained whenever necessary? | | |
| Are there written standard operating procedures for the selection and use of respirators where needed? | | |
| If you have a respirator protection program, are your employees instructed on the correct usage and limitations of the respirators? Are the respirators NIOSH–approved for this particular application? | | |
| Are they regularly inspected and cleaned, sanitized and maintained? | | |
| If hazardous substances are used in your processes, do you have a medical or biological monitoring system in operation? | | |
| Are you familiar with the Threshold Limit Values or Permissible Exposure Limits of airborne contaminants and physical agents used in your workplace? | | |
| Have control procedures been instituted for hazardous materials, where appropriate, such as respirators, ventilation systems, and handling practices? | | |

| | | |
|---|---|---|
| Whenever possible, are hazardous substances handled in properly designed and exhausted booths or similar locations? | | |
| Do you use general dilution or local exhaust ventilation systems to control dusts, vapors, gases, fumes, smoke, solvents or mists which may be generated in your workplace? | | |
| Is ventilation equipment provided for removal of contaminants from such operations as production grinding, buffing, spray painting, and/or vapor degreasing, and is it operating properly? | | |
| Do employees complain about dizziness, headaches, nausea, irritation, or other factors of discomfort when they use solvents or other chemicals? | | |
| Is there a dermatitis problem? Do employees complain about dryness, irritation, or sensitization of the skin? | | |
| Have you considered the use of an industrial hygienist or environmental health specialist to evaluate your operation? | | |
| If internal combustion engines are used, is carbon monoxide kept within acceptable levels? | | |
| Is vacuuming used, rather than blowing or sweeping dusts whenever possible for clean up? | | |
| Are materials that give off toxic asphyxiant, suffocating or anesthetic fumes stored in remote or isolated locations when not in use? | | |

| **HAZARDOUS SUBSTANCES COMMUNICATION** | Yes | No |
|---|---|---|
| Is there a list of hazardous substances used in your workplace? | | |
| Is there a current written exposure control plan for occupational exposure to bloodborne pathogens and other potentially infectious materials, where applicable? | | |
| Is there a written hazard communication program dealing with Material Safety Data Sheets (MSDS), labeling, and employee training? | | |

| | | |
|---|---|---|
| Is each container for a hazardous substance (i.e., vats, bottles, storage tanks, etc.) labeled with product identity and a hazard warning (communication of the specific health hazards and physical hazards)? | | |
| Is there a Material Safety Data Sheet readily available for each hazardous substance used? | | |
| Is there an employee training program for hazardous substances? | | |
| Does this program include:<br><br>• An explanation of what an MSDS is and how to use and obtain one?<br>• MSDS contents for each hazardous substance or class of substances?<br>• Explanation of "Right to Know"?<br>• Identification of where an employee can see the employer's written hazard communication program and where hazardous substances are present in their work areas?<br>• The physical and health hazards of substances in the work area, and specific protective measures to be used?<br>• Details of the hazard communication program, including how to use the labeling system and MSDS's? | | |
| Does the employee training program on the bloodborne pathogens standard contain the following elements:<br><br>• An accessible copy of the standard and an explanation of its contents<br>• A general explanation of the epidemiology and symptoms of bloodborne diseases<br>• An explanation of the modes of transmission of blood-borne pathogens<br>• An explanation of the employer's exposure control plan and the means by which employees can obtain a copy of the written plan<br>• An explanation of the appropriate methods for recognizing tasks and the other activities that may involve exposure to blood and other potentially infectious materials | | |

| | | |
|---|---|---|
| • An explanation of the use and limitations of methods that will prevent or reduce exposure including appropriate engineering controls, work practices, and personal protective equipment<br>• Information on the types, proper use, location, removal, handling, decontamination, and disposal of personal protective equipment<br>• An explanation of the basis for selection of personal protective equipment<br>• Information on the hepatitis B vaccine<br>• Information on the appropriate actions to take and persons to contact in an emergency involving blood or other potentially infectious materials<br>• An explanation of the procedure to follow if an exposure incident occurs, including the methods of reporting the incident and the medical follow-up that will be made available Information on post exposure evaluations and follow-up<br>• An explanation of signs, labels, and color-coding? | | |
| Are employees trained in the following:<br><br>• How to recognize tasks that might result in occupational exposure?<br>• How to use work practice and engineering controls and personal protective equipment and to know their limitations?<br>• How to obtain information on the types, selection, proper use, location, removal, handling, decontamination, and disposal of personal protective equipment?<br>• Who to contact and what to do in an emergency? | | |

| ELECTRICAL | Yes | No |
|---|---|---|
| Do you specify compliance with OSHA for all contract electrical work? | | |
| Are all employees required to report as soon as practicable any obvious hazard to life or property observed in connection with electrical equipment or lines? | | |

| | | |
|---|---|---|
| Are employees instructed to make preliminary inspections and/or appropriate tests to determine what conditions exist before starting work on electrical equipment or lines? | | |
| When electrical equipment or lines are to be serviced, maintained or adjusted, are necessary switches opened, locked-out and tagged whenever possible? | | |
| Are portable electrical tools and equipment grounded or of the double insulated type? | | |
| Are electrical appliances such as vacuum cleaners, polishers, and vending machines grounded? | | |
| Do extension cords being used have a grounding conductor? | | |
| Are multiple plug adaptors prohibited? | | |
| Are ground-fault circuit interrupters installed on each temporary 15 or 20 ampere, 120 volt AC circuit at locations where construction, demolition, modifications, alterations or excavations are being performed? | | |
| Are all temporary circuits protected by suitable disconnecting switches or plug connectors at the junction with permanent wiring? | | |
| Do you have electrical installations in hazardous dust or vapor areas? If so, do they meet the National Electrical Code (NEC) for hazardous locations? | | |
| Is exposed wiring and cords with frayed or deteriorated insulation repaired or replaced promptly? | | |
| Are flexible cords and cables free of splices or taps? | | |
| Are clamps or other securing means provided on flexible cords or cables at plugs, receptacles, tools, equipment, etc., and is the cord jacket securely held in place? | | |
| Are all cord, cable and raceway connections intact and secure? | | |
| In wet or damp locations, are electrical tools and equipment appropriate for the use or location or otherwise protected? | | |
| Is the location of electrical power lines and cables (overhead, underground, underfloor, other side of walls) determined before digging, drilling or similar work is begun? | | |

| | | |
|---|---|---|
| Are metal measuring tapes, ropes, handlines or similar devices with metallic thread woven into the fabric prohibited where they could come in contact with energized parts of equipment or circuit conductors? | | |
| Is the use of metal ladders prohibited in areas where the ladder or the person using the ladder could come in contact with energized parts of equipment, fixtures or circuit conductors? | | |
| Are all disconnecting switches and circuit breakers labeled to indicate their use or equipment served? | | |
| Are disconnecting means always opened before fuses are replaced? | | |
| Do all interior wiring systems include provisions for grounding metal parts of electrical raceways, equipment and enclosures? | | |
| Are all electrical raceways and enclosures securely fastened in place? | | |
| Are all energized parts of electrical circuits and equipment guarded against accidental contact by approved cabinets or enclosures? | | |
| Is sufficient access and working space provided and maintained about all electrical equipment to permit ready and safe operations and maintenance? | | |
| Are all unused openings (including conduit knockouts) in electrical enclosures and fittings closed with appropriate covers, plugs or plates? | | |
| Are electrical enclosures such as switches, receptacles, and junction boxes, provided with tightfitting covers or plates? | | |
| Are disconnecting switches for electrical motors in excess of two horsepower, capable of opening the circuit when the motor is in a stalled condition, without exploding? (Switches must be horsepower rated equal to or in excess of the motor hp rating.) | | |
| Is low voltage protection provided in the control device of motors driving machines or equipment that could cause probable injury from inadvertent starting? | | |
| Is each motor disconnecting switch or circuit breaker located within sight of the motor control device? | | |

| | | |
|---|---|---|
| Is each motor located within sight of its controller or the controller disconnecting means capable of being locked in the open position or is a separate disconnecting means installed in the circuit within sight of the motor? | | |
| Is the controller for each motor in excess of two horsepower, rated in horsepower equal to or in excess of the rating of the motor it serves? | | |
| Are employees who regularly work on or around energized electrical equipment or lines instructed in the cardiopulmonary resuscitation (CPR) methods? | | |
| Are employees prohibited from working alone on energized lines or equipment over 600 volts? | | |

| NOISE | Yes | No |
|---|---|---|
| Are there areas in the workplace where continuous noise levels exceed 85 dBA? | | |
| Is there an ongoing preventive health program to educate employees in: safe levels of noise, exposures; effects of noise on their health; and the use of personal protection? | | |
| Have work areas where noise levels make voice communication between employees difficult been identified and posted? | | |
| Are noise levels being measured using a sound level meter or an octave band analyzer and are records being kept? | | |
| Have engineering controls been used to reduce excessive noise levels? Where engineering controls are determined not feasible, are administrative controls (i.e., worker rotation) being used to minimize individual employee exposure to noise? | | |
| Is approved hearing protective equipment (noise attenuating devices) available to every employee working in noisy areas? | | |

| | | |
|---|---|---|
| Have you tried isolating noisy machinery from the rest of your operation? | | |
| If you use ear protectors, are employees properly fitted and instructed in their use? | | |
| Are employees in high noise areas given periodic audiometric testing to ensure that you have an effective hearing protection system? | | |

| **FUELING** | Yes | No |
|---|---|---|
| Is it prohibited to fuel an internal combustion engine with a flammable liquid while the engine is running? | | |
| Are fueling operations done in such a manner that likelihood of spillage will be minimal? | | |
| When spillage occurs during fueling operations, is the spilled fuel washed away completely, evaporated, or other measures taken to control vapors before restarting the engine? | | |
| Are fuel tank caps replaced and secured before starting the engine? | | |
| In fueling operations, is there always metal contact between the container and the fuel tank? | | |
| Are fueling hoses of a type designed to handle the specific type of fuel? | | |
| Is it prohibited to handle or transfer gasoline in open containers? | | |
| Are open lights, open flames, sparking, or arcing equipment prohibited near fueling or transfer of fuel operations? | | |
| Is smoking prohibited in the vicinity of fueling operations? | | |
| Are fueling operators prohibited in buildings or other enclosed areas that are not specifically ventilated for this purpose? | | |
| Where fueling or transfer of fuel is done through a gravity flow system, are the nozzles of the selfclosing type? | | |

| IDENTIFICATION OF PIPING SYSTEMS | Yes | No |
|---|---|---|
| When nonpotable water is piped through a facility, are outlets or taps posted to alert employees that it is unsafe and not to be used for drinking, washing or other personal use? | | |
| When hazardous substances are transported through above ground piping, is each pipeline identified at points where confusion could introduce hazards to employees? | | |
| When pipelines are identified by color painting, are all visible parts of the line so identified? | | |
| When pipelines are identified by color painted bands or tapes, are the bands or tapes located at reasonable intervals and at each outlet, valve or connection? | | |
| When pipelines are identified by color, is the color code posted at all locations where confusion could introduce hazards to employees? | | |
| When the contents of pipelines are identified by name or name abbreviation, is the information readily visible on the pipe near each valve or outlet? | | |
| When pipelines carrying hazardous substances are identified by tags, are the tags constructed of durable materials, the message carried clearly and permanently distinguishable and are tags installed at each valve or outlet? | | |
| When pipelines are heated by electricity, steam or other external source, are suitable warning signs or tags placed at unions, valves, or other serviceable parts of the system? | | |

| MATERIAL HANDLING | Yes | No |
|---|---|---|
| Is there safe clearance for equipment through aisles and doorways? | | |
| Are aisleways designated, permanently marked, and kept clear to allow unhindered passage? | | |
| Are motorized vehicles and mechanized equipment inspected daily or prior to use? | | |

| | | |
|---|---|---|
| Are vehicles shut off and brakes set prior to loading or unloading? | | |
| Are containers of combustibles or flammables, when stacked while being moved, always separated by dunnage sufficient to provide stability? | | |
| Are dock boards (bridge plates) used when loading or unloading operations are taking place between vehicles and docks? | | |
| Are trucks and trailers secured from movement during loading and unloading operations? | | |
| Are dock plates and loading ramps constructed and maintained with sufficient strength to support imposed loading? | | |
| Are hand trucks maintained in safe operating condition? | | |
| Are chutes equipped with sideboards of sufficient height to prevent the materials being handled from falling off? | | |
| Are chutes and gravity roller sections firmly placed or secured to prevent displacement? | | |
| At the delivery end of the rollers or chutes, are provisions made to brake the movement of the handled materials? | | |
| Are pallets usually inspected before being loaded or moved | | |
| Are hooks with safety latches or other arrangements used when hoisting materials so that slings or load attachments won't accidentally slip off the hoist hooks? | | |
| Are securing chains, ropes, chockers or slings adequate for the job to be performed? | | |
| When hoisting material or equipment, are provisions made to assure no one will be passing under the suspended loads? | | |
| Are material safety data sheets available to employees handling hazardous substances? | | |

| TRANSPORTING EMPLOYEES AND MATERIALS | Yes | No |
|---|---|---|
| Do employees who operate vehicles on public thoroughfares have valid operator's licenses? | | |
| When seven or more employees are regularly transported in a van, bus or truck, is the operator's license appropriate for the class of vehicle being driven? | | |
| Is each van, bus or truck used regularly to transport employees equipped with an adequate number of seats? | | |
| When employees are transported by truck, are provisions provided to prevent their falling from the vehicle? | | |
| Are vehicles used to transport employees equipped with lamps, brakes, horns, mirrors, windshields and turn signals and are they in good repair? | | |
| Are transport vehicles provided with handrails, steps, stirrups or similar devices, so placed and arranged that employees can safely mount or dismount? | | |
| Are employee transport vehicles equipped at all times with at least two reflective type flares? | | |
| Is a full charged fire extinguisher, in good condition, with at least 4 B:C rating maintained in each employee transport vehicle? | | |
| When cutting tools or tools with sharp edges are carried in passenger compartments of employee transport vehicles, are they placed in closed boxes or containers which are secured in place? | | |
| Are employees prohibited from riding on top of any load which can shift, topple, or otherwise become unstable? | | |

| CONTROL OF HARMFUL SUBSTANCES BY VENTILATION | Yes | No |
|---|---|---|
| Is the volume and velocity of air in each exhaust system sufficient to gather the dusts, fumes, mists, vapors or gases to be controlled, and to convey them to a suitable point of disposal? | | |
| Are exhaust inlets, ducts and plenums designed, constructed, and supported to prevent collapse or failure of any part of the system? | | |
| Are clean-out ports or doors provided at intervals not to exceed 12 feet (3.6576 meters) in all horizontal runs of exhaust ducts? | | |
| Where two or more different type of operations are being controlled through the same exhaust system, will the combination of substances being controlled, constitute a fire, explosion or chemical reaction hazard in the duct? | | |
| Is adequate makeup air provided to areas where exhaust systems are operating? | | |
| Is the source point for makeup air located so that only clean, fresh air, which is free of contaminates, will enter the work environment? | | |
| Where two or more ventilation systems are serving a work area, is their operation such that one will not offset the functions of the other? | | |

| SANITIZING EQUIPMENT AND CLOTHING | Yes | No |
|---|---|---|
| Is personal protective clothing or equipment that employees are required to wear or use, of a type capable of being cleaned easily and disinfected? | | |
| Are employees prohibited from interchanging personal protective clothing or equipment, unless it has been properly cleaned? | | |
| Are machines and equipment, which process, handle or apply materials that could be injurious to employees, cleaned and/or decontaminated before being overhauled or placed in storage? | | |

| | | |
|---|---|---|
| Are employees prohibited from smoking or eating in any area where contaminates that could be injurious if ingested are present? | | |
| When employees are required to change from street clothing into protective clothing, is a clean change room with separate storage facility for street and protective clothing provided? | | |
| Are employees required to shower and wash their hair as soon as possible after a known contact has occurred with a carcinogen? | | |
| When equipment, materials, or other items are taken into or removed from a carcinogen regulated area, is it done in a manner that will contaminate non-regulated areas or the external environment? | | |

| **TIRE INFLATION** | Yes | No |
|---|---|---|
| Where tires are mounted and/or inflated on drop center wheels, is a safe practice procedure posted and enforced? | | |
| Where tires are mounted and/or inflated on wheels with split rims and/or retainer rings, is a safe practice procedure posted and enforced? | | |
| Does each tire inflation hose have a clip-on chuck with at least 24 inches (6.9 centimeters) of hose between the chuck and an in-line hand valve and gauge? | | |
| Does the tire inflation control valve automatically shut off the airflow when the valve is released? | | |
| Is a tire-restraining device such as a cage, rack or other effective means used while inflating tires mounted on split rims, or rims using retainer rings? | | |
| Are employees strictly forbidden from taking a position directly over or in front of a tire while it's being inflated? | | |

Adapted from OSHA Web site, OSHA Office of Training and Education May 1997, http://www.osha.gov/SLTC/smallbusiness/chklist.html#Walking, public domain, OSHA's Small Business Outreach, Training Program, http://www.osha.gov/SLTC/smallbusiness/index.html, public domain

# 3

# Developing Systems to Manage Hazards

Once the operational and hazard risks and hazards data has been identified, the process of determining the types of controls must begin. This chapter outlines the use of the *hierarchy of controls*—its selection, benefits and limitations. At the end of the chapter you will be able to:

- Define how a JHA process can provide continuous improvement
- Cite the hierarchy of controls and its use in determining the type of actions to be taken to control hazards
- Identify why personal protective equipment is the last choice of control whenever possible
- Define the use of *change analysis* and where to gather information for it.

An effective management system (discussed in Chapter 12) allows for the continual analysis of the workplace and anticipates changes that may be needed to modify or develop policies and/or procedures regarding new, existing, or reoccurring hazards. The JHA closes the gap between safety and operational initiatives by providing a method to effectively assess the hazards and associated risk of specific tasks. This chapter reviews systems used to manage hazards and provides a detailed discussion of the hierarchy of controls.

To set the stage for this chapter and provide a road map for managing hazards, refer to Figures 3-1, 3-2, and 3-3 for an overview of employee reporting hazards. Figure 3-1 provides a flow diagram for evaluating hazardous conditions; the flow diagram in Figure 3-2 shows how to evaluate procedures; and Figure 3-3 is a flow diagram on defining hazard analysis.

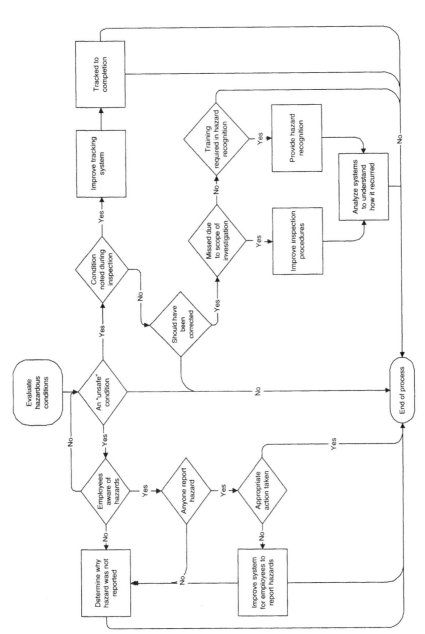

**Figure 3-1** Overview of Employee Reporting Hazards

Evaluating hazardous conditions, Part 1, Adapted from the U.S. Department of Labor, Office of Cooperative Programs, Occupational Safety and Administration (OSHA), Managing Worker Safety and Health, November 1994, http://www.dolir.mo.gov/ls/safetyconsultation/ccp/appendix_9_4.html, public domain

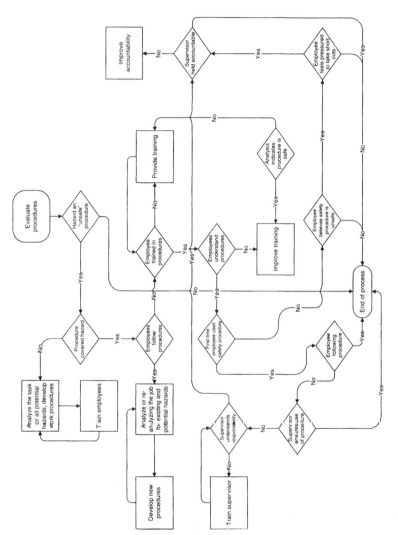

**Figure 3-2** Overview of Employee Reporting Hazards

Evaluating Procedures, Part 2, Adapted from the U.S. Department of Labor, Office of Cooperative Programs, Occupational Safety and Administration (OSHA), Managing Worker Safety and Health, November 1994, http://www.dolir.mo.gov/ls/safetyconsultation/ ccp/appendix_9_4.html, public domain

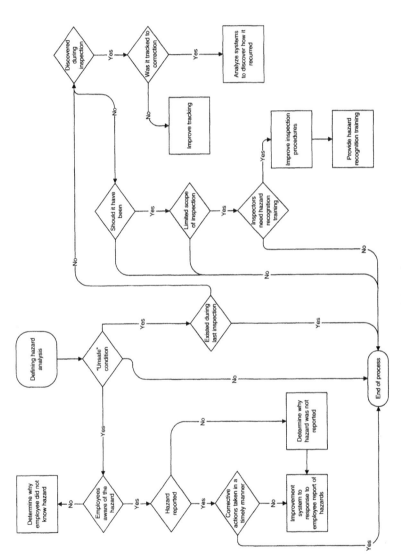

**Figure 3-3** Overview of Employee Reporting Hazards

Defining Hazard Analysis, Part 3, Adapted from the U.S. Department of Labor, Office of Cooperative Programs, Occupational Safety and Administration (OSHA), Managing Worker Safety and Health, November 1994, http://www.dolir.mo.gov/ls/safetyconsultation/ccp/appendix_9_4.html, public domain

# 3.1 HIERARCHY OF CONTROLS

To ensure that proper controls are in place that will minimize hazards, a ranking of potential controls is used, known as the *hierarchy of controls*. Using this hierarchy of controls provides a tiered approach to evaluating and reinforcing your safety system [2]. The order of control preference is:

- Eliminating the hazard from the operation, the method, the material, and/or the structure of the facility. If at all possible, you want to avoid the hazard.
- Abating the hazard by limiting exposure or reducing its risks by substituting a less hazardous material, methods, etc.
- Controlling the hazard by redesigning and engineering the hazard out of the process. The hazard still exists, but mechanisms are in place to contain it.
- Using administrative methods to advise, warn, train, alert, etc., that the hazard exists and that designated safety controls are to be used.
- Utilize PPE only as a last resort to shield employees against hazards.

Refer to Table 3-1 for a summary of the hierarchy of controls and the associated rankings.

## Table 3-1
## Summary of Hierarchy of Controls and Associated Rankings

| |
|---|
| Elimination: Completely avoiding the risk. |
| Substitution: Replacing a high hazard material with a less hazardous element that can be controlled. |
| Engineering Controls: Hazards exist but can be controlled by a process change and/or equipment design:<br>  • Work station design, tool selection, and design.<br>  • Process modification.<br>  • Mechanical assist, machine guarding, ventilation, etc. |
| Administrative controls: Used to communicate actions necessary to reduce exposure to hazards:<br>  • Use of safety rules, procedures, protocols, policies, and/or warning devices.<br>  • Training programs.<br>  • Job rotation/enlargement, pacing, etc. |
| PPE: Direct barriers worn by the employee, such as hard hats, gloves, respirator, etc. |

Adapted from US CDC NIOSH Programs Portfolio, Engineering Controls, http://www.cdc.gov/niosh/programs/eng/, public domain

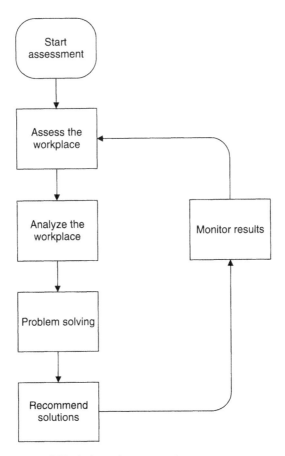

**Figure 3-4** Overview of Workplace Assessment

Oklahoma Department of Labor, Safety and Health Management: Safety Pays, 2000, http://www.state. ok.us/~okdol/, Hazard Identification Flowchart, Chapter 7, pp. 41, public domain

Refer to Figure 3-4 for an overview of workplace assessment and Figure 3-5 for an overview of hazard assessment survey guidelines. These flow diagrams provide a road map for hazard control.

## 3.2 WHY ENGINEERING CONTROLS?

The effective use of engineering controls not only reduces or eliminates the hazard; this type of control can also reduce or eliminate the need for specific employee behaviors through administrative controls [2, 3, 6].

Engineering controls focus on the direct source of the hazard. The basic concept of engineering controls is that the work environment and required tasks

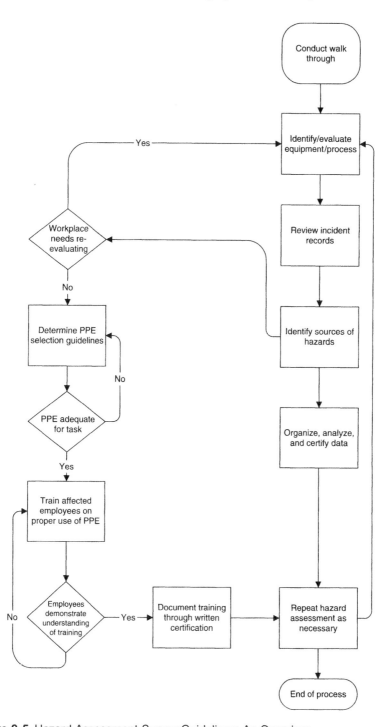

**Figure 3-5** Hazard Assessment Survey Guidelines: An Overview

Adapted from Roughton, James, *Personal Protection Equipment: Compiling with the Standard, Professional Safety*, pp. 30, 1995

should be designed in a manner that eliminates hazards or at least reduces the exposure to hazards. While called "engineering controls," solutions can be as simple as adding a guard to a piece of equipment to prevent someone from sticking their hand into the moving parts, adding an interlock to a guard to shut down the equipment when the guard is open, etc. [6].

---

The best strategy is to control the hazard at its source. The basic concept behind engineering controls is for the work environment and the job to be designed to avoid hazards, eliminate hazards, or reduce exposure to hazards. Engineering controls can be simple in some cases and are based on the following principles:

- Design the facility, equipment, or process to remove the hazard or substitute something that is less hazardous.
- If removal is not feasible, enclose the hazard to prevent exposure in normal operations.

Where complete enclosure is not feasible, establish barriers to reduce exposure in normal operations.

---

Source: Adapted from the OSHA Website http://www.osha-slc.gov/SLTC/safetyhealth_ecat/comp3.htm#SafeWorkPractices, public domain

Roughton, James, J. Mercurio, *Developing an Effective Safety Culture: A Leadership Approach*, Butterworth-Heinemann, 2002, Chapter 11, pp. 196–197

The bottom line for engineering controls is: "No hazard, no exposure, no risk, no injury." Engineering controls may be a simple "quick fix" based on a few basic principles, or they may involve a complete redesign requiring a capital investment. We will discuss a risk matrix in Chapter 10 that will help you to assign a priority to hazards, but for now we will only discuss the "hierarchy of controls."

*NO HAZARD = NO RISK = NO EXPOSURE = NO INJURY OR HARM!!*

## 3.3 ADMINISTRATIVE CONTROLS

Workplace rules and safe work practices are considered administrative controls and are used to communicate actions necessary to minimize any exposure to hazards. As an example, ergonomic-related rules and practices might include the following:

- Longer or more rest breaks.
- Additional substitute employees.

- Job rotation of employees through different jobs to reduce or "even out" exposure to hazards.
- Exercise or stretching that alternates and varies the body motions.
- Limits on the length and number of required work shifts [2].
- Selection of employees that are physically able to do the work.

The challenge in administrative control lies in the fact that it is a "paper-based" function that tells someone what to do to be safe but cannot monitor activities. What happens a couple of days or a couple of weeks after implementing an administrative control? The learning curve of the human mind is such that employees tend to forget to do the task as outlined in the written procedure unless a positive reinforcement and refresher emphasis is provided. To determine whether employees retain the required training, a "gage study" should be conducted to measure how much they actually retain. This will be covered in detail in Chapter 7. Supervisors may also forget to enforce the rules of the workplace for reasons such as pressures to produce, time constraints, or other more pressing problems. Refer to Table 3-2 for administrative control considerations.

### Table 3-2
### Administrative Control Considerations

Administrative Controls:

- Reduce the duration, frequency, location, etc. of exposure but not the hazard.
- Use safe work procedures that consider the scope and severity of the hazards.

Key elements that must be managed:

- Training
- Compliance reporting
- Direct observation of actions of persons working in or around the hazards
- Communication, warnings, signage

Major weakness: Relies on specific actions and appropriate behavior of the worker at all times

Adapted from the OR OSHA website, OR OSHA 100w, *Hierarchy of Hazard Control Strategies*, pp. 16, http://www.cbs.state.or.us/external/osha/educate/training/pages/materials.html, public domain
Roughton, James, J. Mercurio, *Developing an Effective Safety Culture: A Leadership Approach*, Butterworth-Heinemann, 2002, Chapter 11, pp. 194, 198

Administrative controls are "paper-based" actions that tell employees how to be safe, but do not and cannot monitor the activities.

Adapted from the OSHA Website http://www.osha-slc.gov/SLTC/safetyhealth_ecat/comp3.htm#SafeWorkPractices, public domain

Roughton, James, J. Mercurio, *Developing an Effective Safety Culture: A Leadership Approach*, Butterworth-Heinemann, 2002, Chapter 11, pp. 194, 198

## 3.4 PPE

The last and least effective method of control is the use of personal protective equipment (PPE). PPE can consist of a wide array of equipment and clothing such as steel-toed shoes and boots, safety glasses with side shields, goggles, face shields, dusk masks, and respirators, etc. The nature of the work and other factors may prevent the use of engineering, avoidance, or substitution, leaving the need for a combination of administrative and PPE as the primary tools remaining [3]. We have found in our experience that a mismatch can easily occur between the scope and nature of the hazard and the use of the right PPE.

It is important to remember that, as with administrative controls, the use of PPE does not control the hazard. PPE limits exposure to the hazard by setting up a barrier between the employee and the hazard [2, 6]. PPE is typically used:

- When hazards cannot be engineered completely out of the normal process or maintenance work, and
- When safe work practices and other forms of administrative control cannot provide sufficient additional protection.

### 3.4.1 PPE Limitations

It is important to ensure that all PPE is selected and designed for the specific function and hazard. For example, no one type of glove or apron, safety glasses, safety shoes, etc. will protect against all associated hazards. Proper use of PPE is a combination of proper selection and use by exposed individuals. Employees must have in-depth training and a full understanding of the proper PPE, its use, maintenance, cleaning and storage of the equipment. They should understand *why* they must wear the PPE, and not just blindly follow a rule that it is required!

Selecting PPE is not as simple as ordering from a catalogue; you must complete a PPE risk assessment and match the PPE to the hazard and to the employee [5]. Do not assume that the purchasing department or supervisors understand PPE selection. In one company, we found military CS (riot gas) masks in place for emergency use. The recommendation was for a specific respirator for a certain area and the purchasing manager thought that "one mask was the same as any other." Unfortunately, linking the purchase to the hazard was not a part of their buying process.

Administration of PPE requires a positive and fair reinforcement of the rules governing use. Employees may resist wearing PPE, because it may be uncomfortable. We have found that a conflict can easily develop when the administrative controls demand the use of PPE without understanding all of the considerations of proper PPE selection, fitting, use, and so forth. *"One size does not fit all!"*

When faced with completing a task or stopping because of defective equipment, many employees will keep working, putting themselves at risk. Unless the hazard and associated risk are made very clear and the process allows them to stop work without penalty, their behavior will tend toward working at-risk.

## 3.4.2 PPE Hazard Assessment

The fundamental element of the management program for PPE is the in-depth assessment of the job steps and tasks in order to determine how to properly protect all employees against the hazards of the task. The JHA provides the foundation for establishing a standard operating procedure (SOP) for training on the protective limitations of the PPE, its proper use, maintenance, cleaning and storage. Refer to Table 3-3 for an overview of PPE assessment. The JHA is essential and should be mandatory for all jobs requiring PPE.

### Table 3-3
### Overview of PPE Assessment

When hazards cannot be engineered out of the process, environmental conditions or maintenance work, and when safe work practices and other forms of administrative controls cannot provide sufficient protection or separation from the exposures, the last method of control is the use of PPE. PPE may also be appropriate for controlling hazards while engineering and work practice controls are being implemented. For specific OSHA requirements on personal protective equipment, refer to OSHA's standard 1910, Subpart I.

Adapted from the OSHA Website http://www.osha-slc.gov/SLTC/safetyhealth_ecat/comp3.htm# SafeWorkPractices, public domain
Roughton, James, J. Mercurio, *Developing an Effective Safety Culture: A Leadership Approach,* Butterworth-Heinemann, 2002, Chapter 11, pp. 194, 200

Refer to Appendix G for details on the complete process of PPE assessment, along with a sample PPE assessment.

## 3.5 WORK PRACTICES AND SAFETY RULES

Work practices and safety rules help reduce intended exposure to hazards by designing safe work practices and integrating them into job procedures and general safety rules.

Safe work practices and safety rules must be designed from a baseline survey (inventory) of hazards, a comprehensive JHA that is accompanied by solid employee training and reinforcement, with consistent and reasonable enforcement. As an administrative control, work practices are used in conjunction with other controls that minimize or control exposure to hazard and are an important part of a job procedure [2, 6].

### 3.5.1 General Safety Rules

It is important to note that "any policy or rule ignored is a policy or rule rewritten." As an example, if a supervisor or manager ignores a safety rule that states: "You must lock out your equipment before clearing a jam," and if lockout is not used when performing this task, then the rule has been essentially rewritten to: "Lockout is not required when clearing jams from equipment." As the work rules has now been changed, employees who do not properly lock out equipment are, in effect, still complying with the standard set by the manager. When management does not follow its own policies, employees are meeting "established standards" and discipline is not an option [6], since the employee is only following a practice that has been allowed by management. This now becomes the "written policy." In many cases, management must sit back and ask the question, "Did my behavior create this condition?"

> "Any policy or rule ignored is a policy or rule rewritten"

You must understand that safety rules usually begin as generic copies from many sources and must be customized based on your site-specific requirements. What do you want to accomplish by setting your work rules? If you ask any supervisor what is their main safety concern, the typical response is: "I cannot get my employees to wear their PPE." This response indicates that a problem exists with your PPE process. The use of the JHA would assess the exposure and employee participation could pinpoint the underlying hidden issues driving the behavior of "not wearing PPE."

Safety rules are most effective when they are written, posted, and discussed with all employees, particularly in small groups as well as a large plant-wide meeting. The key is to emphasize the linkage between the company policy/vision, the safety rules and the injury consequences of not following the rules. The employees should sign a statement to the effect, "I have read and understand these safety rules and why they are to be followed. I have received an explanation of the injury consequences of not following the rules regarding both possible discipline and potential for injury and damage." The employer and the employee both keep a copy of this signed statement [2].

Developing or revising safe work rules should include employee participation and buy-in as discussed in Chapter 5. When employees play a role in formulating guidelines and rules, they are more likely to understand and follow them [3].

Safe work practices are used in conjunction with and not as a substitute for more effective or reliable engineering controls. As serious injuries may occur while employees are performing nonroutine tasks, general safety rules should identify and document these types of tasks. Refer to Appendix F for a sample set of safety rules.

## 3.5.2 Limitations of Work Practices and Safety Rules

When developing a hazard control system, you must anticipate the natural human trait to resist change. Resistance to change begins when employees are instructed on new job practices and procedures, particularly if they have performed a job for a long period of time without special precautions. Employees are most likely to feel unconcerned about hazards because their perception of risk may be low if no serious injury has occurred. They may even have that perspective even if an injury has occurred, believing it was just a "freak accident" and therefore not controllable. Your process must consider concepts and techniques designed to change the traditional definition of an "accident" into the concept that it is similar to a product defect that can be eliminated [2].

## 3.6 CHANGE ANALYSIS

Any time something new is introduced into the workplace, whether a piece of equipment, modified or relocated equipment, different materials, a new process, or a new building, etc., new hazards may come into play. Some hazards will be obvious while other hazards are hidden in the interrelationships of the new elements. If you are considering any change in the workplace, a thorough analysis of the change is essential before implementation in order to head off a problem. Many safety professionals have learned to hate the phrase "Oh, by the way, . . . ." Coming from a member of management or an employee, the phrase is usually followed by a serious notification of change that brings in a new raft of exposures to the process.

---

"Oh, by the way, I have just come from this meeting and I need you to . . . . . . . . ."

---

Preparing for any workplace change is crucial and must include considering the potential effect on all areas of the operation. Individuals respond differently to change. Even a clearly beneficial change can throw an employee temporarily off-guard and increase the risk of an injury. All affected employees should be consulted on what is being changed to gather their input. Once the change is complete, employees need to be provided with the appropriate education and training. Management must pay attention to the employee responses and monitor the change until everyone has adapted. Figure 3-6 provides a snapshot of the impact of change and an overview of change analysis.

To simplify, large change is composed of a number of several smaller changes. When you begin producing a new product, chances are you will have new equipment, materials, and processes to monitor. Ensure that each new change is analyzed not only individually, but also in relation to all of the other changes [2, 5].

---

You need to consider a change analysis process when:

- Expanding or modifying existing materials.
- Moving operations to different geographic areas.
- Moving into new or leased facilities.
- Using new, relocated, or modified equipment or processes.
- Using new tools, materials, or chemicals.
- Employees, management or staffing changes occur.
- Reorganizing and restructuring any operation.

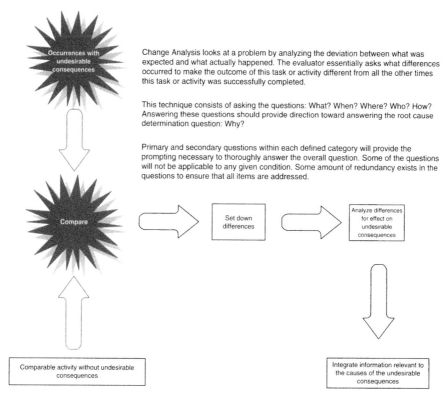

Change Analysis looks at a problem by analyzing the deviation between what was expected and what actually happened. The evaluator essentially asks what differences occurred to make the outcome of this task or activity different from all the other times this task or activity was successfully completed.

This technique consists of asking the questions: What? When? Where? Who? How? Answering these questions should provide direction toward answering the root cause determination question: Why?

Primary and secondary questions within each defined category will provide the prompting necessary to thoroughly answer the overall question. Some of the questions will not be applicable to any given condition. Some amount of redundancy exists in the questions to ensure that all items are addressed.

**Figure 3-6** Change Analysis
DOE Guideline Root Cause Analysis Guidance Document, DOE-NE-STD-1004-92, February 1992, U.S. Department of Energy Office of Nuclear Energy, Office of Nuclear Safety Policy and Standards, Appendix E, Figure E-1, pp. E-1, Six Steps Involved in Change Analysis, modified public domain version

## 3.6.1 A Change in the Process

New processes require employees to perform differently. Consequently, new hazards may develop even when employees are using familiar materials, equipment, and facilities. You must carefully develop safe work procedures for new processes. The JHA must be performed to discover any hidden hazards and to ensure that everyone is familiar with the new procedures [5].

## 3.6.2 Building or Leasing a New Facility

The JHA can assist in clarifying hazards at any level of the operation and is not strictly limited to employee safety. A new building or even a leased facility should be reviewed carefully to identify existing and potential hazards and/or consequences of exposures. A design that appears to enhance productivity and appears delightful to the management may be a poor management decision when reviewed from a risk point of view. One case we know of is that of a retail company that made a great real estate deal on the purchase of a very large warehouse facility as its "just-in-time" center. It met all their financial and location criteria for the support of their many stores. It was later found that the facility was located next to a propane tank farm with a history of fires and serious incidents.

When leasing a facility that was built for a different purpose at an earlier time, the risk of acquiring safety or environmental issues is even greater. You should make a thorough review of the actual facility, plus review the blueprints or plans for any renovations made [5].

## 3.6.3 New Equipment Installation

You cannot blindly rely on equipment manufacturers to analyze and design guarding, controls, or safe procedures to fit your specific requirements. If the equipment is manufactured in a foreign country, it may not meet U.S. requirements. One steel company used a European standard for equipment lockouts. However, these controls did not meet OSHA criteria of the time [4].

These types of cases help the safety professional to be more involved in the purchasing process, in the installation plans, and start-up review of any equipment [2, 5].

Testing and "breaking-in" periods for newly installed equipment should be a part of the process. The most experienced employees should be assigned to assist in the identification of new hazards in the operations before full production begins. As with new facilities, the sooner hazards are detected, the easier and cheaper any corrective action will be [2].

Using the JHA process provides a structure to follow and use in presenting the findings to management. The JHA coupled with the risk matrix (which we will present in Chapter 10), cause-and-effect diagram, and hierarchy of controls will walk you through a logical sequence of events to identify hazards associated with the new process and its specific job hazards. Part 3 of this book will provide the details for this risk assessment.

Refer to Figure 3-7 for a sign-off process flow diagram that can help control new installations. This sign-off procedure can apply to any type of installation, whether it is new, modified, relocated equipment, a chemical process, etc. Refer to Appendix H for an example of a form that can be used for equipment review. This form and an associated checklist will help to identify hazards on any type of equipment and will make the process consistent [1].

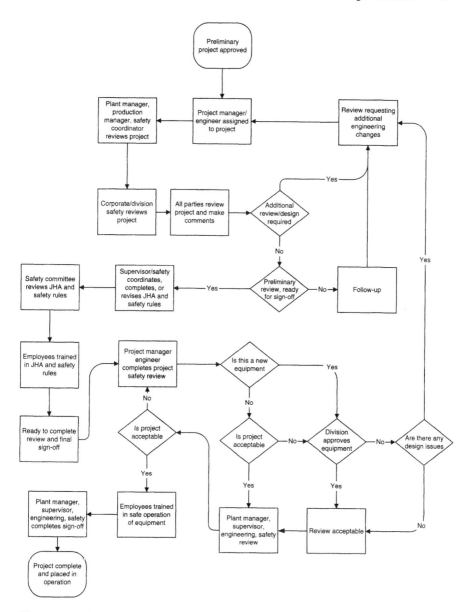

**Figure 3-7** Suggested Safety Review of New/Relocated Equipment or Major Modification Sign-Off Process

## 3.6.4 Using New Materials

Before introducing any new materials into your processes, research the hazards that the materials may present. Try to determine any hazards that may appear due to the processes you plan to use with the materials [2, 5].

When reviewing new material, the best place to start is the manufacturer's Material Safety Data Sheet (MSDS). An MSDS is required for all materials containing hazardous substances and should be provided with each shipment. The MSDS must provide specific hazard information. A person who is well trained in hazard recognition should analyze potential hazards whenever a new substance may be involved, and recommend any change to prevent or control any associated hazard [2, 5]. If the new material is considered a hazard, then it must be listed on the JHA.

## 3.6.5 Employee Changes

When an employee changes job functions, other unknown safety concerns may develop. Employee changes can be divided into three basic areas:

*Changes in staffing.* A new employee performs a task previously completed by another employee or a nonroutine task. The new employee may bring to the position a different level of skill. The employee may possess a different degree of experience when performing the tasks, following specific work rules and procedures, and/or interacting with other employees, especially in high hazard situations. In addition, their skills may have been on-job training and passed down from employee to employee along with undesired at-risk events. These differences can be examined and specific steps should be taken to minimize any increased risk through the JHA [2, 5].

*Individual change.* The second category of employee change is the sometimes sudden, sometimes gradual, change that can occur with the individual employee. The change may be related to temporary or chronic medical problems, a partially disabling condition, family responsibilities, health problems, family crisis and other personal issues, alcohol or drug abuse, aging, or the worker's response to workplace changes. The analysis of personal change may require physical and/or administrative accommodations to ensure that safe conditions remain consistent. Administrative controls must be fully coordinated with human resources to make sure labor and health laws and regulations are followed. Accommodations for individual skill and physical changes may be mandated by law and the impact of such laws on the control of the individual exposure to hazards, use of PPE and other criteria must be considered and responded to.

*Temporary employees.* The last category is temporary employees. This group of employees is usually not treated the same as regular employees and is not, in many cases, trained appropriately for the task being performed. It is important that these employees are trained appropriately.

## 3.6.6 Adapting to Change

Change analysis is most useful when there is a side-by-side comparison of the old process against the changes in the new process. This will highlight subtle changes that may be hidden. Techniques should include:

- Developing a process map detailing the old process versus the new process, noting the differences.
- Developing cause-and-effect diagrams of the current vs. proposed changes.
- Comparing JHAs of tasks currently being performed and the proposed change. Jobs at high levels may look the same but outlining the steps and tasks will highlight the differences.
- Comparing the old work environment against the proposed changes, what environmental elements will change: i.e., ventilation, lighting, air quality, temperatures, environmental, etc.
- Comparing old policies, procedures, and protocols against new requirements: What is no longer relevant or needs updates/modifications? Are new regulatory requirements introduced? New technology may make current administrative controls obsolete.
- Comparing the current organizational structure changes. Is supervision still adequate? Will additional or different employees be needed?
- Completing a risk assessment using the process maps and JHA comparisons.

## 3.6.7 Other Analytical Tools for Consideration

A number of analytical tools such as a "What if" checklist, Hazard and Operability studies (HAZOP), Failure Mode and Effect Analysis (FMEA), and "Fault-tree" analysis can be used to determine possible process breakdowns. These tools can be used to design prevention/controls for the likely causes of unwanted events. Refer to Appendix I for a brief summary of each tool.

## 3.7 SUMMARY

Workplace hazards do not exist in a vacuum. The workplace is a dynamic place. Things and elements are always in some state of flux. Not only physical needs must be monitored; the human element is always present, and the human condition is also one of continuous change. An effective manager will be sensitive to change and its potential effect on the safety of the employees and the organization as a whole.

To complement hazard controls, you must have a good management system that involves preventive maintenance; hazard correction tracking; and a fair and consistent enforcement of rules, procedures, work practices, and PPE [2].

Effective planning and design of the workplace or task can minimize related hazards. Where eliminating hazards is not feasible, planning and design can minimize the exposure to unsafe conditions or situations.

Even after you have followed the process, conducted a comprehensive hazard survey, analyzed the possibility of changes in the workplace, routinely analyzed jobs and/or processes for hazards, and developed a program of hazard prevention and control, residual risk may still be present. Hazards may have been missed because their presence, characteristics, and routine appearance are not clearly visible at the time of the surveys and analysis.

As we move to the next chapter, we will discuss how to handle hazards and the associated risk. A great hazard control program is worthless if the individuals will not or cannot follow its criteria. We will discuss a detailed analysis of behavior-related concepts to provide some background information on how the process works and its value in providing for continuous improvement of the JHA process.

## CHAPTER REVIEW QUESTIONS

1. When should a hazard identification and assessment analysis be conducted?
2. What are the hierarchy of controls?
3. Why are engineering controls important?
4. How can you use engineering controls to minimize hazards through design?
5. What are administrative controls?
6. What are examples of administrative controls?
7. When are safety rules most effective?
8. What are the limitations of PPE?
9. What are the limitations of work practices?
10. Why is employee education and training important?

## REFERENCES

1. Oklahoma Department of Labor, Safety and Health Management: *Safety Pays*, 2000, http://www.state.ok.us/~okdol/, public domain
2. U.S. Department of Labor, Office of Cooperative Programs, Occupational Safety and Administration (OSHA), *Managing Worker Safety and Health*, November 1994, public domain

3. OSHA Web Site, http://www.osha-slc.gov/SLTC/etools/safetyhealth/index.html, public domain
4. Oregon Website, *Hazard Identification and Control*, OR-OSHA 104, http://www. cbs.state.or.us/osha/educate/training/pages/courses.html, public domain
5. Missouri Department of Labor, *Establishing Complete Hazard Inventories*, Chapter 7, http://www.dolir.mo.gov/ls/safetyconsultation/ccp/, public domain

# Appendix F

## F.1 SAMPLE SAFETY RULES

### F.1.1 Codes of Safe Practice (Example)

Work shall be well-planned and supervised to prevent injuries when working with equipment and handling heavy materials. Employees will be trained in proper material handling.

Employees shall not handle or tamper with any energy sources (electrical, hydraulic, thermal, mechanical, etc.) in a manner not within the scope of their duties, unless they have received instructions from their supervisor/employer.

All injuries shall be reported promptly to the supervisor so that arrangements can be made for medical and/or first-aid treatment. First aid materials are located in _____; emergency, fire, ambulance, rescue squad, and doctor's telephone numbers are located on _____; and fire extinguishers are located at _____.

### F.1.2 Suggested Safety Rules

- Do not throw material, tools or other objects from heights (whether structures or buildings) until proper precautions are taken to protect others from the falling objects.
- Wash your hands thoroughly after handling hazardous substances.
- Flammable liquids shall not be used for cleaning purposes.
- Arrange work so that you are able to face ladder and use both hands while climbing.

### F.1.3 Use of Tools and Equipment

- Keep hammers in good condition to avoid flying nails and bruised fingers.
- Files shall be equipped with handles; never use a file as a punch or pry.

- Do not use a screwdriver as a chisel.
- Do not lift or lower portable electric tools by the power cords; use a rope.
- Do not leave the cords of these tools where cars or trucks will run over them.

## F.1.4 Machinery and Vehicles

- Do not attempt to operate machinery or equipment without special permission, unless it is one of your regular duties.
- Loose or frayed clothing, dangling ties, finger rings, and similar items must not be worn around moving machinery or other places where they can get caught.
- Machinery shall not be repaired or adjusted while in operation.

Source: Adapted from OSHA Handbook for Small Businesses, Safety Management Series, U.S. Department of Labor, Occupational Safety and Health Administration, OSHA 2209, 1996 (Revised), Appendix C, pp. 60, public domain.

# Appendix G

## G.1 PERSONAL PROTECTIVE EQUIPMENT (PPE) ASSESSMENT

Information in this appendix was adapted from Washington State Department of Labor and Industries website, http://www.lni.wa.gov/Safety/Basics/Programs/Protective/default.asp, public domain.

## G.2 GUIDELINES FOR COMPLYING WITH PPE REQUIREMENTS

Use this checklist to help determine the PPE requirements at your work place. You can use the available tools in the far right column to help you accomplish the step. Check off the boxes in the far left column as you complete each step.

| Done | STEP | Tools |
|---|---|---|
| ☐ | Do a work place walk-through and look for hazards (including potential hazards) in all employees' work spaces and work place operating procedures. | Checklist #1, Option 1: PPE Hazard Assessment **or** Checklist #2, Option 2: JHA PPE Hazard Assessment |
| ☐ | Consider engineering, administrative, and/or work practice methods to control the hazards first. Identify those existing/potential hazards and tasks that require PPE. | |
| ☐ | Select the appropriate PPE to match the hazards and protect employees. | |

| | | |
|---|---|---|
| ☐ | Communicate PPE selection to each at-risk employee. Provide properly fitting PPE to each employee required to use it. | |
| ☐ | Train employees on the use of PPE and document it. | PPE Training Certification Form |
| ☐ | Test employees to make sure they understand the elements of the PPE training. | Sample PPE Training Quiz *(required)* |
| ☐ | Follow up to evaluate effectiveness of PPE use, training, policies, etc. against the hazards at your work place. <br> ☐Yes ☐No    All employees have been trained <br> ☐Yes ☐No    Employees are using their PPE properly and following PPE policies and procedures <br> ☐Yes ☐No    Supervisors are enforcing use of required PPE <br> *(If you checked any No boxes, go back through the steps and correct the deficiencies.)* <br> ☐Yes ☐No    Have things changed at your work place? (e.g., fewer injuries/illnesses) | |

## G.3 HAZARD ASSESSMENT FOR PPE

This tool can help you conduct a hazard assessment to determine if your employees need to use personal protective equipment (PPE) by identifying activities that may create hazards for your employees. The activities are grouped according to what part of the body might need PPE.

This tool can serve as written documentation that you have completed a hazard assessment. Make sure that the blank fields at the beginning of the checklist (indicated by *) are filled out (see below, Instructions #4).

## G.3.1 Instructions

1. Do a walk-through survey of each work area and job/task. Read through the list of work activities in the first column, putting a check next to the activities performed in that work area or job.

2. Read through the list of hazards in the second column, putting a check next to the hazards to which employees may be exposed while performing the work activities or while present in the work area. (e.g., work activity: chopping wood; work-related exposure: flying particles).

3. Decide how you are going to control the hazards. Consider engineering, work place, and/or administrative controls to eliminate or reduce the hazards before resorting to using PPE. If the hazard cannot be eliminated without using PPE, indicate which type(s) of PPE will be required to protect your employee from the hazard.

4. Make sure that you complete the following fields on the form (indicated by *) to certify that a hazard assessment was done:
   *Name of your work place
   *Address of the work place where you are doing the hazard assessment
   *Name of person certifying that a workplace hazard assessment was done
   *Date the hazard assessment was done

# PPE Hazard Assessment Certification Form

**\*Name of work place:** _____

**\*Work place address:** _____

Work area(s): _____

**\*Assessment conducted by:** _____

**\*Date of assessment:** _____

Job/Task(s): _____

\*Required for certifying the hazard assessment.    Use a separate sheet for each job/task or work area

## EYES

| Work activities, such as: | | Work-related exposure to: | Can hazard be eliminated without the use of PPE? |
|---|---|---|---|
| ☐ abrasive blasting | ☐ sanding | ☐ airborne dust | Yes ☐  No ☐ |
| ☐ chopping | ☐ sawing | ☐ flying particles | If no, use: |
| ☐ cutting | ☐ grinding | ☐ blood splashes | ☐ Safety glasses      ☐ Side shields |
| ☐ drilling | ☐ hammering | ☐ hazardous liquid chemicals | ☐ Safety goggles      ☐ Dust-tight |
| ☐ welding | | ☐ intense light | ☐ Shading/Filter (# )      goggles |
| ☐ punch press operations | | ☐ other: | ☐ Welding shield |
| ☐ other: | | | ☐ Other: |

## FACE

| Work activities, such as: | Work-related exposure to: | Can hazard be eliminated without the use of PPE? |
|---|---|---|
| ☐ cleaning ☐ foundry work<br>☐ cooking ☐ welding<br>☐ siphoning ☐ mixing<br>☐ painting ☐ pouring molten<br>☐ dip tank operations      metal<br>☐ chemical handling<br>☐ other | ☐ hazardous liquid chemicals<br>☐ extreme heat/cold<br>☐ potential irritants:<br>☐ other: | Yes ☐   No ☐<br>If no, use:<br>☐ Face shield<br>☐ Shading/Filter (# )<br>☐ Welding shield<br>☐ Other: |

## HEAD

| Work activities, such as: | Work-related exposure to: | Can hazard be eliminated without the use of PPE? |
|---|---|---|
| ☐ building maintenance<br>☐ confined space operations<br>☐ construction<br>☐ electrical wiring<br>☐ walking/working under catwalks<br>☐ walking/working under conveyor belts<br>☐ walking/working under crane loads<br>☐ utility work<br>☐ other: | ☐ beams/structures<br>☐ pipes<br>☐ exposed electrical wiring<br>   or components<br>☐ falling objects<br>☐ machine parts<br>☐ other: | Yes ☐   No ☐<br>If no, use:<br>☐ Protective Helmet<br>☐ Type A (low voltage)<br>☐ Type B (high voltage)<br>☐ Type C<br>☐ Bump cap (not ANSI-approved)<br>☐ Hair net or soft cap<br>☐ Other: |

## HANDS/ARMS

| Work activities, such as: | Work-related exposure to: | Can hazard be eliminated without the use of PPE? |
|---|---|---|
| ☐ baking ☐ material handling<br>☐ cooking ☐ sanding<br>☐ grinding ☐ sawing<br>☐ welding ☐ hammering<br>☐ working with glass ☐ Unjamming equipment<br>☐ using computers<br>☐ using knives<br>☐ dental and health care services<br>☐ other: | ☐ blood<br>☐ irritating chemicals<br>☐ tools or materials that could scrape, bruise, or cut<br>☐ extreme heat/cold<br>☐ other: | Yes ☐ No ☐<br>If no, use:<br>☐ Gloves<br>☐ Chemical resistance<br>☐ Liquid/leak resistance<br>☐ Temperature resistance<br>☐ Abrasion/cut resistance<br>☐ Slip resistance<br>☐ Protective sleeves<br>☐ Other: |

## FEET/LEGS

| Work activities, such as: | Work-related exposure to: | Can hazard be eliminated without the use of PPE? |
|---|---|---|
| ☐ building maintenance<br>☐ construction<br>☐ demolition<br>☐ food processing<br>☐ foundry work<br>☐ logging<br>☐ plumbing<br>☐ trenching<br>☐ use of highly flammable materials<br>☐ welding<br>☐ material handling<br>☐ other: | ☐ explosive atmospheres<br>☐ explosives<br>☐ exposed electrical wiring or components<br>☐ heavy equipment<br>☐ slippery surfaces<br>☐ tools<br>☐ dropped tools, material, etc.<br>☐ other: | Yes ☐ No ☐<br>If no, use:<br>☐ Safety shoes or boots    ☐ Metatarsal protection<br>☐ Toe protection    ☐ Heat/cold protection<br>☐ Electrical protection    ☐ Chemical resistance<br>☐ Puncture resistance<br>☐ Anti-slip soles<br>☐ Leggings or chaps<br>☐ Foot-Leg guards<br>☐ Other: |

## BODY/SKIN

Work activities such as:
- [ ] baking or frying
- [ ] battery charging
- [ ] dip tank operations
- [ ] fiberglass installation
- [ ] irritating chemicals
- [ ] sawing
- [ ] knife or other sharp tools
- [ ] other:

Work-related exposure to:
- [ ] chemical splashes
- [ ] extreme heat/cold
- [ ] sharp or rough edges
- [ ] other:

Can hazard be eliminated without the use of PPE?
Yes [ ]   No [ ]

If no, use:
- [ ] Vest, Jacket
- [ ] Coveralls, Body suit
- [ ] Raingear
- [ ] Apron
- [ ] Welding leathers
- [ ] Abrasion/cut resistance
- [ ] Other:

## BODY/WHOLE

Work activities such as:
- [ ] building maintenance
- [ ] construction
- [ ] utility work
- [ ] order pulling
- [ ] warehouse operation
- [ ] other:

Work-related exposure to:
- [ ] working from heights of 4 feet or more
- [ ] working near water
- [ ] other:

Can hazard be eliminated without the use of PPE?
Yes [ ]   No [ ]

If no, use:
- [ ] Fall Arrest protection/Restraint:   Type:
- [ ] PFD:   Type:
- [ ] Other:

*(See Footnote 1)

## LUNGS/RESPIRATORY

| Work activities such as: | | Work-related exposure to: | Can hazard be eliminated without the use of PPE? |
|---|---|---|---|
| ☐ cleaning | ☐ pouring | ☐ irritating dust or particulate | Yes ☐   No ☐ |
| ☐ mixing | ☐ sawing | ☐ irritating or toxic gas/vapor | |
| ☐ painting | | ☐ other: | ☐ hearing protection, Type |
| ☐ fiberglass installation | | | |
| ☐ compressed air or gas operations | | | |
| ☐ other: | | | *(See Footnote 1) |

## EARS/HEARING

| Work activities such as: | | Work-related exposure to: | Can hazard be eliminated without the use of PPE? |
|---|---|---|---|
| ☐ generator | ☐ grinding | ☐ loud noises | Yes ☐   No ☐ |
| ☐ ventilation fans | ☐ machining | ☐ loud work environment | |
| ☐ motors | ☐ routers | ☐ noisy machines/tools | |
| ☐ sanding | ☐ sawing | ☐ punch or brake presses | |
| ☐ pneumatic equipment | | ☐ other: | |
| ☐ punch or brake presses | | | |
| ☐ use of conveyors | | | |
| ☐ other: | | | *(See Footnote 1) |

# G.4 JOB HAZARD ANALYSIS ASSESSMENT FOR PPE

The Job Hazard Analysis (JHA) approach to doing a hazard assessment for PPE is useful in larger businesses with many hazards and/or complex safety issues. It also helps you assign a *Risk Priority Code* to the hazard to determine the course of actions you need to take to control the hazard.

Follow the instructions as you conduct your hazard assessment and fill in the hazard assessment form. Customize the form to fit the needs of your work place.

This tool can also serve as written documentation that you have done a hazard assessment in order to document your hazard assessment for PPE. Make sure that the blank fields at the bottom of the form (indicated by *) are filled out.

- Name of your work place
- Address of the work place where you are doing the hazard assessment
- Name of person certifying that a workplace hazard assessment was done
- Date the hazard assessment was done

## G.4.1 Job Hazard Analysis Assessment for PPE

### G.4.1.1 Instructions

1. **Conduct a walk-through survey of your business.** For each job/task step, note the presence of any of the following hazard types (see table below), their sources, and the body parts at risk. Fill out the left side of the hazard assessment form. Gather all the information you can.
   - Look at all steps and task in a job and ask the employee if there are any variations in the job steps and task that are infrequently done and that you might have missed during your observation.
   - For purposes of the assessment, assume that no PPE is being worn by the affected employees even though they may actually be wearing what they need to do the job safely.
   - Note all observed hazards. *This list does not cover all possible hazards that employees may face or for which personal protective equipment may be required.* Noisy environments or those which may require respirators must be evaluated with appropriate test equipment to quantify the exposure level when overexposure is suspected.

| Hazard Type | General Description of Hazard Type |
|---|---|
| Impact | Person can strike an object or be struck by a moving or flying or falling object. |
| Penetration | Person can strike, be struck by, or fall upon an object or tool that would break the skin. |
| Crush or pinch | An object(s) or machine may crush or pinch a body or body part. |
| Harmful Dust | Presence of dust that may cause irritation, or breathing or vision difficulty. May also have ignition potential. |
| Chemical | Exposure from spills, splashing, or other contact with chemical substances or harmful dusts that could cause illness, irritation, burns, asphyxiation, breathing or vision difficulty, or other toxic health effects. May also have ignition potential. |
| Heat | Exposure to radiant heat sources, splashes or spills of hot material, or work in hot environments. |
| Light (optical) Radiation | Exposure to strong light sources, glare, or intense light exposure which is a byproduct of a process. |
| Electrical Contact | Exposure to contact with or proximity to live or potentially live electrical objects. |
| Ergonomic hazards | Repetitive movements, awkward postures, vibration, heavy lifting, etc. |
| Environmental hazards | Conditions in the work place that could cause discomfort or negative health effects. |

2. **Analyze the hazard.** For each job task with a hazard source identified, use the Job Hazard Analysis Matrix table and discuss the hazard with the affected employee and supervisor. Fill out the right side of the hazard assessment form:
   - Rate the SEVERITY of injury that would *reasonably* be expected to result from exposure to the hazard.
   - Rate the PROBABILITY of an accident actually happening.
   - Assign a RISK CODE based upon the intersection of the SEVERITY and PROBABILITY ratings on the matrix.

## Job Hazard Analysis Matrix

### Severity of Injury

| | | Probability of an Accident Occurring | | | | |
|---|---|---|---|---|---|---|
| Level | Description | A Frequent | B Several Times | C Occasional | D Possible | E Extremely Improbable |
| I | Fatal or Permanent Disability | 1 | 1 | 1 | 2 | 3 |
| II | Severe Illness or Injury | 1 | 1 | 2 | 2 | 3 |
| III | Minor Injury or Illness | 2 | 2 | 2–3 | 3 | 3 |
| IV | No Injury or Illness | 3 | 3 | 3 | 3 | 3 |

### Risk Priority

| Code | Risk Level | Action Required |
|---|---|---|
| 1 | **High** | Work activities must be suspended immediately until hazard can be eliminated or controlled or reduced to a lower level. |
| 2 | **Medium** | Job hazards are unacceptable and must be controlled by engineering, administrative, or personal protective equipment methods as soon as possible. |
| 3 | **Low** | No real or significant hazard exists. Controls are not required but may increase the comfort level of employees. |

3. **Take action on the assessment**. Depending on the assigned Risk Level/Code (or Risk priority), take the corresponding action according to the table above:
   - If Risk priority is LOW (3) for a task step → requires no further action.
   - If Risk priority is MEDIUM (2) → select and implement appropriate controls.
   - If Risk priority is HIGH (1) → immediately stop the task step until appropriate controls can be implemented. Note: We will cover risk assessment in detail in Chapter 6, Defining Associated Risk.

A high risk priority means that there is a reasonable to high probability that an employee will be severely disabled doing this task and/or a high probability that the employee will suffer severe illness or injury!

4. **Select PPE:**
   - Try to reduce employee exposure to the hazard by first implementing engineering, work practice, and/or administrative controls. If PPE is supplied, it must be appropriately matched to the hazard to provide effective protection, durability, and proper fit to the worker. Note the control method to be implemented in the far right column.
5. **Certify the hazard assessment:**
   - Certify on the hazard assessment form that you have done the hazard assessment and implemented the needed controls.
   - Incorporate any new PPE requirements that you have developed into your written injury prevention program.

**Job Hazard Analysis for Personal Protective Equipment (PPE) Assessment**

Job/Task: _____    Location: _____

| Job/Task Step | Hazard Type | Hazard Source | Body Parts At Risk | Severity | Probability | Control Method[1] |
|---|---|---|---|---|---|---|
|  |  |  |  |  |  |  |
|  |  |  |  |  |  |  |
|  |  |  |  |  |  |  |
|  |  |  |  |  |  |  |
|  |  |  |  |  |  |  |
|  |  |  |  |  |  |  |
|  |  |  |  |  |  |  |
|  |  |  |  |  |  |  |

(1) Note: Engineering, work practice, and/or administrative hazard controls such as guarding must be used, if feasible, before requiring employees to use personal protective equipment.

Certification of Assessment

**Name of work place:** _____    **Address** _____

**Assessment Conducted By:** _____    Title: _____    **Date(s) of Assessment** _____

**Page**

Implementation of Controls Approved By: _____    Title: _____    Date: _____

## G.5 EXAMPLE PERSONAL PROTECTIVE EQUIPMENT TRAINING CERTIFICATION FORM

Employee's Name: _____ Employee ID No. _____
Job Title/Work area: _____
Employer: _____
Trainer's Name (person completing this form): _____
Date of Training: _____

Types of PPE employee is being trained to use:

_____        _____

_____        _____

_____        _____

The following information and training on the personal protective equipment (PPE) listed above were covered in the training session:

The limitations of personal protective equipment: PPE alone cannot protect the employee from on-the-job hazards.

What work place hazards the employee faces, the types of personal protective equipment that the employee must use to be protected from these hazards, and how the PPE will protect the employee while doing his/her tasks.

When the employee must wear or use the personal protective equipment.

How to use the personal protective equipment properly on-the-job, including putting it on, taking it off, and wearing and adjusting it (if applicable) for a comfortable and effective fit.

How to properly care for and maintain the personal protective equipment: look for signs of wear, clean and disinfect, and dispose of PPE.

*Note to employee*: *This form will be made a part of your personal file. Please read and understand its contents before signing.*

(Employee) I understand the training I have received, and I can use PPE properly.

_____        _____

Employee's signature                          Date

(Trainer must check off)
Employee has shown an understanding of the training.
Employee has shown the ability to use the PPE properly.

_____        _____

Trainer's signature                            Date

# G.6 EXAMPLE PERSONAL PROTECTIVE EQUIPMENT TRAINING QUIZ

(REQUIRED)

(This is a sample quiz that you can use to make sure an employee has understood the training and can demonstrate the proper use and care of personal protective equipment. Also quiz any employee who has been retrained due to improper use of the PPE in performing his/her job tasks. You can keep this form in the employee's file with the PPE Certification Form.)

1. What are the limitations of personal protective equipment?
2. List the types of personal protective equipment you must use when doing your work/tasks.
3. What are the hazards in your job for which you must use each type of PPE, and when must you use your personal protective equipment?
4. What are the procedures for the proper use, care, and maintenance of your PPE?
5. What should you look for to determine that your PPE is in good working condition?
6. What do you do when your PPE is no longer usable?
7. (Trainer/Supervisor:) Have the employee demonstrate putting on, wearing and adjusting, and taking off each PPE properly. Also have employee demonstrate how to clean and disinfect each PPE.

Has employee demonstrated proper use and care of each PPE?

PPE #1: _____   Yes _____   No _____
PPE #2: _____   Yes _____   No _____
PPE #3: _____   Yes _____   No _____
PPE #4: _____   Yes _____   No _____

The employee has answered all the questions adequately and has demonstrated the ability to properly use and care for the PPE needed to do his/her job.

_____          _____

Trainer's/Supervisor's signature          Date

_____          _____

Employee's signature          Date

# G.7 SAMPLE PPE POLICIES

## G.7.1 Instructions

In addition, the Consultation Section of the Department of Labor and Industries may be called on for assistance at any time.

THE FOLLOWING PERSONAL PROTECTIVE EQUIPMENT (PPE) POLICIES PROVIDE A POSSIBLE FORMAT THAT CAN BE CUSTOMIZED for YOUR WORK PLACE.

REMEMBER: YOUR SAFETY AND HEALTH PROGRAM CAN ONLY BE EFFECTIVE IF IT IS PUT INTO PRACTICE!

## G.7.2 Personal Protective Equipment Policies
**(Customize by adding the name of your business)**

### Introduction

The purpose of the Personal Protective Equipment Policies is to protect the employees of (**Name of your business**) from exposure to work place hazards and the risk of injury through the use of personal protective equipment (PPE). PPE is not a substitute for more effective control methods and its use will be considered only when other means of protection against hazards are not adequate or feasible. It will be used in conjunction with other controls unless no other means of hazard control exist.

Personal protective equipment will be provided, used, and maintained when it has been determined that its use is required to ensure the safety and health of our employees and that such use will lessen the likelihood of occupational injury and/or illness.

This section addresses general PPE requirements, including eye and face, head, foot and leg, hand and arm, body (torso) protection, and protection from drowning. Separate programs exist for respiratory protection and hearing protection as the need for participation in these programs is established through industrial hygiene monitoring. **(List other programs or policies requiring PPE, such as Hearing Protection, Respiratory Protection, Fall Protection, etc., that you may have at your work place)** are also addressed in **(State the section or location in your Accident Prevention Program where they are found)**

The **(Name of your business)** Personal Protective Equipment Policies includes:

- Responsibilities of supervisors and employees
- Hazard assessment and PPE selection
- Employee training
- Cleaning and maintenance of PPE

### Responsibilities

**(Customize this page by modifying or adding any additional responsibilities and deleting those that may not apply to your company.)**

### Safety Person (or designated person responsible for your work place safety and health program.)

*Note: Depending on your business and the number of employees you have, you may simply have a "designated safety coordinator" (who may be a supervisor/lead worker) or a larger organized safety and health unit. Customize this section to fit the needs of your plant*

**(Safety Coordinator or designated person)** is responsible for the development, implementation, and administration of **(Name of your business)**'s PPE policies. This involves

1. Conducting workplace hazard assessments to determine the presence of hazards which necessitate the use of PPE.
2. Selecting and purchasing PPE.
3. Reviewing, updating, and conducting PPE hazard assessments whenever
   - a job changes
   - new equipment is used

- there has been an accident
- a supervisor or employee requests it
- or at least every year
4. Maintaining records on hazard assessments.
5. Maintaining records on PPE assignments and training.
6. Providing training, guidance, and assistance to supervisors and employees on the proper use, care, and cleaning of approved PPE.
7. Periodically re-evaluating the suitability of previously selected PPE.
8. Reviewing, updating, and evaluating the overall effectiveness of PPE use, training, and policies.

## Supervisors (leads, etc., and/or designated persons)

Supervisors **(leads, etc., and/or designated persons)** have the primary responsibility for implementing and enforcing PPE use and policies in their work area. This involves

1. Providing appropriate PPE and making it available to employees.
2. Ensuring that employees are trained on the proper use, care, and cleaning of PPE.
3. Ensuring that PPE training certification and evaluation forms are signed and given to **(Safety Person or designated person responsible for your work place safety and health program)**.
4. Ensuring that employees properly use and maintain their PPE, and follow **(Name of your business)** PPE policies and rules.
5. Notifying **(Name of your business)** management and the Safety Person when new hazards are introduced or when processes are added or changed.
6. Ensuring that defective or damaged PPE is immediately disposed of and replaced.

## Employees

The PPE user is responsible for following the requirements of the PPE policies. This involves:

1. Properly wearing PPE as required.
2. Attending required training sessions.
3. Properly caring for, cleaning, maintaining, and inspecting PPE as required.
4. Following **(Name of your business)** PPE policies and rules.
5. Informing the supervisor of the need to repair or replace PPE.

Employees who repeatedly disregard and do not follow PPE policies and rules will be **(Write in the actions management will take concerning this matter.)**

> (Customize this page by modifying or adding any additional responsibilities and deleting those that may not apply to your company.)

## Procedures

### a. Hazard Assessment for PPE

**(Safety Person or designated person)**, in conjunction with Supervisors, will conduct a walk-through survey of each work area to identify sources of work hazards. Each survey will be documented using the Hazard Assessment Certification Form, which identifies the work area surveyed, the person conducting the survey, findings of potential hazards, and date of the survey. **(Safety Person or designated person)** will keep the forms in the **(Specify exact location, e.g., your company's business files).**

**(Safety Person or designated person)** will conduct, review, and update the hazard assessment for PPE whenever

- a job changes
- new equipment or process is installed
- there has been an accident
- whenever a supervisor or employee requests it
- or at least every year

Any new PPE requirements that are developed will be added into **(Name of your business)**'s written accident prevention program.

### b. Selection of PPE

Once the hazards of a workplace have been identified, **(Safety Person or designated person)** will determine if the hazards can first be eliminated or reduced by methods other than PPE, i.e., methods that do not rely on employee behavior, such as engineering controls (refer to Appendix B – Controlling Hazards).

If such methods are not adequate or feasible, then (**Safety Person or designated person**) will determine the suitability of the PPE presently available; and as necessary, will select new or additional equipment which ensures a level of protection greater than the minimum required to protect our employees from the hazards (refer to Appendix C – Selection of PPE). Care will be taken to recognize the possibility of multiple and simultaneous exposure to a variety of hazards. Adequate protection against the highest level of each of the hazards will be recommended for purchase.

All personal protective clothing and equipment will be of safe design and construction for the work to be performed and will be maintained in a sanitary and reliable condition. Only those items of protective clothing and equipment that meet NIOSH or ANSI (American National Standards Institute) standards will be procured or accepted for use. Newly purchased PPE must conform to the updated ANSI standards which have been incorporated into the PPE regulations, as follows:

- Eye and Face Protection ANSI Z87.1-1989
- Head Protection ANSI Z89.1-1986
- Foot Protection ANSI Z41.1-1991
- Hand Protection (There are no ANSI standards for gloves, however, selection must be based on the performance characteristics of the glove in relation to the tasks to be performed.)

Affected employees whose jobs require the use of PPE will be informed of the PPE selection and will be provided PPE by (**Name of your business**) at no charge. Careful consideration will be given to the comfort and proper fit of PPE in order to ensure that the right size is selected and that it will be used.

## A. Training

Any worker required to wear PPE will receive training in the proper use and care of PPE before being allowed to perform work requiring the use of PPE. Periodic retraining will be offered to PPE users as needed. The training will include, but not necessarily be limited to, the following subjects:

- When PPE is necessary to be worn
- What PPE is necessary
- How to properly don, doff, adjust, and wear PPE
- The limitations of the PPE
- The proper care, maintenance, useful life, and disposal of the PPE

After the training, the employees will demonstrate that they understand how to use PPE properly, or they will be retrained.

Training of each employee will be documented using the Personal Protective Equipment Training Documentation Form *(or **whatever form your company uses)*** and kept on file. The document certifies that the employee has received and understood the required training on the specific PPE he/she will be using.

The PPE Training Quiz will be used to evaluate employees' understanding and will be kept in the employee training records.

*Retraining*
The need for retraining will be indicated when

- an employee's work habits or knowledge indicates a lack of the necessary understanding, motivation, and skills required to use the PPE (i.e., uses PPE improperly)
- new equipment is installed
- changes in the work place make previous training out-of-date
- changes in the types of PPE to be used make previous training out-of-date

## B. Cleaning and Maintenance of PPE

It is important that all PPE be kept clean and properly maintained. Cleaning is particularly important for eye and face protection where dirty or fogged lenses could impair vision. Employees must inspect, clean, and maintain their PPE according to the manufacturers' instructions before and after each use (see attached). *(**Attach a copy of the manufacturers' cleaning and care instructions for all PPE provided to your employees)*.** Supervisors are responsible for ensuring that users properly maintain their PPE in good condition.

Personal protective equipment must not be shared between employees until it has been properly cleaned and sanitized. PPE will be distributed for individual use whenever possible.

Defective or damaged PPE will not be used and will be immediately discarded and replaced.

*NOTE*: *Defective equipment can be worse than no PPE at all. Employees would avoid a hazardous situation if they knew they were not protected; but they would get closer to the hazard if they erroneously believed they were protected, and therefore would be at greater risk.*

It is also important to ensure that contaminated PPE which cannot be decontaminated is disposed of in a manner that protects employees from exposure to hazards.

## C. Safety Disciplinary Policy

*(Customize by adding your company name here)* believes that a safety and health Accident Prevention Program is unenforceable without some type of disciplinary policy. Our company believes that in order to maintain a safe and healthful workplace, the employees must be cognizant and aware of all company, State, and Federal safety and health regulations as they apply to the specific job duties required. The following disciplinary policy is in effect and will be applied to all safety and health violations.

The following steps will be followed unless the seriousness of the violation would dictate going directly to Step 2 or Step 3.

1. A first time violation will be discussed orally between company supervision and the employee. This will be done as soon as possible.
2. A second time offense will be followed up in written form and a copy of this written documentation will be entered into the employee's personnel folder.
3. A third time violation will result in time off or possible termination, depending on the seriousness of the violation.

---

**(Customize this page by adding any additional disciplinary actions and deleting those that may not apply to your company.)**

# Appendix H

## H.1 SAFETY REVIEW OF NEW/RELOCATED EQUIPMENT MAJOR MODIFICATION SIGN-OFF FORM

**Do not Operate Equipment Until this Tag is Completed**

| Project Title | |
|---|---|
| Type of Equipment | Catalog # |
| MFG Serial # | Plant/facility Location |
| Comments: (as applicable) | |

**Installation Completed**

| This equipment has been properly installed and the proper guarding and safety equipment has been put in place. Have discussed installation with Division. | | |
|---|---|---|
| Project Manager | | |
| | Signature | Date |
| Comments | | |

| This equipment has been properly installed and the proper guarding and safety equipment has been put in place. | | |
|---|---|---|
| Maintenance Manager | | |
| | Signature | Date |
| Comments | | |

| I have reviewed this equipment from a safety standpoint. This equipment is accepted. | | |
|---|---|---|
| Operation Manager | | |
| | Signature | Date |
| Comments | | |

| I have reviewed this equipment from a safety standpoint. This equipment is accepted. | | |
|---|---|---|
| Production Supervisor | | |
| | Signature | Date |
| Comments | | |

| I have reviewed this equipment from a safety standpoint. This equipment is accepted. | | |
|---|---|---|
| Safety    Coordinator/ Manager | | |
| | Signature | Date |
| Comments | | |

# Appendix I

## I.1 OTHER ANALYTICAL TOOLS FOR CONSIDERATION

**A number of other analytical tools, such as "What if," checklist, hazard and operability study (HAZOP), failure mode and effect analysis (FMEA), or "fault-tree" analysis can be used to determine possible process breakdowns. You then can design prevention/controls for the likely causes of these unwanted events.**

*—Process Hazard Analysis*

A process can be defined as any series of actions or operations that convert raw material into a product. A process terminates in a finished product ready for consumption or in a product that is the raw material for subsequent processes. The process hazard analysis is used to assess the types of hazards found within a process.

OSHA's Process Safety Management (PSM) of Highly Hazardous Chemicals defines process as any activity involving a highly hazardous chemical, including any use, storage, manufacturing, handling, or on-site movement of such chemicals, or a combination of these activities. The objective of this standard is to protect employees by preventing or minimizing the consequences of chemical incidents involving very hazardous chemicals.

Refer to the appendices to determine what is covered by the standard and to take advantage of the standard's greater detail regarding requirements for establishing a Process Safety Management (PSM) system. The concept of PSM is relevant and useful to the full range of workplaces, not only those subject to the standard's requirements [7].

The PHA is a detailed study of the actions and operations to identify to the degree possible hazards that may develop. Every element of the process must be studied. Each action of every piece of equipment, each substance present, and every move made by an employee must be assumed initially to pose a hazard to employees. PHA is applied to show that the element either poses no hazard, poses a hazard that is controlled in every foreseeable circumstance, or poses an uncontrolled hazard [7].

## I.1.1 "What if" Analysis

The analysis starts with points in the process where something could go wrong. You determine "what else" could happen, and assess all possible outcomes. You must plan additional prevention and controls for those possible unplanned events that could contribute to an undesirable outcome [7].

For more complex processes, the "what if" study is best organized through the use of a "checklist." Aspects of the process are assigned to analysis team members with the greatest experience or skill in those areas. Operator practices and job knowledge are audited, the suitability of equipment and materials of construction is studied, the chemistry of the process and the control systems are reviewed, and the operating and maintenance records are audited [7].

### What if

For a relatively uncomplicated process, review the process from raw materials to finished product. At each handling or processing step, formulate and answer "what if" questions to evaluate the effects of component failures or procedural errors on the process (OSHA 3071, p. 18).

### What-if/Checklist

The what-if/checklist is a broadly based hazard assessment technique that combines the creative thinking of a selected team of specialists with the methodical focus of a prepared checklist. The result is a comprehensive hazard analysis useful in training operating personnel on the hazards of the particular operation.

The review team is selected to represent a wide range of disciplines, such as production, mechanical, technical, and safety. Each employee is given a basic information packet regarding the operation to be studied. This packet typically includes information on known hazards of materials, process technology, procedures, equipment design, instrumentation control, incident experience, and previous hazard reviews. A tour of the operation is conducted. The review team methodically examines the operation from receipt of raw materials to delivery of the finished product to the customer's site. At each step, the group collectively generates a listing of "what-if" questions regarding the hazards and safety of the operation. When the review team has completed listing its spontaneously generated questions, it systematically goes through a prepared checklist to stimulate additional questions.

Subsequently, the review team develops answers for each question. They work to achieve a consensus for each question and answer. From these answers, a listing of recommendations is developed, specifying the need for additional action or study. The recommendations, along with the list of questions and answers, become the key elements of the hazard assessment report (OSHA 3071, p. 19–20).

## I.1.2 Hazard and Operability Study (HAZOP)

Hazard and operability (HAZOP) is a method for systematically comparing each element of a system against potential for critical parameters deviation from the intended design conditions that could create hazards and operability problems. Typically, an analysis team studies the piping and instrument diagrams or plant model. The team analyzes the effects of potential deviations from design conditions in, for example, flow, temperature, pressure, and time. Then the effect of deviations from design conditions of each parameter is examined. A list of keywords, such as "more of," "less of," "part of," is selected for use in describing each potential deviation.

The team then assesses the system's existing safeguards, the causes of and potential for system failure, and the requirements for protection.

The system is evaluated as designed and with deviations noted. All causes of failure are identified. Existing safeguards and protection are identified. An assessment is made weighing the consequences, causes, and protection requirements involved (OSHA 3071, p. 20).

## I.1.3 Failure Mode and Effect Analysis (FMEA)

The FMEA is a methodical study of component failures. This review starts with a diagram of the operations, and includes all components that could fail and conceivably affect the safety of the operation. Typical examples of components that fail are instrument transmitters, controllers, valves, pumps, and rotometers. These components are listed on a data tabulation sheet and individually analyzed for the following:

- Potential mode of failure (i.e., open, closed, on, off, leaks).
- Consequence of the failure; effect on other components and effects on whole system.
- Hazard class (i.e., high, moderate, low).
- Probability of failure.
- Detection methods.
- Compensating provision/remarks.

Multiple concurrent failures are also included in the analysis. The last step in the analysis is to analyze the data for each component or multiple component failure and develop a series of recommendations appropriate to risk management (OSHA 3071, p. 20–21).

## I.1.4 Fault Tree Analysis (FTA)

The FTA begins with the definition of an undesirable outcome and pulls together all of the components necessary for that occurrence.

A fault tree analysis can be either a qualitative or a quantitative model of the undesirable outcomes, such as a toxic gas release or explosion, which could result from a specific initiating event. It begins with a graphic representation (using logic symbols) of all possible sequences of events that could result in an incident. The resulting diagram looks like a tree with many branches. The diagram lists the sequential events (failures) for different independent paths to the top or undesired event. Probabilities (using failure rate data) are assigned to each event and then used to calculate the probability of occurrence of the undesired event.

The technique is particularly useful in evaluating the effect of alternative actions on reducing the probability of occurrence of the undesired event (OSHA 3071, p. 20–21) (Insert Figure 9-2).

## I.1.5 Phase or Activity Hazard Analysis

Phase hazard analysis is a useful tool for construction and other industries that involve a rapidly changing work environment, different contractors, and widely different operations. A phase is defined as an operation involving a type of work that presents hazards not experienced in previous operations, or an operation where a new subcontractor or work crew is to perform work. In this type of hazard analysis, before beginning each major phase of work, the contractor or site manager should assess the hazards in the new phase. Appropriate supplies and support are coordinated as well as preparation for hazards expected through a plan to eliminate or control them [7].

To find these evolving hazards for elimination or control, you will use many of the same techniques that you use in routine hazard analysis, change analysis, process analysis, and job analysis. The major additional challenge will be to find those hazards that develop when combinations of activities occur in close proximity. Employees for several contractors with differing expertise may be intermingled. They will need to learn how to protect themselves from the hazards associated with the work of nearby colleagues as well as the hazards connected to their own work and the hazards presented by combinations of the two kinds of work [7].

The Corps of Engineers uses an Activity Hazard Analysis that must be developed before each activity. According to EM 385-1-1 before beginning each activity involving a type of work presenting hazard not experienced in previous project operations, or where a new work crew or subcontractor is to perform the work, activity hazard analyses shall be performed by the contractors performing the work activity. Refer to the appendix in the referenced document for an example Activity Hazard Analysis [8] EM385-1-1, pp. 3, 4.

# Part 2

## Developing Systems that Support Hazard Recognition

# 4

## Understanding the Human Role in the Safety Process

Control of hazards must be supported by management systems that take into account basic concepts of human behavior. This chapter provides an overview of the basic elements used to evaluate aspects of human behavior in the workplace. At the end of the chapter you will be able to:

- Discuss a basic overview of the behavioral process
- Define elements that contribute to at-risk events
- Identify some reasoning as to why people put themselves at risk
- Cite the seven guiding principles and five core functions of integrated safety management
- Determine if your organization is ready for a behavioral approach to safety.

"The culture is the way that it is around here."

—Dan Peterson

"There are two primary choices in life: to accept conditions as they exist, or accept the responsibility for changing them."

—Dr. Denis Waitley

An enormous body of literature from psychology to management science attempts to explain and map what can be expected when humans work together. Social norms, laws, rules, and procedures are all an attempt to funnel human activities toward some goal or objective.

The shift from one individual to a work team to a department, and ultimately to the organization as a whole, alters at each level how humans behave in their environment.

In evaluating organizations with the intent of improving the safety process, various levels of activities must be considered. A misconception often exists that unsafe actions are only restricted to certain selected individuals ("employees") who need to change their behaviors. So, when individuals hear about safety performance, many still believe that efforts must be aimed only at the "employees"—trying to change their behaviors by conducting observations of their work activities. In reality, *everyone* in an organization is an employee, as both management and workers are influenced by peer pressure and other factors that result in positive and/or negative feedback. The feedback comes from many sources and in various strengths, depths, and scope.

Management has to stop and think about what it must do day-by-day to change the culture of an organization in order to shape the actions of employees. Even different departments within an organization have unique behavioral patterns. The overall organizational actions, and the norms in the industry, establish patterns of behavior that are expected from the employee. These include adherence to facility rules, dress codes, etiquette, performance goals, etc.

For the purpose of JHA development, we are targeting existing and potential hazards and looking for physical consequences of exposures that are associated with the risk and hazards. We are trying to stop at-risk events by developing preventive measures.

The ability to stop at-risk events really rests on how well we manage to create the proper consequences of events that lead to a desired behavior. For example, are predetermined consequences driving actions towards safe behaviors and use of hazard controls? Or is the system generating unintended consequences that create an environment in which employees attempt to hide safety issues, because they believe that their actions must be done in a certain way in order to keep on production schedules, no matter what the outcome.

---

The Governor of Georgia has introduced a bill to curb "excessive speeders." The problem is that the speed limit is 55 and "excessive" speeding is defined as 85 mph, which is 30 mph over the existing speed limit. The question in the author's mind is: are we trying to change behaviors in the correct manner? In our sense, this is driving the behavior in the wrong direction.

---

## 4.1 HOW ARE AT-RISK EVENTS DEVELOPED?

Each industry responds differently to the challenges it faces. Members of the news media will knowingly enter an unsafe area, such as a war zone or hurricane, in search of a BIG story and consider this an acceptable risk. An acceptable risk in the construction industry is different from that in general industry, and also from that of an office environment [2]. Our attitudes are driven by an array of management, co-workers, family members, friends, peer groups, television, and other environmental factors that surround us.

The term culture has many definitions: "the ideas, customs, skills, arts, perceptions, pre-judgments, etc. of a given person, in a given period." Dan Peterson said it best: "The culture is the way that it is around here." [3]. Every organization develops a unique culture derived from its industry, its leadership, and individuals in the organization.

We see use of psychological or behavioral theory, for example, when popular TV hosts, such as Dr. Phil, talk to guests about their behaviors and how they react to situations. Reality television shows like "Big Brother," Donald Trump's "The Apprentice," "Top Chef," and "Super Nanny," just to name a few, all display various types of responses when individuals are put in specific situations.

Humans repeat activities that their culture reinforces through direct or an implied recognition and/or punishment system. If you use that view and relate it to hazard recognition, you might say that "Our actions are predictable based on our surroundings." We can say that in the work environment, our actions are a function of the management system and the perceived safety culture. Because of the relatively low probability that any one at-risk event will result in an immediate injury, an employee's perception that risks can be taken and no harm or injury will occur is reinforced. By recognizing that an array of social and other consequences drives our actions and by establishing a process of specific structured quality feedback, an organization can move towards an improved culture.

The safety culture in an organization is driven by the leadership. Management must have a clear understanding of its operations and must make full use of active employee participation in operational hazard identification. With both management and employees, the process of active involvement in identifying and encouraging safe behavior, identifying existing, potential hazards and consequences of exposures, is enhanced.

## 4.2 WHAT CONTRIBUTES TO AN AT RISK EVENT?

What are at-risk events? How do at-risk events differ from observed employee actions during the safety process? In theory, during the enhanced safety process, each employee should be observed and/or coached until an at-risk behavior has changed. Data is collected to ensure that the behavior is either changed or an action plan is developed to target specific behaviors. So the question is: If you do not see the undesirable behavior again, has it actually been changed? Or did the employee stop that behavior only when an observation was conducted?

Let's look at an example. When an observation of an employee is performed and the employee is noted performing the task under an at-risk condition, this observation would count as one observation. Under a typical behavioral process, if an employee is trying to clear a jam from a piece of equipment in an at-risk manner, we count that observation as one observation. Using the at-risk event concept, we would count the number of time that the employee actually performed the event, i.e., sticking their hands into the piece of equipment to clear the jam. In either case the employee would be coached about what could be done differently, or be shown the safe method to accomplish the task.

Therefore, the safety structure brings together the aspects of behavioral science with general management criteria to help create a process. This process, in turn, will help to promote safety as an important value within an organization. It is sometimes forgotten that behavioral observations are only one part of a process and must interact within the safety management structure in place and not be considered as an independent action.

To put all of this into perspective, we will discuss the Safety Pyramid, as it is generally used to document proportions of incidents with serious injuries, minor injuries, property damage, or no injury or damage. Refer to Figure 4-1 for an example of how the Safety Pyramid is used in many safety programs. While the validity of the pyramid has been questioned, it does provide a framework that shows a relative spectrum of loss-producing situations or potential situations. Given each situation, the elements may or may not be related, and are dependent on hazards and associated risk in a particular workplace. The pyramid is only of value if one is trying to explore the potential proportions of loss-producing events. Sometimes an incident or injury happens based on the interaction of many other things going on in both the person and the workplace [2].

As employees use the enhanced safety process and make working safely a habit, injuries should decrease. The strength of the safety process is in the measurement of "safe behaviors" shown as a percentage of specific defined

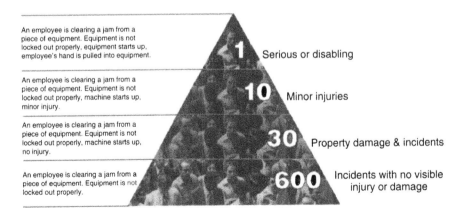

**Figure 4-1** Safety pyramid modified

Bird Frank E., Germain George L., *Loss Control Management: Practical Loss Control Leadership*, revised edition, Figure 1-3, p. 5, Det Norske Veritas (U.S.A), Inc., 1996, Adapted for use – designed by Damon Carter

observed actions as a *leading* safety indicator. Many safety metrics are lagging measures that are recorded after an injury has occurred—i.e., Total Case Incident Rates (TCIR) as offered by OSHA's recordability standard (29 CFR 1904), audit scores, etc. [1]. The behavioral approach provides a method for defining and establishing a clear safety criteria, as it is being used in real time and addresses actual actions or behaviors.

According to the Department of Energy (DOE), anecdotal evidence exists to indicate that measurement of "percentage of safe behaviors" is predictive. If used correctly, this means that the employee observation and feedback techniques may be used to predict safety issues that exist in a facility. Intensifying the safety process cycle to increase the contact rate (number of employees contacted each month or in a given short time frame) can strengthen behaviors and can help to prevent an injury [1].

As we discussed in Chapter 3, engineering and management approaches are designed to counter at-risk events by use of automation equipment, procedure compliance, administrative controls, and OSHA-type enforcement standards and rules. The engineering of hazards from the workplace has been successful in reducing the number of injuries significantly. However, injuries still continue to plague organizations, which implies that a natural lower limit may exist to injury reduction [1].

Proportions of various levels of injury and damage compared to total number of incidents, as shown in the Safety Pyramid, are estimated and are not fixed but vary widely based on the operation. In addition to the pyramid, an hourglass concept can also be used. Both graphical techniques emphasize the probability of at-risk events and the near-miss potential. Refer to Figure 4-2 for the hourglass concept [2].

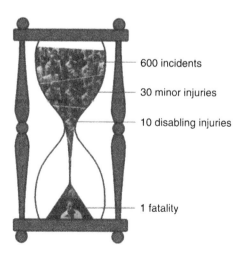

600 incidents

30 minor injuries

10 disabling injuries

1 fatality

**Figure 4-2** Hourglass concept

Adapted, Roughton, James, James Mercurio, *Developing an Effective Safety Culture: A Leadership Approach,* Butterworth-Heinemann, 2002, Safety accident pyramid, Figure 12-10, designed by Damon Carter

## 4.3 THE FEEDBACK LOOP

The "ladder of inference" indicates that we become so skilled in the way that we perform our jobs, we may operate from our prejudgments, perception, thinking, and habits. We tend to jump up a mental decision ladder without recognizing that we have made the leap [6]:

- Our mental models of how the world works and our repertoire of actions (prejudgments, perceptions, thinking, and habits) influence the data (decisions) that we make each day, the judgments that we make, and the conclusions we draw.

We tacitly register some preliminary decisions, such as from our individual day-to-day thoughts, prejudgments, perceptions, thinking, and habits, while ignoring other decisions that may indicate interaction with at-risk events.

- As we have performed our task one, two, three, one thousand, ten thousand times, etc., we tend to lose sight of how we infer what is actually happening, because we do not think about our actions or consequences. Our actions just become "natural." "Nothing risky has happened and nothing can and will ever happen," we think.

"Can't see the forest for the trees."

We feel comfortable with our conclusions because they are so obvious to us. We see no need to retrace our steps (or change our ways). We feel comfortable with what we are doing and, therefore, take unnecessary risk.

Our reasoning skill is both essential and also tends to get us in trouble (such as when an injury occurs):

- If we thought about each and every inference we made, life would pass us by and we would be incapable of making even small decisions; we would be overwhelmed by all the sensory input.
- People can and do reach different conclusions about reality and react accordingly. When we view our conclusions as "obvious," we do still need to know how we reached that conclusion.
- When people disagree, they unconsciously throw conclusions at each other from the tops of their respective Ladders of Inference.
- From this position, it is hard for individuals to resolve differences and to learn from mistakes and from each other. This slows or prevents the potential for effective team development [6].

Adapted from Actiondesign.com, The Ladder of Inference, Fifth Discipline, Peter Senge http://www.actiondesign.com/resources/concepts/ladder_intro.htm

Refer to Figure 4-3 for an overview of the Feedback Loop and the Ladder of Inference concept.

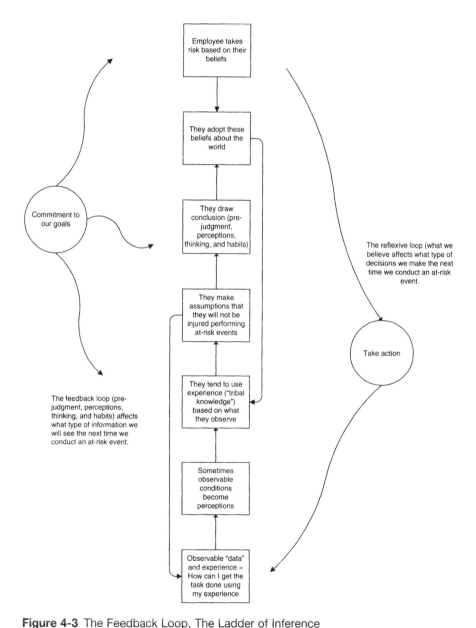

**Figure 4-3** The Feedback Loop, The Ladder of Inference

Adapted from Senge, Peter, *5th Discipline Fieldbook*, 1994 and Adapted from Acton Design.com, The Ladder of Inference, http://www.actiondesign.com/resources/concepts/ladder_ intro.htm

## 4.4 BEHAVIORAL APPROACH

The core philosophies of the behavioral approach complement and are usually integrated into other programs, such as Voluntary Protection Process (VPP), ANSI/AIHA, Z10-2005, the Occupational Health and Safety Management System [4]. The behavioral approach provides the safety process with a systematic approach for identifying and correcting at-risk events and conditions that can be immediately corrected.

The behavioral approach can be applied across a broad range of safety areas, from the production environment to the office environment, and is applicable in off-the-job safety as well. The behavioral approach enhances traditional safety tools, such as management reviews, housekeeping audits, safety meetings, etc., thereby reducing the overall safety cost. The behavioral approach provides a method to shift the focus of safety from theory or program management only, to focus on the operational process.

## 4.5 CHANGING BEHAVIOR

When you attempt to implement a behavior-based safety (BBS) process without serious discussions with management and without developing a solid action plan, it can seriously harm the safety process. Attempting to change an individual's behavior is complex. On the surface, there is the appearance of manipulation, as the employee is involved in the planning process. As humans we prefer to have a stable, generally predictable environment. As a result, when an employee who has performed a task over time both routinely and nonroutinely and has been exposed to risk without suffering any negative consequences, the need for change is not apparent and the perception of associated risk is low. This is the case of someone doing something at-risk for so long that the at-risk event has become the "right way" to do things. After all, what could go wrong?

As noted in the Department of Energy Action Plan, *Lessons Learned from the Columbia Space Shuttle Accident and Davis-Besse Reactor Pressure-Vessel Head Corrosion Event*: "When the space shuttle Columbia launched on January 16, 2003, there were 3,233 Criticality 1/1R critical item list hazards that were waived. Hazards that result in Criticality 1/1R component failures are defined as those that will result in loss of the orbiter and crew. In both the Challenger and Columbia accidents: 'The machine was talking to us, but nobody was listening.' " [5]

"Safety must be a value to the organization and not a Priority or Number One."

An organization is a complex interactive real-time-event-developing system. Changing any part of a system can create cascading effects that may or may not be desired or anticipated. Changes in the workplace can and will create new and exotic hazards. For example, the introduction of the personal computer initiated a severe epidemic of carpal tunnel and other ergonomic-related issues, resulting in a new industry dedicated to workstation and equipment design. New audio devices, while bringing joy to many, may also in the long run result in hearing loss to serious users of such devices. Studies have shown that cell-phone usage while driving may be increasing automobile crashes. The use of new technologies has developed greater potential for at-risk events and has resulted in new behaviors based on the consequences of their use—i.e., immediate entertainment is a positive consequence while the loss of hearing is a remote negative consequence (we cannot see it because it is not immediate). We tend to move from one behavior to another in a manner that may limit our ability to weigh the impact of the new behavior, as we may or may not recognize the change we are undergoing.

As we discussed in Part 1, hazard-recognition techniques and other actions can be used to locate and map hazards and associated risk. A behavioral approach does not identify the interrelationships of specific hazards since it focuses only on employee's behaviors. The JHA provides the mechanism that helps bridge the gap between how the job is completed and the observation of defined at-risk events using a structured data collection method. Refer to Figure 4-4 for an overview of how the JHA and at-risk events can be integrated into the JHA process.

## 4.5.1 Understanding Why Employees Put Themselves at Risk

All human actions invoke an element of risk. Each day, we are engaged in thousands of at-risk events, many times putting ourselves at risk. Think about what you do on a daily basis. Try to become aware and notice the level and severity of risk that you actually create for yourself and others. A daily example would be driving on a major highway, moving from one lane to another in heavy traffic while on a cell phone, eating, or even reading. What about stopping too close to the car in front of you, slowing down but not stopping at a stop sign (known as a rolling stop), not observing street signs, or learning an unfamiliar dashboard layout of a new car? Then reflect on the other drivers around you who are doing similar things. And we wonder why accidents happen!

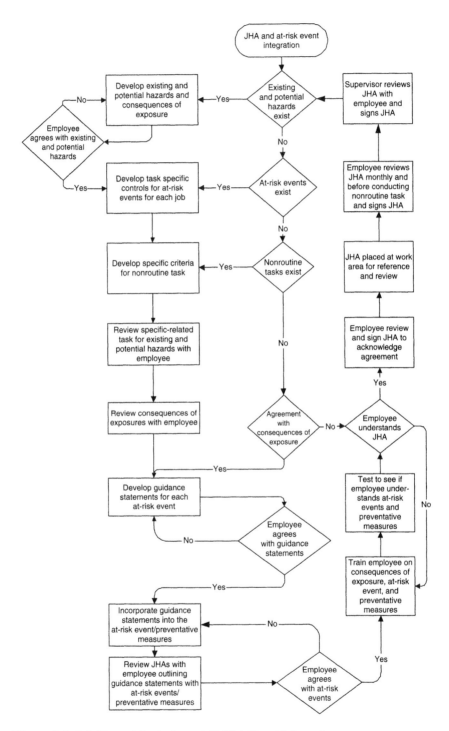

**Figure 4-4** Job Hazard Analysis and At-Risk Event Integration

One of the best way to describe behaviors is by using the ABC model: *antecedents* (activators), *behaviors* (at-risk events), and *consequences* (of exposure). While this model has been around for many years, it is not adequately or fully used to aid in assessing the workplace. Refer to Figure 4-5 for the ABC model that explains why we do what we do. This figure illustrates in a simple overview how we react in specific situations, in this case answering the telephone with a salesperson on the other side of the call.

Safety rules and operating procedures, investigating injuries, conducting safety meetings, training employees are not "behaviors"—rather, they are the "antecedents" that should drive the desired safe behavior. Refer to Chapter 7 for suggested training methods. The development of the JHA addresses the necessary antecedents and combines knowledge of the hazards and associated risk of the task with a structured way to communicate the required knowledge consistently.

Consequences can be stronger than the antecedent. To assess behavior, as many consequences of a behavior as possible must be evaluated.

As a test, develop a list of your personal behaviors during the day, along with a list of potential and existing hazards you encounter. Make an estimate of the potential consequences of those exposures. As you recognize the at-risk events you engage in without thinking, you create a better awareness of your hazard exposures and can begin to take action to minimize your chance of injury. For example, think about a motivational speaker that you heard. The speaker touched on things that you wanted to do and you got excited listening to the speaker. However, what happened when you left the seminar and tried to maintain that enthusiasm on your own? You must begin to routinely practice what you see and hear in order to become successful and to understand that changing behavior begins with your own behavior.

Consequences from your changed behavior will determine whether you stay enthusiastic or drift back into old patterns.

## 4.6 UNDERSTANDING THE OTHER SIDE OF SAFETY

To be successful in program implementation, there must be an understanding of the operational "linking pins" of management commitment, leadership and employee participation [2]. The workplace is not linear, where one action or decision leads directly to another. Having good intentions is not enough, as you are surrounded by other professional disciplines that are also demanding the attention, time and budget of management.

An overview of this type of thinking is summarized in Table 4-1. In comparison, we will discuss two scenarios with possible reasons why humans act the way they do [2].

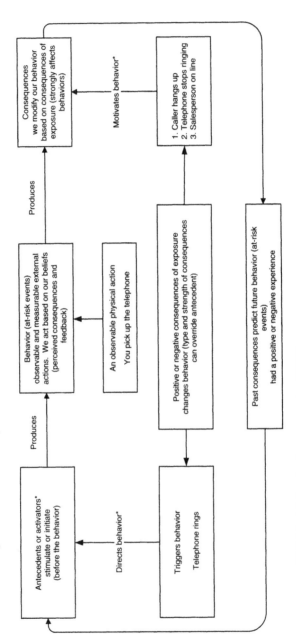

Understanding what drives behavior

Our behavior is caused by thoughts. The body responds to thoughts by producing feeling. Behavior (at-risk events) represents the outward expression of internal thoughts and beliefs.

Will you continue to answer your telephone each time it rings, or only when you want to talk to someone?

**Figure 4-5** ABC Model, Look at an example of BBS

*Adapted from the U.S. Department of Energy, Environmental, Safety, Health, Behavior Based Safety, Total Safety Culture, Worker Applied Safety Program (WASP) Training, Web Site http://www.eh.doe.gov/bbs/wasp/ineel_training2.pdf, public domain
Adapted from OR-OSHA 113, Train The Trainer II: Conducting Classroom Safety Training, Presented by the Public Education Section Oregon OSHA, Department of Consumer and Business services, public domain

## Table 4-1
## Self-Imposed Behavior vs. Traffic Cop Mentality

| Employee attitude, prejudgments, thinking, perceptions | Reason for shortcut (possible consequences) |
| --- | --- |
| Time | |
| An employee would rather be somewhere else and/or doing other things more important than working. | Taking shortcuts is usually due to a "perceived" time gain to do other things: Increase production, take more breaks or talk to co-workers. |
| Employee Perception | |
| Low Risk: better employee than anyone else; "me against them" mentality; "It will not happen to me"; Nothing has ever happened before. Therefore, I can take as many chances (at-risk events) as I want and I will not get injured. | Increases "free" time; increases production that increases bonuses; medium risk; at-risk events accepted and/or ignored by management and others, maybe even expected by management; "It can't happen to me". |
| Feelings/Emotions | |
| Excitement/thrill: pleasure; challenge (may be lessened without enforcers unless perceived as a physical ability challenge); control; freedom; choice/right. | Excitement/Thrill; pleasure; challenge (employee against machine); control; freedom; choice/right; (perceived, pre-judgments, thinking) acceptance from management/co-workers; resistance to change. |
| Beliefs | |
| This is the way that it has always been; Trained; at-risk events; proven safer technology (seatbelts, etc.) reactive | I have done this thousands of times before and have never been hurt. Trained; learned practice; at-risk events; resistance to change; reactive |
| Hazard Recognition and Design | |
| Equipment runs faster. "If they could not, they would not." | Allows archaic-unforeseen danger, hazard risk. "If they could not, they would not." |

Adapted from Roughton, James, James Mercurio, *Developing an Effective Safety Culture: A Leadership Approach*, Butterworth-Heinemann, 2002

These are addressed through acceptance and an overall culture change through an Activity-Based Safety System (ABSS), such as a safety meeting before the shift starts to set the stage for the day, one-on-one contacts, safety meetings before the shift, etc. [2].

## 4.7 BENEFITS OF BEHAVIOR-BASED SAFETY

The behavioral approach is a process that provides organizations with the opportunity to move to a higher level of safety excellence by using an understanding of antecedents and consequences and statistically valid data to build ownership, trust, and unity with employees and developing empowerment opportunities which relate to employee safety. Equally important to organizational culture, this behavioral approach provides line management with the opportunity to develop and demonstrate core values, and improve coaching and leadership skills.

A properly designed behavioral process involves employees at every level. The atmosphere of trust that results from nonpunitive observation and the feedback process leads to creating more employee participation. If a positive approach and communication is developed, employees frequently start asking to be observed and use the feedback given to modify their activity to make themselves and their fellow employees safer. A rapport develops between the observers and the employees being observed, leading to more open discussions. As trust increases, the reporting of minor injuries increases, allowing root causes to be determined.

Variation refers to the fluctuations that can occur based on inconsistent environmental factors, human actions, equipment, tools, etc. Fluctuations of injuries and behaviors occur due to these variations in actions and conditions in the workplace. The Statistical Process Control (SPC) (used in quality control) system can aid in the review of injuries and observations, and the organization can show the variations in its system. The SPC shows whether the system is out of control and needs specific issues addressed or if it is "in control" but requires major across-the-board safety program improvements.

## 4.8 BEHAVIOR-BASED SAFETY AND INTEGRATED SAFETY MANAGEMENT FUNCTIONS

The Department of Energy (DOE) BBS process outlines "Seven Guiding Principles" of a successful management system and "Five Core Functions."

Information that is broadcast across the entire organization on a day-to-day basis does not restrict the process to the actual performance of work activities. Many injuries occur when employees are involved in non-task-related activities such as walking from point A to point B, performing non-routine tasks, etc. BBS processes also provide the footprints to show that a safety management system is at work around the clock [1]. The following section has been adapted from the DOE public domain manual, *Seven Guiding Principles of a successful management system and Five Core Functions.* For more details, refer to this document.

## 4.8.1 Seven Guiding Principles of Integrated Safety Management

- Management commitment and leadership responsibility for safety. The responsibility for safety and the behavioral process is led by management with a shared involvement from knowledgeable employees. All levels of the organization are involved in an effective behavioral process.
- Clearly defined roles and responsibilities must be in place with job functions defined within the management process. These responsibilities must be performed at the proper level and must be integrated and adapted to fit the organization.
- Competence commensurate with responsibilities. An effective behavioral process ensures that the skills needed to perform the tasks and functions associated with the job (steps and tasks) in a timely manner are present and provides the opportunity to use those skills on a regular basis. It provides for coaching and interaction with other people and organizations.
- Effective use is made of Balanced Safety Data. The behavioral process provides a stream of safety data that enables managers to balance safety effectively within production and other operational needs.
- Safety standards and requirements are identified and followed. Existing safety standards and requirements aid in developing the list of behaviors and definitions used in the behavioral process.
- Hazard controls are tailored to work being performed via a JHA. The observation process along with observation data provides ongoing monitoring of processes so that hazard controls reflect the risks associated with work being performed in changing environments and conditions.
- Operations authorization. The behavioral process helps provide the behavior-related safety information necessary to make informed decisions prior to initiating operations [1].

## 4.9 WILL A BEHAVIORAL PROCESS WORK FOR YOU?

For a behavioral process to succeed, a careful assessment must be made as to whether your company is ready and whether the necessary communications, time availability, and budgets are in place. Management commitment and support, effective management systems as we already discussed, the AIHA/ANSI, VPP, and the company's culture are all keys to determining if a company is ready for the behavioral process [1].

The behavioral process requires learning not just by trial and error, but by getting things to work right the first time whenever possible. To avoid "reinventing the wheel," the management system must adapt to conditions by providing continuous improvement through the management of risk. By use of the JHA, you establish the foundation for the observation criteria [1].

The basic concept of behavior modification is the systematic use of specific targeted reinforcement. The results of using defined specific reinforcement are improved performance in the area where the reinforcement is connected closely with the behavior [3]. If a person does something and immediately following the act something pleasurable happens, he/she will be more likely to repeat that act. If a person does something and immediately following the event something painful occurs, he/she will be less likely to repeat that event again, or at least will make sure that he or she is not caught in that the same event the next time [3]. Positive reinforcement may or may not drive safe behavior. In some cases, positive consequences may exist that drive unsafe behavior—the good feeling from accomplishing a task, the thrill of doing something dangerous and getting away without injury, etc.

## 4.10 SUMMARY

This chapter has provided a basic overview of the BBS process. It was the intent of this chapter to introduce and highlight a concept as it will apply to the JHA and show how it fits into the process.

Safe behavior reinforcement requires understanding what happens after a specific behavior and recognizing employees when they do a job safely at the time they are doing it [3].

We have been involved in behavior-based safety for many years and agree 100% with the behavioral process. As we get more into its core concepts, there are many aspects of at-risk events to which we can relate. Everyone in life makes choices and, depending on those choices, the consequences can be positive or negative.

As an example, one case involved an employee getting cut with a knife. The organization's focus was on its OSHA incident rate and management tried to find fault with the employee. The supervisor was set to discipline the employee because of the injury. The possibility of a different root cause was discussed but the supervisor did not want to consider anything except that the employee should have known that this was a hazard. The supervisor was asked to focus on only one thing when discussing the injury with the employee: he was told to think about what he, as a leader, could have done differently to prevent the injury. During the review with the employee, the supervisor finally heard one thing that changed his mind:

> Supervisor: "How and why did you cut your hand? You know that you are supposed to use the proper glove and the correct tool."
>
> Employee: "I was using what was provided to me. The tool that I was using was worn, and I had to do my job. The tool slipped and I cut my finger on the knife. I have asked for a new tool for several weeks."

The employee was willing to continue to work with the worn tool in order to get the task done. He was willing to accept the risk of injury, because the environment set the tone. The consequence of continuing to work was stronger than the consequences of working under unsafe conditions.

The process can be mapped showing the consequences expected, and the expected behavior reviewed for potential acceptance of risk. A hazard was present and the control (proper tool and PPE) was allowed to deteriorate. Management did not respond in a timely fashion. The company culture dictated that the employee was to continue the work even under unsafe conditions.

In the next chapter we will discuss the importance of having the employee involved in the process. The success of any business depends on the total involvement of each employee. Without the involvement of employees, the potential for developing a full understanding of the job and how it is completed cannot be achieved.

The objective of employee participation is to encourage everyone to help in the structuring and effective functioning of the safety process and help with the decisions that directly affect their personal safety. If done correctly, the increased trust will provide everyone a means to use their insight and energy for achieving the safety goals and objectives.

## CHAPTER REVIEW QUESTIONS

1. After the risks and hazards have been identified and assessed, what needs to be assessed next?
2. What are the basic concepts of human behaviors?

3. What is the concept behind the hourglass that we presented?
4. What are the four main components of human behaviors?
5. How would you define the difference between priority and value?
6. How is the word culture defined?
7. What are some examples of measuring safe behavior responsibility, authority, and accountability?
8. A management system with a behavioral component can consist of what elements?
9. How would you compare self-imposed behavior to traffic cop mentality?
10. What are some reasons why programs and efforts are not working?

## REFERENCES

1. DOE Handbook, *Good Practices for the Behavior-Based Safety Process*, U.S. Department of Energy, Washington, D.C. October 23, 2003, Website http://www.eh.doe.gov/bbs/BBS102403.pdf, Public Domain
2. Roughton, James, James Mercurio, *Developing an Effective Safety Culture: A Leadership Approach*, Butterworth-Heinemann, 2002
3. Peterson, Dan, *The Challenge of Change*: *Creating a New Safety Culture*, Implementation Guide, CoreMedia, Development, Inc., 1993, Recognition for Performance, pp. 57–58
4. American National Standard – Occupational Health and Safety Management Systems, ANSI Z10-2005, July 25, 2005
5. *Lessons Learned from the Columbia Space Shuttle Accident and Davis-Besse Reactor Pressure-Vessel Head Corrosion Event*, U.S. Department of Energy, July 2005, page 14
6. Senge, Peter, 5th Discipline Fieldbook, 1994 and Adapted from Acton Design.com, The Ladder of Inference, http://www.actiondesign.com/resources/concepts/ladder_intro.htm

# Appendix J

## J.1 SAMPLE BEHAVIOR (AT-RISK EVENTS) LIST

The following is a sample of generic behaviors adapted from the U.S. Department of Energy, Environmental, Safety, Health, and "A Primer for Local Safety Improvement Teams in the Behavioral Based Process," developed by employees of the Westinghouse Savannah River Company.

### J.1.1 Body Use/Position

- Walking/Ascending/Descending
- Stabilized Work/Surfaces
- Proper Lifting Technique
- Ease of Movement
- Entry/Exit from Job Site
- Proximity Hazard Awareness

### J.1.2 PPE

- Protecting Hands
- Protecting Eyes/Face
- Using Fall Protection
- Protecting Head
- Protecting Feet
- Protecting Against Radiation Exposure or Contamination

### J.1.3 Work Environment

- Work Area Maintained Free of Clutter
- Waste Disposed of, Properly Tagged and Segregated

- Barricades/Warning Signs Used Properly
- Aware of Moving/rotating Equipment
- Protecting Against Heat Stress/Cold Exposure
- Working with Adequate Lighting
- Housekeeping

## J.1.4 Tools/Equipment

- Using Right Tool
- Tool Use
- Using Tool as Intended/designed
- Tool Condition
- Powered Equip/Vehicle Used as Intended
- Operating Vehicle Safely (Driving/Parking)
- Pedestrian Right of Way Honored

Adapted from the U.S. Department of Energy, Environmental, Safety, Health, "A Primer for Local Safety Improvement Teams in the Behavioral Based Process," developed by employees of the Westinghouse Savannah River Company, Website, http://www.eh.doe.gov/bbs/primer.pdf, public domain.

# 5

# Effective Use of Employee Participation

Much has been written in business literature about the need for not just management support, but for the total involvement of all personnel in continuous improvement programs. This chapter provides insights into the importance of employee involvement in the JHA process and methods that can be used to increase participation. At the end of the chapter you will be able to:

- Define why employees should be involved in the JHA process
- Identify ways to gain employee support
- Develop guidelines for employee participation
- Identify the types of safety committees that can be developed
- Identify what management can do to gather support.

"The bottom line is that people sharing their expertise in any area and in any sector leads to improvement. When you factor in the natural drive that leaders have to improve their business, then ultimately performance improvement is very possible. Good safety performance is good business as far as I am concerned."

—Duncan Hawthorne

The success of any business depends on the total involvement of each and every employee in the operation. Without employee involvement, potential for developing a full understanding of the job steps and tasks and how the job is currently completed is limited [2].

"In organizations, real power and energy is generated through relationships. The patterns of relationships and the capacities to form them are more important than tasks, functions, roles, and position."

—Margaret Wheatly

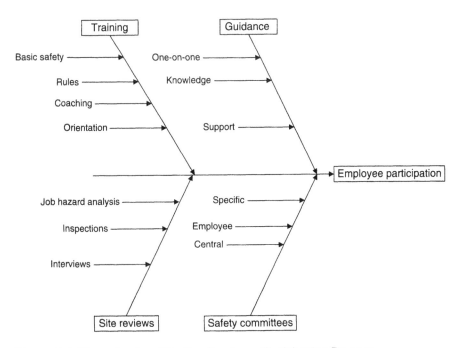

**Figure 5-1** Elements of an Effective Employee Participation Process

The objective of employee participation is to encourage everyone in the organization to help in the structuring and effective functioning of the safety process and help with the decisions that directly affect personal safety. If done consistently, the increased trust will provide everyone with a means to use their insight and energy for achieving the safety goals and objectives [2].

These activities require effective training and clear communication to ensure that each employee can perform their functions without risk of an injury. Training does not need to be elaborate classroom sessions but can be short, concise activities conducted at the workplace by employees who are appropriately trained and have an understanding of the JHA concept [2]. Refer to Figure 5-1 for the elements that can be used in an effective employee participation process.

The success of any safety process depends on the participation of employees. As part of management, you do not have to face this task alone. In this chapter, we will outline how employee participation can strengthen your management system [4].

## 5.1 WHY SHOULD EMPLOYEES BE INVOLVED?

"Without involvement, there is no commitment. Mark it down, asterisk it, circle it, underline it. No involvement, no commitment!"

—Steven Covey

This chapter takes a look at the reasons behind employee participation, and the ways you can implement a successful process [4]. Andrew Carnegie understood well the importance of quality employees as a competitive weapon. At one point in his career he suggested an epitaph he felt would be appropriate for him, which captured his management philosophy: "Here lies a man who was able to surround himself with men far cleverer than himself." [10] However, Carnegie apparently did not understand well how to include the employee as part of the team. This is the essential difference between safety efforts and all other priority initiatives in building a successful organization culture [10]. Refer to Table 5-1 for an overview of guidelines for employee participation.

As mentioned, many of these activities require a level of training to ensure that each employee can perform these functions proficiently. The training does not need to be elaborate and can be conducted at the workplace by employees who are appropriately trained [4].

## Table 5-1
## Guidelines for Employee Participation

Employee participation provides the means for employees to develop and express their own commitment to safety, for both themselves and their co-workers.

You need to recognize the value of employee participation and how the increasing number and variety of employee participation arrangements can raise legal concerns. It makes good sense to consult your human resources department to ensure that your employee participation program conforms to any legal requirements [5].

Why should employees be involved? The obvious answer is that it is the right and smart thing to do because:

- Employees are the individuals most in contact with potential hazards. They have a vested interest in an effective management system that supports safety.
- Group decisions have the advantage of the group's wider range of experience.
- Employees are more likely to support and use programs in which they have input (buy-in).
- Employees who are encouraged to offer their ideas and whose contributions are taken seriously are more satisfied and productive on the job.

## Table 5-1
### *Continued*

What can employees do to be involved? The following are examples of employee participation:

- Participating on joint labor-management committees and other advisory or specific purpose committees
- Conducting workplace inspections
- Analyzing routine hazards (JHA) in each step of a job or process, and preparing safe work practices or controls to eliminate or reduce exposure
- Developing and revising the workplace safety rules
- Training both current and newly hired/transferred and seasoned employees
- Providing programs and presentations at safety meetings
- Conducting and participating in incident investigations
- Reporting hazards and fixing hazards under their control
- Supporting co-workers by providing feedback on risks and assisting them in eliminating hazards

Performing a pre-use or change analysis for new equipment or processes in order to identify hazards up front before use.

OSHA Web site, http://www.oshaslc.gov/SLTC/safetyhealth_ecat/comp1_empl_envolv.htm, public domain

## 5.2 INVOLVING EMPLOYEES IN THE SAFETY MANAGEMENT SYSTEM

Involving all employees in the safety management system is one of the most effective approaches you can take in developing a positive safety culture. The advantage to this participation is that it promotes general awareness, instills an understanding of the comprehensive nature of a management system, and allows employees to "own" a part of the safety system. Employees make valuable problem-solvers because they are closest to the action. No one knows the job better than the employees [6]. The following outline provides some ideas on how to deal with employee participation:

- Employees are the individuals that are in contact most with potential physical hazards every day. They have a vested interest in ensuring that there is an effective management system.
- Group decisions have the advantage of a wider range of experience.

- Employees who participate in the safety process are more likely to support and use the programs.
- Employees who are encouraged to offer their ideas and whose contributions are taken seriously are more satisfied and more productive [4].

"Quality is the most important factor in business."

—Andrew Carnegie

## 5.2.1 Close Contact with Hazards

As a manager, you have an understanding of how your overall business operates. You provide the vision. However, employees who work in the organization have a more detailed knowledge of each operation and task because they do these tasks day in and day out. Unless you have recently performed a particular task over and over, your employees will know the job much better than you do [7].

Employees who understand the hazards that are associated with the workplace will realize that they have the most to gain from preventing or controlling exposure to identified hazards. Employees who are knowledgeable and aware tend to be safe employees and also good sources of ideas for better control [4].

## 5.2.2 Improved Support

Supervision and managers must understand all aspects of the safety management system [9]. Given the separation of management from day-to-day activities, management can become short-sighted or complacent with regard to the operational hazards. They may complain that they cannot get employees to use or follow safety guidelines. Chapter 4 addressed how they may have established, knowingly or unknowingly, conditions or consequences that are driving that behavior.

Management may not have the correct perception of risk or the scope of hazards that employees face daily. Financial, purchasing, or other commitments of sales or production may obscure what conditions really are. Just as the quality management process stresses total participation, a broader approach benefits the safety management process. Decisions made in the confines of an office using a spreadsheet may or may not be effective when the actions are implemented.

We as humans naturally resist change, but we have a tendency to support ideas that we help to develop and implement. Employees who are allowed to participate in the development and implementation of safety rules and procedures have more of a personal stake in making sure that these rules are followed [4].

### 5.2.3 More Participation, More Awareness

Using safety committees may not always be the best approach in reaching a decision. However, these group decisions often can establish the best buy-in from employees. Benefits are also derived from many viewpoints and varied experiences, and these can help produce better decisions.

Employee participation can be used to help identify and solve safety issues. Employees who participate enjoy their work more than those who simply do what they are told. When employees enjoy their work, they take a greater responsibility in their jobs and tend to produce a better quality product. Reduced employee turnover is a potential benefit of increased employee participation [4].

## 5.3 HAWTHORNE STUDY

In the early years of management science, the Hawthorne Study came about as the result of experiments conducted at Western Electric and involved changes in workplace conditions that produced unexpected results in employee performance. Two teams of employees took part in these experiments, in which the lighting conditions for one team were changed. Production for that group rose dramatically [3]. The interesting thing that happened is that production also improved in the group for which the lighting remained unchanged.

The study was undertaken in an effort to determine what effect environmental factors such as hours of work and periods of rest might have on employee fatigue and productivity. As the study progressed it was discovered that the social environment could have an equal if not a greater effect upon productivity than the physical environment [2].

The study also revealed the influence that informal work groups can have on the productivity of employees and on their response to such factors as supervision and financial incentives [2]. In the author's opinion, this study is a clear example of paying more attention to employees as outlined in this chapter. When employees are involved in the safety process, they can help to provide a more effective and safer environment by identifying hazard that they are exposed.

## Table 5-2
## Summary of Hawthorne Study

| Prejudgments | Finding | Safety Culture |
|---|---|---|
| Job performance depends on the individual employee. | The group is the key factor to employee's job performance. | Being involved in activities and providing input to management. |
| Fatigue is the main factor affecting output. | Perceived meaning and importance of the work determine output. | Daily leadership from management, employee (buy-in) ownership of process elements, several keys to employee safety. |
| Management sets production standards. | Workplace culture sets its own production standards. | Top management shows employee visible commitment to safety and is the driving force. |

Adapted and modified from Louis E., David L. Kurtz, *Management*, 3rd edition, Figure 2-4, pp. 41

This study emphasized the impact of human motivation on production and output. When they began the first phase of experiments it was believed that every social problem was "ultimately individual." The results found that group rather than individual psychology was a key factor in the production performance of the workers [1]. Refer to Table 5-2 for a summary of the Hawthorne Study.

The contribution to management philosophy was important to the field of managing employees by revealing the importance of human emotions, reactions, and response to managing others. It also pioneered the concept of good communication between management and employees [3].

The importance of the relationship of working groups to management is one of the fundamental problems of organizations. Organizing teamwork, developing, and sustaining cooperation has to be a major occupation of management [3].

An organization is a formal arrangement of functions, as well as a social system with different cultures.

Since the Hawthorne Study, many studies have added much to our knowledge of human behavior. Sincere participatory support of all levels of the operation in solving organization problems and can to some degree foster a more open and trusting environment and a greater emphasis on the groups [8].

## 5.4 COMMITTEE PARTICIPATION

Joint labor-management committees are a popular method of employee participation. Other types of committees have been used successfully to allow employee participation. At many unionized facilities, employee safety committees (with members selected by the union or elected by employees) work alone, with little direct management participation, on various tasks. In other workplaces, employees participate on a central safety committee. Some worksites use employees or joint committees for specific purposes, such as conducting workplace surveys, investigating incidents, and training new employees, etc. [4].

### 5.4.1 Getting Employee Participation Started

A journey begins with the first step, so one of the keys in getting employees involved is to meet with employees. The following list provides some techniques of employee participation:

- Meet with employees in small groups, as appropriate, (by shift, department, one-on-one, etc.) depending on the nature of the business.
- Explain the safety policy and the JHA goals and objectives that you want to achieve.
- Explain how you want their help with the safety efforts. Ask for their input, suggestions, etc. Implement as many of the reasonable suggestions as possible in some visible way [4].

### 5.4.2 Form a Committee

The next step is actually forming the safety committee. The following list provides some suggestions:

- Form a committee that is suited for your business. It should be large enough to represent different parts of the organization. For example, a committee designed for one part of a facility should integrate with other parts of the facility. If there is no cross-departmental or cross-discipline structure, then there is division between different departments.
- Balance management and non-supervisory employees on the committee. One of the authors has found the following technique to be effective: the author calls this "the safety committee chain letter." Have several supervisors pick one person that he/she can trust. Then that employee will

pick someone that they can trust, that employee picks someone that they can trust, etc. This continues until the committee is the desired size. With this method, employees get a diverse set of members and it is now harder for anyone to dominate the committee. The author has discovered that this helps to define the safety committee in a more effective manner.

- Make sure that the safety professional serves as a resource for the committee and is not the leader of the committee. Choose middle management members who can get things done.
- If there is a union, your agreement defines the selection process.

If your workplace is not unionized, you may want to consult with a qualified labor relations professional on the best way to obtain employee participation if you decide to use a committee [4].

A critical item in forming a safety committee is the development of a "charter." Charters are recommended by quality and Six-Sigma projects and provide a reason for the committee's existence. Each committee must have a written description that defines what is to be accomplished. In addition, a charter will allow you to perform some level of measurement on work that is accomplished. An example charter is provided in Appendix K.

## 5.4.3 How to Use Employees in the Process

You can involve employees in the management system and safety program by having them conduct regularly scheduled routine physical surveys using appropriate tools and methods, checklists, guidelines, etc.

- Ensure that employees have adequate and appropriate training.
- Employees should be expected to help with decisions about hazard correction as well as hazard identification.

Once the committee is established and functioning successfully, it will be in a position to suggest other ways to involve your employees in the safety program.

Always remember that it is the employer who has ultimate legal responsibility for making sure that the workplace is safe for all employees [4].

### 5.4.3.1 Joint Labor-Management Committees

This type of committee usually has equal representation of labor and management. The chairperson may alternate between an employee representative and a management representative as appropriate. The power of this type of committee is worked out through negotiation. Although tasks depend on the outcome of these negotiations, the committees typically conduct the following activities [4]:

- Site evaluations with oversight of hazard recognition.
- Investigating hazards reported by employees.
- Incident investigations.
- Safety awareness program development [4].

### 5.4.3.2 Other Joint Committees

While their usual functions are similar to the joint labor management committees, in other joint committees, there may be either more employee participants (for example, at construction sites where several different trade unions represent employees) or more management participants (especially where safety, medical, and industrial hygiene professionals are counted as management). The top management at the workplace frequently chairs these committees. In other cases, these committees are led by employees who are elected by the committee [4]. They work by consensus and do not take formal votes.

### 5.4.3.3 Employee Safety Committees

These committees are usually union safety committees, with membership determined by the union. Some workplaces with more than one union will have more than one union safety committee. The committee operates without management and meets regularly with management to discuss safety issues. At these meetings, the committee raises concerns, and management provides the necessary responses to help fix the identified hazard. The committee may conduct inspections, investigate hazards reported by employees, and bring safety issues to management for corrective actions. The committee also may design and present safety awareness programs to employees [4].

### 5.4.3.4 Central Safety Committee

The central safety committee is an oversight committee with an interest in every part of the safety process. It sometimes serves as the hazard correction tracking management system. The central committee receives reports from all other committees and reviews inspection reports, incident investigations, and hazards reported by employees, to ensure that all reported hazards are tracked until resolved [4].

Some sites rotate employees on this committee so that all employees can participate in safety planning. At other sites, management selects the employees for their experience and achievements in other safety management systems [4].

### 5.4.3.5 Function-Specific Committees

The special function committee has a single responsibility, such as incident investigation, site safety inspections, development of safety rules, providing training, creating safety awareness programs, etc. The company provides committee members with the necessary training and resources. These committees may consist of employees only, with management support; or there may be a joint membership with some management and/or the safety professional (including on-site nurse/doctor or medical provider, as applicable) [4]. Refer to Figure 5-2 for a sample committee structure.

## 5.5 AREAS OF EMPLOYEE PARTICIPATION

### 5.5.1 Conducting Site Inspections

Whatever method you choose, employees must be trained to recognize hazards. They should have access to a safety professional. For meaningful participation, the committee or safety observer should be able to suggest methods of correcting hazards and to track corrections to completion [4].

### 5.5.2 Routine Hazard Analysis

Employees can be helpful in analyzing jobs, processes, or activities for hidden hazards and in helping to design improved hazard controls. Employees and supervisors are teamed up to accomplish these activities. For complicated processes, an engineer should probably lead the team [4].

Employees are more likely to accept the changes that result from these analyses if they are involved in the decisions that revise practices and processes. For more information on routine hazard analysis and job hazard analysis in particular, refer to Chapters 8 and 10.

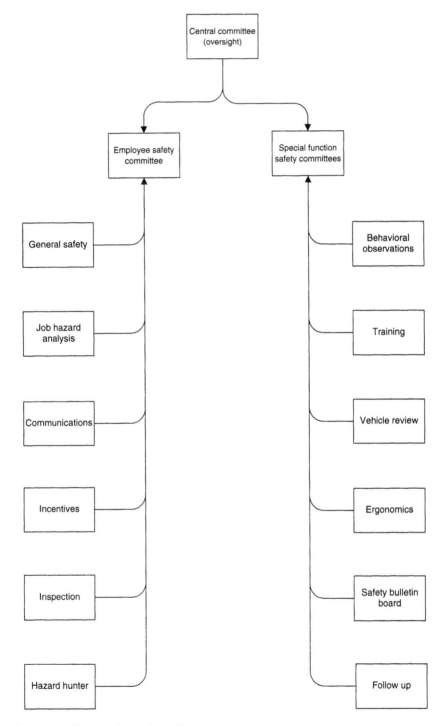

**Figure 5-2** Sample Committee Structure

### 5.5.3 Developing or Revising Site-Specific Safety Rules

Giving employees responsibility for developing or updating the site's safety rules can be profitable. Employees who help make the rules are more likely to adhere to them and to remind others to adhere to them [4].

### 5.5.4 Training Other Employees

Use of qualified employees to train other employees on safety matters—for example, rules, procedures, and other topics—can be effective and can improve your ongoing training efforts. These employees bring direct experience to the training. For those qualified employees who have the skills and enjoy training, it also offers them an area of professional development. For more information on safety training refer to Chapter 7 [4].

### 5.5.5 Employee Orientation

Qualified employees can make excellent safety trainers for new employees. The trainer who provides this introduction to the job can follow up by acting as "buddy" or "mentor" and watching over the new employee, providing advice and answering questions that a new employee might be afraid to ask a supervisor [4]. A problem that must be addressed is that "seasoned" employees may train new employees with old habits or behaviors; their procedures may not be correct and could in effect cause injuries. You need to be careful that you are not allowing new employees to be trained in bad habits that may exist in your workplace.

If the JHA is used in the training and its process followed, the correct information can be transmitted to the employee.

### 5.5.6 Different Approaches: Union and Non-Union Sites

Employee participation at a unionized facility is usually achieved differently than at non-union facilities. Neither type of workplace is more conducive than the other to successful safety awareness through employee participation. At both union and non-union worksites, employee participation is characterized by a commitment to cooperative problem-solving. Just as important, employee participation relies on respect among employees and representatives of organizations [4].

### 5.5.6.1 Unionized Work Sites

A reduction in injuries is clearly beneficial to both management and the union. This goal should lend itself to joint union-management effort. The most common form of cooperative, participatory effort is the joint labor-management safety committee, as discussed. Sometimes an all-employee safety committee will be used. These committee duties can range from reviewing hazards reported

**Table 5-3**
**Sample Employee Participation Case Studies for Union Sites**

| |
|---|
| Case 1: Employee participation at a paint manufacturing facility with 42 employees works primarily through the Safety Committee. Three members of the committee are hourly bargaining unit employees selected by the union, and three are salaried nonbargaining unit employees selected by management. Members participate in committee meetings, hold monthly plant inspections, and recommend safety and health related improvements to management. |
| Case 2: An oil refinery with almost 400 employees involves its employees in a variety of ways. Employees act as safety and health monitors assigned to preventive maintenance contractors. Employees develop and revise safe work procedures and are part of the team that develops and reviews job hazard analyses. They serve as work group safety and health auditors. |
| Case 3: A chemical company with 1,200 employees has found numerous ways to include its employees in the site's safety and health program. For example, the Safety and Health Committee, which includes equal labor and management representatives, has responsibility for a variety of activities: for example, monthly plant inspections, incident investigations, and examination of any unsafe conditions in the plant. Employees are also involved in process and operations review, safety inspection, and quality teams. Two hourly employees work full-time at monitoring the safety and health performance of on-site contractors. |
| Case 4: An electronics manufacturer with almost 5,800 employees has established a joint committee consisting of seven management and eight hourly employees. They conduct monthly inspections of preselected areas of the facility, maintain records of these inspections, and follow up to make sure that identified hazards are properly abated. They investigate all incidents that occur in the facility. All committee members have been trained extensively in hazard recognition and incident investigation. |
| U.S. Department of Labor, Office of Cooperative Programs, Occupational Safety and Administration (OSHA), Managing Worker Safety and Health, November 1994, public domain |

and suggesting corrections to conducting site surveys to handling incident investigations and follow-up of hazards. Some committees are advisory, while others have specific powers to correct hazards and, in some circumstances, to shut down unsafe operations [4]. Refer to Table 5-3 for several case studies for union sites.

### 5.5.6.2 Non-Union Work Sites

Employee hesitation (resistance) may develop when trying to implement employee participation in a non-union work environment. The key is to be careful not to force or impose "voluntary" employee participation on any employees who do not want to get involved. One method is to conduct buy-in meetings where employees, with management support, present the benefits of participation. Once employees feel comfortable with management support, then the program will have an improved chance of being successful. This includes protecting employees from reprisal when they get involved in safety activities [4]. Refer to Table 5-4 for examples of case studies for non-union sites.

**Table 5-4**
**Sample Employee Participation Case Studies for Non-Union Sites**

| |
| --- |
| Case 1: A textile manufacturer with more than 50 plants and employee populations of 18 to more than 1,200 has established joint safety and health committees on all shifts at its facilities. All members are trained in hazard recognition and conduct monthly inspections of their facilities. |
| Case 2: A small chemical plant with 85 employees involves employees in safety activities through an Accident Investigation Team and a Safety and Communications Committee consisting of four hourly and three management employees. The team investigates all incidents that occur at the facility. The committee conducts routine site inspections, reviews all accident and incident investigations, and advises management on a full range of safety and health matters. |
| Case 3: Employee participation at a farm machinery manufacturer of 645 employees includes active membership on several committees and sub-committees. Members change on a voluntary, rotational basis. Committees conduct routine plant-wide inspections and incident investigations. Employees also are involved in conducting training on a variety of safety and health topics. Maintenance employees are revising the preventive maintenance program. |

**Table 5-4**
*Continued*

| |
|---|
| Case 4: A large chemical company with 2,300 employees implemented a dynamic safety and health program that encourages 100 percent employee participation. Its safety and health committee is broad and complex, with each department having its own committee structure. Subcommittees deal with specific issues, for example, off-site safety, training, contractors, communication, process hazard analysis, management, and emergency response. The plant-wide committee, which includes representatives from all departmental committees, is responsible for coordination. All committee members are heavily involved in safety and health activities, for example, area inspections and incident investigations. They also act as channels for other employees to express their concerns. Members receive training in incident investigations, area assessments, and interpersonal skills. |

U.S. Department of Labor, Office of Cooperative Programs, Occupational Safety and Administration (OSHA), Managing Worker Safety and Health, November 1994, public domain

## 5.5.7 Forms of Employee Participation

Employee participation can be rotated through the entire employee population. Programs then receive the benefit of a broad range of employee experience. Other employees benefit from increased safety knowledge and awareness. At some non-union facilities, employee participation relies on volunteers, while at some facilities supervisors appoint employees to safety committees [4].

The best method for employee participation will depend on what you want to achieve and the direction in which you need your program to go. If improved employee awareness is a major objective, rotational programs are a good choice. If high levels of skill and knowledge are required to achieve your safety objectives, volunteers or appointees who possess this knowledge may be preferable [4].

## 5.5.8 What Can Management Do?

Management sets the tone. Refer to Chapter 12 on management commitment and leadership. If management is not supportive in getting employees involved, and unless your employees believe you want their participation, participation will be limited and will not be successful [4].

Employees may not believe management actually wants their input on serious matters, so the participation is minimal. They may have heard about committees that are formed and then die due to the lack of management support in time and budget. This could be because managers have experienced safety committees that only want to talk about "trivial" things like cafeteria menus. They might decide from this evidence that employees are either unwilling or unable to address the serious issues of the worksite. It is essential that mistrust and miscommunication between management and employees be corrected. You can accomplish this by demonstrating visible management commitment and support and providing positive feedback [4].

The following are possible actions that you can use to encourage employee participation:

- Show your commitment through management support and leadership. This helps your employees to believe that you want a safe workplace, whatever it takes. Support means a commitment of time and resources.
- Communicate clearly to your employees that a safe workplace is a condition of their employment, and your commitment to them.
- Tell your employees what you expect of them. Document those requirements. Refer to Chapter 12 for an overview of a Safety Management Process. Communicate to all employees specific responsibilities in the safety program, and provide appropriate training and adequate resources for performing specific activities that were assigned.
- Get as many employees involved as possible: brainstorming, inspecting, detecting, and correcting.
- Make sure that employee participation is expected as part of the job during normal working hours or as part of their assigned normal jobs.
- Take your employees seriously. Implement their safety suggestions in a timely manner, or take time to explain why they cannot be implemented.
- Make sure co-workers hear about it when other employee ideas are successful.
- Provide any committee with a clear charter that outlines duties, direction, resources, etc. Refer to Figure 5-2 for a sample structure of a safety committee.

## 5.6 SUMMARY

Employee participation has been shown to improve the quality of a safety process. Your employees are equipped to provide assistance in a wide variety of areas. What employees need are opportunities to participate. This can be shown by clear signals through management leadership, training, and resources.

Employee participation differs at unionized and non-union sites. No matter what forms of participation you choose to establish in your program, you have the opportunity and responsibility to set a management tone that communicates your commitment to safety, providing a high-quality response from your employees [4].

The objective of employee participation is to provide for and encourage employees to help in the structure and operation of the safety program and in decisions that affect their safety. If this is done properly, employees commit their insight and energy into achieving the safety program's goals and objectives.

If your workplace is not unionized, you may wish to solicit employee suggestions on how to select non-supervisory members of the committee.

One important consideration is to consult with your human resources professionals before selecting or holding an election for Safety Committee members. If you are not sure where you stand on this issue, it would be advisable to consult with your company attorney. Employees may volunteer and be put on a rotational basis, to provide as much information and knowledge from as many employees as possible.

---

The following are general guidelines for involving employees. The key is to provide employees:

- The opportunity for participation.
- Clear support from management—in both time and budget.
- Training and resources.
- Take employees seriously and communicate that a safe environment is a condition of employment.
- Implement employee suggestions in a timely manner or take the time to explain why they cannot be implemented.
- Include in your policy statement that employees are protected from reprisal resulting from safety program participation.
- Make sure that all employees hear about the success of other employees' ideas.
- Provide opportunities and mechanism(s) for employees to influence safety program design and operation. Make sure that there is evidence of management support of employee safety interventions.
- Employees have a substantial impact on the design and operation of the safety program.

# CHAPTER REVIEW QUESTIONS

1. What is the objective of employee participation?
2. Why should employees be involved in the JHA process?
3. Does employee participation create more awareness?
4. What are methods of getting employees to participate?
5. What are some guidelines for employee participation?
6. Why is it important to form a safety committee?
7. How can you involve employees in the JHA process?
8. What are employee safety committees?
9. What is a central safety committee?
10. Why is employee orientation important?

# REFERENCES

1. Bone, Louis E., David L. Kurtz, *Management*, 3rd edition, Random House Business, New York, 1984, p. 644
2. Chruden, Herbert J., Arthur W. Sherman, Jr., *Personal Management: The Utilization of Human Resources*, Sixth Edition, South-Western Publishing Co. Cincinnati, Ohio, 1980
3. Kennedy, Carol, *Instant Management*, revised 1993, William Morrow and Company, Inc., New York, 1991
4. Oklahoma Department of Labor, *Safety and Health Management: Safety Pays*, 2000, http://www.state.ok.us/~okdol/, Chapter 4, pp. 21–23, public domain
5. Ouchi, William, Theory, Z., *How American Business Can Meet the Japanese Challenge*, Avon Publishers, 1981
6. Rue, Leslie W., Lloyd L. Byers, *Management: Theory and Application*, Fifth Edition, Irwin, Homewood, Illinois, 1989
7. U.S. Department of Labor, Office of Cooperative Programs, Occupational Safety and Administration (OSHA), *Managing Worker Safety and Health*, November 1994, public domain
8. Wertheim, Edward G., *Historical Background of Organizational Behavior*, College of Business Administration, Northeastern University, Boston, MA 02115, Adapted from website: http://www.cba.neu.edu
9. OSHA Web Site, http://www.osha-slc.gov/SLTC/etools/safetyhealth/index.html, public domain
10. George, Michael L., *Lead Six Sigma: Combining Six Sigma Quality with Lean Speed*, McGraw-Hill, 2002

# Appendix K

## K.1 EXAMPLE OF A COMMITTEE TEAM CHARTER

Team Name _____

Date: _____

---

Problem Statement:

(Provide a clear statement describing the improvement opportunity or problem.)




Committee Mission Statement:

(Provide a clear statement describing what the committee is being chartered to do.)




Description:

(Describe the process to be improved or problem to be solved, or identify the steps in the process from beginning to end.)




---

Background:

(Identify what has been happening, the importance of the committee.)

Scope:

(Identify the limits on the project to include whether the committee will be able to pilot improvements/solutions or just make recommendations.)

Time Frame for Committee Tenure:

Date/Time for Committee Launch:

Date/Time for Process Owner to meet with the team to discuss the assignment and agree on the project specifics:

Process Owner:

Sponsor:

Team Leader:

Facilitator:

Committee Members:

List of Resources:

Committee Contract:

We have read and understand this committee's charter, understand our roles and responsibilities, and have come to an agreement with the Sponsor and/or Process Owner on the opportunity being addressed, the actions to be taken, and the limitations of the committee. If at any time it becomes apparent that the committee charter needs to be modified, we will consult the Sponsor and/or Process Owner and come to agreement on the modifications.

(Signatures of Team Members)

# 6

# Defining Associated Risk

*"Dice have no memory, they change all of the time" CSI TV*

"Dice have no memory; they change all of the time."

—CSI TV

A major factor in the JHA process is that the concept of risk must be clearly understood. Using only loss-related data based on injuries and damage does not provide a full understanding and determination of where operational hazards may exist but have not yet created a loss. This chapter provides an overview of the concepts necessary to evaluate risk. At the end of the chapter you will be able to:

- Define risk management and its principles
- Utilize a model to follow in developing the elements of a system
- Classify risk using the risk matrix
- Cite the importance of incorporating risk into the JHA process.

"Trust but verify."

—Ronald Reagan

"There are two primary choices in life: to accept conditions as they exist, or accept the responsibility for changing them."

—Dr. Denis Waitley

Risk can be defined as a measure of the probability of, and probable severity of, adverse effects. Risk management principles have been around for many years and have been used in many high-hazard industries and operations. However, most safety programs typically remain designed around regulatory compliance or reported losses. Risk management concepts are new to many and still are rarely used in many organizations to assess events which could cause an injury. Even safety professionals still go on their "gut" instinct, "I think, I feel," or personal experience to develop safety programs. A shift must take place in which we understand the need to collect risk-related data, truly analyze the data, and make decisions based on risk assessment. Risk management would then be used to address and clarify the role of the injury prevention program [1].

We will provide simple, logical formats to help you understand how to use effective risk management principles. These formats will outline how hazards can be associated with specific job steps and related tasks. In addition, we will provide several qualitative (descriptive) and quantitative (mathematical) models to help you assess the risk of hazards.

We have introduced the concepts of identifying and analyzing workplace hazards; methods on how to verify hazards through inspections and surveys; an overview of the hierarchy of controls; discussions of at-risk events, and why employee participation is an important role in risk analysis. Now we will turn our attention to assessing risks and their impact on specific tasks. If hazard identification is conducted, as in traditional safety, without assessing risk in a quantified way, then the severity of the hazard remains questionable. Therefore the concept of risk assessment must be incorporated into the JHA process, as we will discuss in this chapter.

## 6.1 RISK MANAGEMENT

Risk management is the systematic application of management and engineering principles and uses an array of tools to control inherent hazards of specific tasks.

Risk analysis concepts provide a method of establishing and prioritizing events. The JHA format that we will discuss in Chapter 10 provides a method for the development of an effective risk analysis program that can be used to identify existing and potential hazards and the consequences of exposure. Using the JHA, as we suggest, you will begin to look at the job tasks in different ways, identifying a combination of actions, environments, tools/equipment/materials, and employees. At this point, a series of questions must be addressed:

- How often does the exposure to a specific hazard occur?
- What is the frequency of the task or activity? Can we quantify?
- What is the probability of a loss-producing event during the task or activity? Can we quantify?
- What would be the consequences of the exposure if the wrong conditions came together?
- What is the potential cost of a risk versus the benefit it may produce?
- What is the cost of taking a risk versus the cost of not taking a risk?
- Do the total benefits outweigh the total costs? [1]

The first step in the JHA process is to develop a clear mental picture of the elements of a job, the related steps and specific tasks needed to complete each specific step. Traditionally, JHA development stops at the job step and does not take into account the tasks that are needed to perform the step [2]. The lack of full employee participation limits understanding of the interactions needed to provide adequate control and job improvement.

To be effective, the JHA must identify all aspects of the job tasks and should include a method to provide a risk assessment of each identified task. This process will allow the JHA developer to determine the scope of any associated at-risk events [1, 2]. Consideration must be given to whether the job is routine or nonroutine, as training, guidelines and controls must consider familiarity with the job.

## 6.2 GENERAL RISK MANAGEMENT THEORIES AND MODELS

A number of general risk management theories and models have been developed to provide a graphic presentation of risk complexity. One such model used by NASA involves the implementation of a thorough, disciplined risk management approach required of all NASA programs and projects. NASA uses a process called *Continuous Risk Management* (CRM). Refer to Figure 6-1 for an overview of this process.

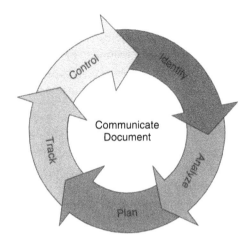

**Figure 6-1** Continuous Risk Management (CRN)

NASA, Safety, Environmental and Mission Assurance, Independent Risk Assessment, http://www. extremeenvironment.com/2003/expedition/sema.htm, public domain

Another model that is used and that we find very valuable is the U.S. Air Force "5M" model used in its Operational Risk Management (ORM). Refer to Figure 6-2 for the 5M model of system engineering. This model is a system that consists of Man, Machine, Media, Management, and Mission [1]. This model

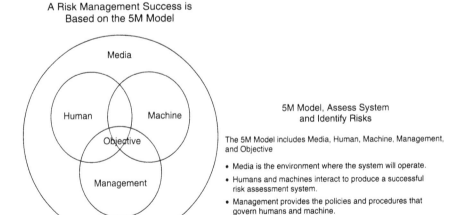

**Figure 6-2** 5M Model of System Engineering

Adapted from FAA System Safety Handbook, Chapter 15: Principles of System Safety, Figure 3-6, pp. 17, 5M Model of Safety Engineering, December 30, 2000, public domain

provides a framework for analyzing systems and determining the relationships between the elements that work together to perform specific tasks. We adapted the 5M model to align with current terminology that we feel allows the method to be used in general industry as a guide to risk assessment.

- Employee Considerations: Used in place of "Human" as this includes both genders and can incorporate all human demographics.
- "Environment Considerations": Used instead of "Media" as it includes all aspects of the environment such as the general conditions (weather, lighting and temperature and other conditions found at a job site), biohazards, area obstructions, animals, terrain and walking surface and/or anything that defines where the job is being performed.
- "Tools/Equipment/Materials Considerations": Used instead of Machine. This category is self-explanatory and provides for the mixture of various items that must be used to complete the job.
- "Management Support and Policy/Procedures Considerations": This provides for Management commitment and the structure, policies, procedures, and rules that govern the interactions between the other elements.
- "Job" (steps and tasks) instead of "Objective": Taking the role of Six Sigma (discussed in Chapter 13), we use: "What is the objective, or what are you trying to achieve?" What is the job to be accomplished? This will be described in Table 6-5.

This model provides a structured approach to assess the interactions that produce a successful "goal," i.e., the completion of a task. The amount of overlap or interaction between the individual elements and components is dynamic as each part of the model can change and evolve [1, 5].

When a process is unsuccessful and an injury occurs, the steps and tasks must be analyzed and the various inputs and interactions of the model thoroughly assessed [1, 5]. The question that must be asked on every case is simple: "What did we do as management to create this set of conditions?" We need to ensure that this is taken into consideration, as there may have been an implied consequence in place or process assumptions that allowed the conditions for an injury or an event to occur.

## 6.2.1 Employee Considerations

The human factor is an area of great variability due to differences in values, personalities, skills, capabilities, etc. of individuals. With differences between individuals, one might expect that the major source of associated risks would be the individual reacting to their environment. In addition, we are relying on employees to do the right thing. Humans are subject to error when it comes to bypassing safety procedures, forgetting rules, avoiding safe procedures, etc., as you may have correctly determined the necessary physical controls for clearly identified risks and hazards. One has to assess the consequences of exposure and apply the appropriate behavioral concepts. However, the selection and ensuring that employees are physically and mentally capable of doing the job must be analyzed. Refer to Table 6-1 for an overview of human factor considerations.

**Table 6-1**
**Employee Considerations**

| Factor | Examples |
|---|---|
| Selection | Individual psychologically and physically trained in event proficiency, procedures, and habit (behavior) patterns |
| Performance | Awareness, pre-judgment, perceptions, task saturation, distraction, channeled attention, stress, peer pressure, confidence, insight, adaptive skills, pressure/workload, fatigue (physical, motivational, sleep deprivation, circadian rhythm) |
| Personal | Expectancies, job satisfaction, values, families/friends, perceived pressure (over tasking) and communication skills |

Adapted from The Federal Aviation Administration (FAA), Office of System Safety, *FAA System Safety Handbook*, Chapter 15, "Operational Risk Management, December 30, 2000, http://www.faa.gov/library/manuals/aviation/risk_management/ss_handbook/, public domain

## 6.2.2 The Environment

Environmental considerations are defined as external environmental and internal operational conditions. Refer to Table 6-2 for an overview of environment considerations.

## Table 6-2
## Environmental Considerations

| Factor | Examples |
| --- | --- |
| Climatic | Visibility, temperature, humidity, wind, precipitation, daylight, darkness, etc. |
| Operational | Human-made obstructions, facility design, equipment configuration, new/modified/relocation equipment reviewed, pace and flow of product or service, machine guarding, etc. |
| Hygienic | Ventilation/air quality, noise/vibration, dust, and contaminants |

Adapted from The Federal Aviation Administration (FAA), Office of System Safety, *FAA System Safety Handbook*, Chapter 15, "Operational Risk Management," December 30, 2000, http://www.faa.gov/library/manuals/aviation/risk_management/ss_handbook/, public domain

## 6.2.3 Tools/Equipment/Material Considerations

Tools/equipment/materials bring the items used or handled with their direct task hazards into contact with employees or others [5]. Refer to Table 6-3 for an overview of tools/equipment/material considerations.

## Table 6-3
## Tools/Equipment/Material Considerations

| Factor | Examples |
| --- | --- |
| Design | Equipment and engineering reliability and performance, ergonomics |
| Maintenance | Availability of time, tools, and parts, ease of access |
| Logistics | Supply, maintenance, repair, hazardous materials and chemicals, tool and equipment purchase and design |
| Technical data | Clear, accurate, useable, and available |

Adapted from The Federal Aviation Administration (FAA), Office of System Safety, FAA System Safety Handbook, Chapter 15, Operational Risk Management, December 30, 2000, http://www.faa.gov/library/manuals/aviation/risk_management/ss_handbook/, public domain

## 6.2.4 Management Support and Policies, and Procedure Considerations

Management directs the process by showing leadership and support, defining and implementing the necessary safety policies, standards, and procedures, along with engineering controls. Although management provides procedures and rules to govern interactions, it cannot completely control all aspects of the system elements. Refer to Table 6-4 for an overview of policy, procedures, and management considerations.

### Table 6-4
### Management Support and Policies and Procedure Considerations

| Factor | Examples |
|---|---|
| Management Commitment and Employee Participation | The level of commitment from management determines the success of support for the supervision and employees. |
| Standards | Regulatory requirements (OSHA, EPA, DOT, etc), company policies and procedures. |
| Procedures | JHA, audit checklists, manuals, policies, rules etc. |
| Controls | Rest, restrictions, training rules/limitations, excessive hours of work. |
| Objective | The desired outcome, task analysis and completion. |

Adapted from The Federal Aviation Administration (FAA), Office of System Safety, *FAA System Safety Handbook*, Chapter 15, "Operational Risk Management, December 30, 2000, http://www.faa.gov/library/manuals/aviation/risk_management/ss_handbook/, public domain

## 6.2.5 Job Steps and Task Considerations

Job "steps" are the specific actions and movements required to achieve the successful completion through the interactions of the parts of the model. Job "tasks" are considered as functions to be performed, an objective of each step. Refer to Table 6-5 for an overview of job step and task considerations.

### Table 6-5
### Job Step and Task Considerations

| Factor | Examples |
|--------|----------|
| Job | An overall activity based on performing specific steps to complete each task. |
| Steps | Clearly defined actions that must be taken; define how each step is completed. Movements, lifting, travel, observation, use of tools, equipment and materials in sequence. One of a series of actions, processes, or measures taken to achieve a goal; a stage in the process. |
| Task | A function, a series of functions, to be performed, based on the job steps, an objective. The finer details required to complete a step. |

Adapted from The Federal Aviation Administration (FAA), Office of System Safety, *FAA System Safety Handbook*, Chapter 15, "Operational Risk Management," December 30, 2000, http://www.faa.gov/library/manuals/aviation/risk_management/ss_handbook/, public domain

## 6.3 THE SYSTEM ENGINEERING MODEL

As risk assessments are developed, subpopulations such as temporary employees, relocated employees, transferred employees, etc., must be identified. These individuals may be particularly susceptible and at significantly more risk than what is faced on a routine basis. Organizations tend to overlook these sub-populations for a number of reasons:

- The company has no financial stake in their safety. Temporary agencies provide their own workers' compensation programs.
- Relocated and transferred workers may be erroneously thought to be "experienced" even through they are the equivalent to a new hire.

System Safety Areas:

Manage and implement the safety system plan.

Identify hazards associated within the management system.

Incorporate safety using behavioral observations, product design, operation, test, and maintenance.

Evaluate identified hazards and design action plans to eliminate or minimize and control the hazards.

Develop safety design criteria to be incorporated into the product design.

Conduct hazard analyses on the job, its steps, and tasks being developed.

Identify characteristics of hazardous materials and energy sources, including explosives, flammables, corrosives, toxics, and methods of control and disposal.

Equipment pre- and post-inspection.

Adapted from Air Force System Safety Handbook, Air Force Safety Agency, Kirtland AFB, NM 87117-5670, Revised July 2000, Types of Risk, Table 15-2, pp. 117, public domain

## 6.4 RISK VERSUS BENEFIT

Risk management is a logical process that weighs the potential costs of losses against the possible benefits of various levels of control ranging from none to total avoidance. We routinely make risk determinations as benefits can be realized through controls as well as prudent risk-taking [5]. Risks below a certain threshold can be considered "acceptable." Unfortunately, when risks are not assessed, decisions are made that can result in harm. However, because in many activities, harm may or may not occur, we begin to believe no risk is present. People make decisions on the false premise that "NO LOSS = NO RISK." Of concern are those employees who may have a low perception of risk, thereby putting themselves in harm's way because they have never been injured.

The benefits of risk assessment and management are not limited to decreased injuries or damage, and may also be realized as increases in efficiency. Risk management provides a reasoned and repeatable process that reduces the reliance on "intuition," past history, sheer luck, or hopes that nothing goes wrong.

"Hope is not a strategy."

—Rick Page

Risk assessments should take into account existing and potential hazards and any potential consequences of exposure in a given situation within a given environment.

The risk assessment should analyze the scope of the job task and the benefits associated with each control selected for the specific task. Reasonable risk management strategies include consideration of both the frequency of exposures to the hazards and the potential consequences of that exposure to those hazards. A combination of techniques are used that look at the extreme possible events through "What if. . . ?" scenarios. Then a reality check based on a review of the specific task is completed. A risk matrix will be introduced later that simplifies the assessment when used in the JHA process [3].

# 6.5 RISK MANAGEMENT COMMUNICATION

To be effective, risk communication must involve an open, two-way exchange of information between all levels of the organization. Risk management goals must be clearly defined and decisions communicated accurately and objectively in a meaningful manner. As part of the communication, responsibilities must be assigned as discussed in the following sections.

## 6.5.1 Risk Management Responsibilities

To have an effective risk management system, as in safety management, levels of responsibility must be defined. The following are suggested management responsibilities for specific roles in risk management:

- Assessing risks, developing risk reduction alternatives, and defining remaining risk issues. Identifying unnecessary or ineffective risk controls for analysis.
- Selecting risk reduction options from recommendations. Accepting or rejecting risk based on the benefit to be derived. Management must define unacceptable risk.
- Training and motivating employees to use risk management techniques. Refer to Chapter 7 for training techniques.
- Elevating risk decisions to a higher management level (i.e., if severe or catastrophic). The higher the risk, the greater the liability and loss potential and the decision to accept must be elevated to upper management.
- Integrating risk controls into the action planning process to ensure that time and money are budgeted for implementation of such a process [4, 5].

## 6.5.2 Supervision Responsibilities

The following are suggested supervision responsibilities for specific roles in risk management:

- Consistently using effective risk management concepts in identifying hazardous events potential. Routinely applying all appropriate risk management elements in the process.
- Ensuring that all employees follow appropriate risk control procedures and protocols as defined by the JHA.
- Elevating risk issues beyond their control or authority to management for resolution [5].

## 6.5.3 Employee Responsibilities

While it does not relieve management of their specific duties, all employees have job responsibilities. The following are employee responsibilities for specific roles in risk management:

- Understanding, accepting, and implementing all risk management principles.
- Maintaining a constant awareness of the changing risks associated with specific procedures, the operation and/or task.
- Making supervision immediately aware of any unrealistic risk reduction measures or high-risk procedures [5].
- Ensuring that all co-workers follow procedures and assist each other in maintaining effective controls.

Using risk management principles, the safety process will move from a "crisis management role" to a proactive decision-making process that will help to manage safety hazards. Anticipating what could go wrong becomes a part of everyday life. As the culture of the organization changes, the management of risks will become an integral part of the overall management system.

## 6.6 RISK ASSESSMENT

To help everyone understand the basis for the methods we have discussed, we will use an eight-step model based on the Federal Aviation Administration (FAA) model. Figure 6-3 provides an overview of eight steps for accomplishing a risk assessment. Refer to Table 6-6 for an explanation of the risk assessment model.

## Table 6-6
## Eight Steps to Accomplish a Risk Assessment

| | Process | Description |
|---|---|---|
| 1 | Define the objective | The first step is to define the objectives of the system and ask the question: What are we trying to accomplish? |
| 2 | Define and describe the system | Describe how the system elements interact: Employee, Tools/Equipment/Materials, Environment, and Policies and Procedures. |
| 3 | Hazard identification: identify hazards and consequences of exposure | Hazards are identified by analyzing the tasks grouped by function. During the identification of each task, the risk analysis assesses the potential consequence of exposure to the hazards. The classic problem-solving format "Who, What, Where, When, Why, and How (How much? How Many?)" Is used to define the root cause. This method is called the "5W1H" method that is used in Six Sigma to get to the root cause. |
| 4 | Risk analysis: analyze hazards and identify the risks | Assessment is the application of quantitative and qualitative measures to determine the level of risk associated with specific hazards. This process uses the estimated probability and potential severity of an injury [5]. |
| | | The risk analysis reviews hazards to determine what can happen. The inability to quantify and/or the lack of historical data on a particular hazard do not exclude the hazard from the need for analysis. |
| 5 | Risk assessment: group steps/tasks and prioritize risks | Risk assessment combines the impact of risks and compares them against defined acceptable level criteria. These criteria can include the consolidation of risks into categories that can be jointly mitigated, combined and used in decision making. Refer to Table 6-7 for an example of Severity Level Categories and Table 6-8 for an example of Hazard Probability Levels. |

**Table 6-6**
*Continued*

| | Process | Description |
|---|---|---|
| 6 | Decision-making: developing action plans | Once a list of tasks has been prioritized, the list is reviewed to determine how to address each risk beginning with the highest priority or most severe risk. Management develops an action plan to apply control methods that have been selected. Management provides the resources and individuals needed to put these measures in place. The "hierarchy of controls" is used during this phase. |
| 7 | Validation and control: evaluate results of action for further planning needs | Evaluate the effectiveness of the action planning process. This evaluation will include identification of data to be collected and the review of data collected. |
| | | "Residual" risk (any remaining risks) can be acceptable, unacceptable, or remain unknown. If acceptable, documentation is required to show the rationale for accepting the risk. If it is unacceptable, an action plan is established for additional actions needed. If listed as unknown, then the process continues the data search. |
| 8 | Modify system/process, as applicable | If the identified risk changes or action plans do not produce the intended effect, a determination must be made as to why. Was the wrong hazard addressed? Was the hazard missed? Does the system/process need to be modified? Re-evaluate the process beginning at the hazard identification step [2]. After controls are in place, the new process must be periodically reevaluated to ensure effectiveness. Management and employees must ensure that the controls are maintained over time. The risk management process continues throughout the life cycle of the system, mission or activity [5]. |

Adapted from The Federal Aviation Administration (FAA), Office of System Safety, *FAA System Safety Handbook*, Chapter 3, "Principles of System Safety," December 30, 2000, Chapter 4, "Pre-Investment Decision Assessment," and document, http://www.asy.faa.gov/risk/ssprocess/ssprocess.htm, public domain

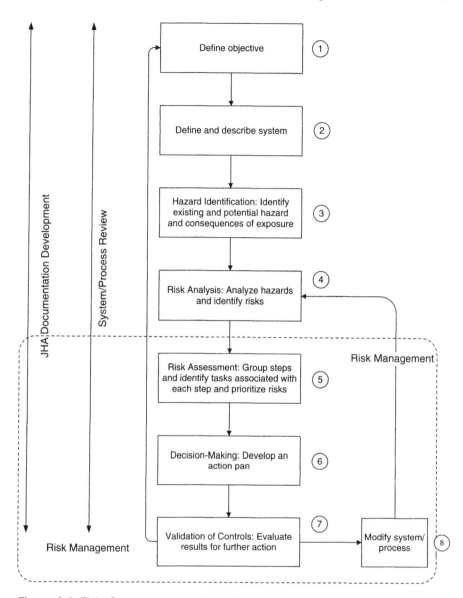

**Figure 6-3** Eight Steps to Accomplish a Successful Risk Assessment

Adapted from The Federal Aviation Administration (FAA), Office of System Safety, FAA System Safety Handbook, January 1, 2005, Risk Management, System Safety Process Step, Order, 8040.4. http://www.faa.gov/library/manuals/aviation/risk_management/ media/ssprocdscrp.pdf, public domain

A continuous loop of assessment, identification, prioritization, action planning/implementation and evaluation provides validation of decisions, desired results and determination of the need for further action as described in Figure 6-3, showing eight steps to accomplish a successful risk assessment [2].

## 6.7 CLASSIFICATION AND RANKING HAZARDS

One of the keys to a successful risk management system is establishing priorities on related hazards to better communicate the various levels of risk. Refer to Figure 6-4 for an overview of establishing task priorities. To categorize and compare the probability and the severity of the risk, the following section provides an overview of the process. We will discuss severity and probability

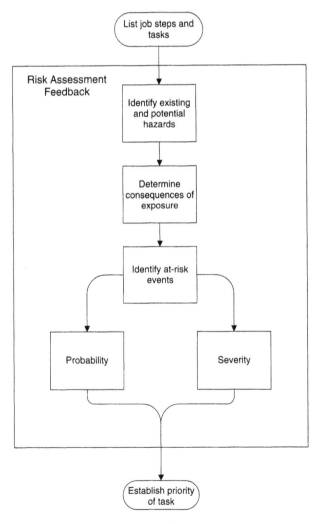

**Figure 6-4** Overview of Establishing Task Priorities

Adapted from Oregon OSHA Website, Job Hazard Analysis, http://www.cbs.state.or.us/external/osha/pdf/workshops/103w.pdf, public domain

of identifying hazards and provide methods that can help to determine ratings for the risk found within a task.

The JHA process begins with a list of all jobs in a facility, with a risk assessment completed on each job. Once the assessment has been conducted, the JHAs can be prioritized in a logical way using verified data.

## 6.7.1 Risk versus Opportunity

"Risk in itself is not bad; risk is essential to progress, and failure is often a key part of learning. But we must learn to balance the possible negative consequences of risk against the potential benefits." [1]

Risk and opportunity go hand in hand. The opportunity for advancement cannot be achieved without taking risk [1]. Risk assessment establishes a disciplined environment for proactive decision-making to:

- Assess continuously what could go wrong (high risk hazards).
- Determine which risks are important.
- Implement strategies for dealing with identified risks [1].

When conducting a risk assessment, one must understand what constitutes unnecessary risk. Accepting risk is a function of both risk assessment and risk management and is not as simple a matter as it first appears. Several principles apply:

- A degree of risk is a fundamental reality. Hazards will exist and realistically some level of risk must always be accepted. We would like for all hazards to be controlled or "fixed" so that there are no risks. It is the way that these hazards are managed that helps to reduce injuries. The criteria for a risk to be accepted or not accepted are the prerogative of management; when the decision is made to accept risk, the decision should be communicated to all affected employees and all other levels of management. The acceptance is documented to record the decision-making criteria, so everyone can know and understand the elements of the decision and why it was made [4, 5].
- Risk management is a process of trade-offs. You may fix one hazard and create another hazard. Quantifying risk does not in itself ensure safety.

It is critical that no risk should be accepted without a full understanding of the consequences. In a litigious climate of lawsuit frenzy, the "how and why" of accepting a risk must clearly demonstrate that the level of risk did not unduly set in motion possible harm to any individual or group, environmental factor, etc. Years ago, the Ford Motor Company made a risk decision with its Pinto design using a cost benefit analysis that showed the savings to Ford would exceed any potential lawsuits. The company paid dearly for that decision. They failed to comprehend the potential impact on their brand and how the public would react to a callous decision concerning a life-threatening design issue [6].

To effectively establish risk control priorities, you must take into account all relevant business and social considerations including the different types of impacts on the business (the brand); prejudgment; individual perceptions of the feasibility of reducing or avoiding risks; the impact on quality of life and the environment; and the magnitude and distribution of both short- and long-term benefits. After you understand the consequences of exposures, then you can start to define the severity and probability factors for each situation. These factors are used to evaluate risk through a simple "decision-making matrix." The risk matrix will allow you to create a Pareto chart (list items in rank order of severity or probability). In addition, we will provide an advanced method that can be used to prioritize hazards in Chapter 13, Appendix R.

To quantify a risk assessment, specific definitions should be designed as shown in Tables 6-7 and 6-8. These definitions are not set in stone and should be modified for your specific requirements. The dollar amounts are for example only.

In example 1, let's assume that the Severity factor is determined to be "Low" and the Probability is determined to be "Seldom." First, draw a line through "Low" on the severity factor, referring to Figure 6-5, and then another line through "Seldom" on the probability factor, referring to Figure 6-6; where the two lines intersect, the combined ranking is shown as a "14". This is the number

## Table 6-7
## Example 1, Severity Rating

| Severity | Description |
|---|---|
| High | Fatality, permanent total disability, major property damage (> $10,000) or more serious consequence, potential lost workday injuries, multiple injuries. |
| Medium | Permanent partial and temp. disability, major property damage (> $2,000–< $10,000), or more serious physical harm. |
| Low | Other than serious, minor injury, lost workday, compensable, minor property damage ($200 < $2,000). |
| Minor | No potential injury to potential first aid, medical treatment, near miss, or minor property damage (< $2,000). |

## Table 6-8
## Example 1, Probability Rating

| Probability | Description |
|---|---|
| Frequent | Steps/Tasks occur very often, i.e., 8, 10, 12 hours per day. Likely to occur frequently. |
| Probable | Steps/Tasks occur often, i.e., less than 8, 10, 12 hours per day. Will occur several times in the life of the task. |
| Non-Routine | Steps/Tasks do occur fairly often. This task may occur each week, month, yearly. Likely to occur sometimes in the life of the task. |
| Seldom | Steps/Tasks occur sometimes, sporadically or several times. Unlikely, but possible. |
| Unlikely | Assume that the steps/tasks will not occur but improbably, very rarely. So unlikely that it can be assumed that it will not occur. |

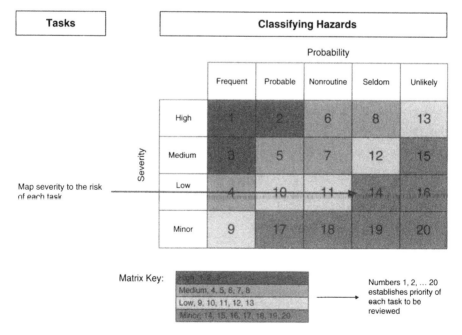

**Figure 6-5** Example 1, Risk Assessment Steps in Hazard Classification Identifying Severity

U.S. Army Communications-Electronics Command, Directorate for Safety, Fort Monmouth, New Jersey, public domain

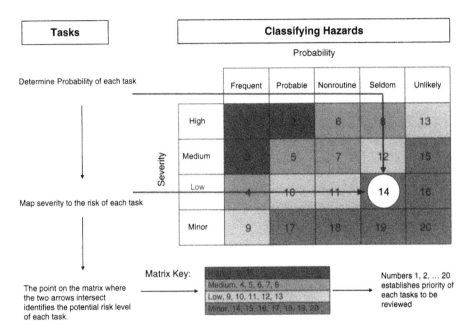

**Figure 6-6** Example 1, Risk Assessment Steps in Hazard Classification Identifying Probability

U.S. Army Communications-Electronics Command, Directorate for Safety, Fort Monmouth, New Jersey, public domain

that will be used to rank the risk in the order of priority. You can now arrange, in a Pareto chart, the risk, from the highest to the lowest.

In example 2, another set of definitions basically works the same way. Refer to Tables 6-9 and 6-10.

## Table 6-9
### Example 2, Probability Rating

| Probability | Occurrence | Exposure |
|---|---|---|
| Frequent | Highly likely to occur | Exposure at all times |
| Probable | Probable to occur | Exposure > once per shift |
| Occasional | Possible to occur | Exposure < once per shift |
| Remote | Unlikely to occur | Exposure < once per year |
| Improbable | Highly unlikely to occur | Exposure so improbable that it can't be reasonably quantified |

### Table 6-10
### Example 2, Severity Rating

| Severity | Description |
|---|---|
| Catastrophic | Fatality, major permanent impairment |
| Critical | Disability in excess of 3 months or some permanent impairment |
| Marginal | Medical treatment case or lost time with full recovery |
| Negligible | First aid or minor medical treatment |

In this example, we have changed our terminology slightly. Now let's assume that the probability is determined to be "Probable" and the severity is determined to be "Critical." Just as we did in example 1, first draw a line through "Probable," referring to Figure 6-7, and then another line through "Critical," referring to Figure 6-8. Where the two lines intersect, the combined ranking is shown as "High." Again, you can now range your risk of each task in the correct order, from high to low.

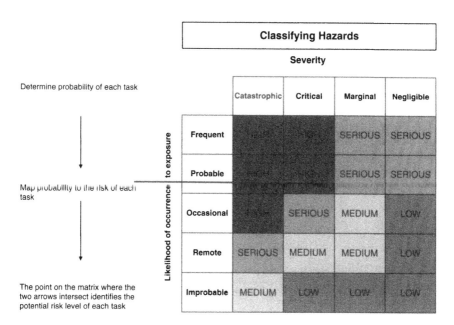

**Figure 6-7** Example 2, Risk Assessment Steps in Hazard Classification Identifying Probability

U.S. Army Communications-Electronics Command, Directorate for Safety, Fort Monmouth, New Jersey, public domain

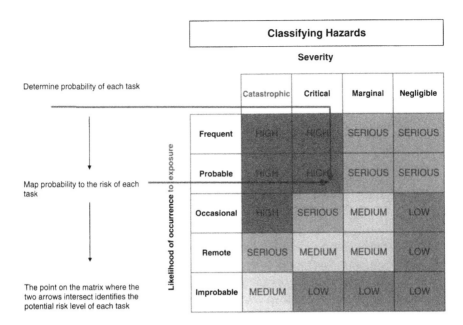

**Figure 6-8** Example 2, Risk Assessment Steps in Hazard Classification Identifying Severity
U.S. Army Communications-Electronics Command, Directorate for Safety, Fort Monmouth, New Jersey, public domain

Example 3 is the simplest method that can be used to prioritize hazardous tasks. Refer to Figure 6-9. This example can be found on the Oregon OSHA website in their JHA workbook.

Severity

| | | Minor | Serious | Fatality |
|---|---|---|---|---|
| Probability | Very likely | 3 | 6 | 9 |
| | Likely | 2 | 5 | 8 |
| | Unlikely | 1 | 4 | 7 |
| | | The intersection of the Probability and Severity is the Risk Assessment Code | | |

**Use the risk matrix to prioritize task**
(1 = lowest priority, 9 = highest priority)

**Figure 6-9** Example 3, Risk Assessment, Prioritize Hazardous Jobs
Oregon OSHA, http://www.cbs.state.or.us/osha/pdf/workshops/103w.pdf, pp. 8, public domain

In this example, once you have identified tasks you believe might be hazardous, you need to determine which tasks:

* Are most likely to cause injury.
* Will cause the most severe injury.


As we have discussed, analyze the "worst first"—i.e., the hazards presenting the most risk need to be analyzed first. To determine risk objectively, use a structured method to prioritize hazards. Example 3 offers another way to prioritize tasks, using a risk matrix. The matrix works on the principle of 1 = lowest priority, 9 = highest priority.

Risk assessment provides a systematic approach that allows ranking of risks so that opportunities to improve the process can be targeted to tasks requiring immediate corrective actions.

Since we have discussed Six Sigma tools in this book, a more comprehensive (advanced version) of developing a risk assessment can be developed using the XY matrix. Refer to Chapter 13, Appendix S for a complete discussion of this tool.


## 6.8 SAFETY SIGNIFICANCE


A research paper titled "Construction Safety Risk: Improving the Level of Hazard Identification," studied the possible relationship between "safety significance" and "risk assessment" [4]. The article noted that risk assessments are often treated as "attachments" to a safety program and must be completed only to comply with regulatory requirements. A process is needed that will link the safety system more closely with risk assessment. This can be accomplished by using the hazards identified during risk assessment in the development of a successful JHA [4].

To illustrate the boundaries between risk and injury, we need to look at two factors, the "visible factors" (what we can see) and the "hidden factors" (what we cannot see) [4]. The visible factors represent what the user actually sees in hazard identification while the hidden factors represent the at-risk events that are usually hidden within interrelationships of the various elements required to complete the task. In the traditional JHA, a generic list of hazards is generated as they relate to the job (limited to the steps of a job). Hazards are selected that are thought to be relevant to the particular job. Refer to

Unacceptable/Eliminate

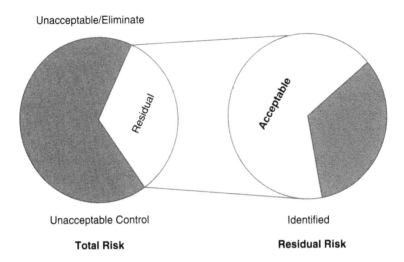

| Unacceptable Control | Identified |
|:---:|:---:|
| **Total Risk** | **Residual Risk** |

**Figure 6-10** Types of Risk

Air Force System Safety Handbook, Air Force Safety Agency, Kirtland AFB, NM 87117-5670, Revised July 2000, Types of Risk, Figure 3-1, pp. 20, Public Domain

Figure 6-10 for an overview of the types of risk, going from the total risk to the residual risk.

"Safety significance" as outlined by Carter and Smith can help to identify and help focus on the tasks that have the greatest potential for harm to employees [4]. Using a method that applies the Pareto law (the 80/20 rule), the safety significance of tasks using risk as the measure of significance can be graphically displayed.

> 80 percent of the observed defects in a product or in a process can be attributed to 20 percent of the possible causes.

We can apply this Pareto's law to the data collected to determine the safety significance of tasks. Each task listed now has a risk value assigned to it. These values are used to determine the order of risk that needs to be considered first. Refer to Figure 6-11 for an overview of the safety-significance concept.

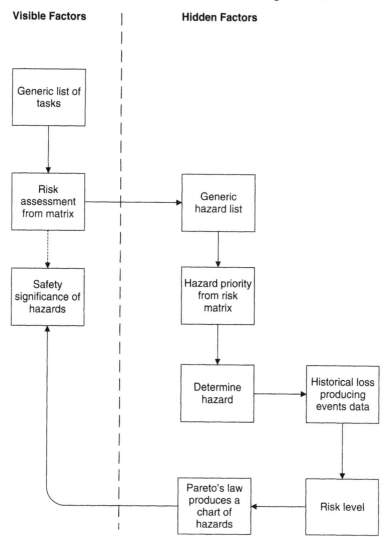

**Visible Factors**                     **Hidden Factors**

**Figure 6-11** Safety Significance

Adapted from Construction Safety Risk, Improving The Level Of Hazard Identification, ESREL 2001 European Safety & Reliability International, Conference, Torino, Italy, September, 2001, Gregory Carter (Postgaduate Student) and Simon Smith (Lecturer in Project Management, School of Civil and Environmental Engineering) The University of Edinburgh, The King's Buildings, Edinburgh, EH9 3JN, UK, Process for Establishing Safety-Significance, pp. 6

## 6.9 WHAT DOES SUCCESS LOOK LIKE?

A successful risk management system is one in which risks are continuously identified and analyzed for relative importance and impact. Risks are mitigated, controlled and tracked by effectively using tools and resources such as the

JHA, inspections, employee participation, Six Sigma, etc. Hazards are avoided, reduced or prevented by use of the Hierarchy of Controls before they create a loss. Employees consciously focus on what conditions and actions could affect safety and feel free to report and work with supervision to fix the issues [1].

## 6.10 SUMMARY

Hazard identification and assessment are fundamental to good safety management. Risk analysis principles provide a structure for the assessment, management, and communication of operational risk. Risk analysis is an important tool as it provides a methodology for evaluating and implementing controls designed for the reduction of risks. One must recognize that risk analysis is an ongoing, continuously evolving and improving process [1].

Risk management does not have to be a complex process. It does require individuals to support and implement basic principles. Risk management offers individuals and organizations a powerful tool for increasing effectiveness and improving JHAs. The risk management process must be accessible and usable by everyone in every conceivable setting or scenario [5]. It ensures that all individuals will have a voice (as in Six Sigma, the author refers to Voice of the Customer (VOC)) in the critical decisions that determine success or failure in all operations and activities.

In setting priorities, gather and review as much information as possible about the risks posed by the various jobs, events or activities in order to consider the degree of the severity and probability of loss-producing events.

If a risk management strategy has been well designed, it will favorably change both the physical conditions and employee behavior. The challenge is to determine the extent and direction of what change is taking place. In assessing possible effective solutions, it may appear that you need to only determine if the number of injuries or other losses has decreased. It may be a year or more before enough data is gathered to determine if significant changes actually occur.

Program and procedural risk control initiatives such as revisions to standard operating procedures can be assessed through use of a standard set of questions or statements reflecting desirable standards of performance against which actual operating situations are compared [5].

We will continue our discussion on how to identify hazards in Chapter 7 and also introduce safety training. One thing that we must all remember is that training does not occur in a vacuum. To be successful, training efforts must have the support of all levels of the organization. We will demonstrate to the reader that in the right culture supervisors should see themselves as coaches who reinforce training.

# CHAPTER REVIEW QUESTIONS

1. How do you define risk?
2. What is risk management?
3. What are some human factor considerations?
4. What are some of the environmental considerations?
5. What are some of the tools/equipment/materials considerations?
6. What are some of the policy and procedures and management considerations?
7. What are job step and task considerations?
8. Define risk versus benefit.
9. What are the principles of risk management?
10. Why is risk management communication important?

# REFERENCES

1. U.S. Army Communications-Electronics Command, Directorate for Safety, Fort Monmouth, New Jersey, public domain
2. Adapted from The Federal Aviation Administration (FAA), Office of System Safety, *FAA System Safety Handbook*, Chapter 3, "Principles of System Safety," December 30, 2000; Chapter 4, "Pre-Investment Decision Assessment," and document, http://www.faa.gov/library/manuals/aviation/risk_management/ss_handbook/, public domain
3. Adapted from The Federal Aviation Administration (FAA), Office of System Safety, *FAA System Safety Handbook*, Chapter 15, "Operational Risk Management," December 30, 2000, Chapter 4, "Pre-Investment Decision Assessment," and document, http://www.faa.gov/library/manuals/aviation/ risk_management/ss_handbook/, public domain
4. *Construction Safety Risk: Improving the Level of Hazard Identification*, ESREL 2001 European Safety & Reliability International, Conference, Torino, Italy, September, 16–20, 2001, Gregory Carter (Postgraduate Student) and Simon Smith (Lecturer in Project Management, School of Civil and Environmental Engineering), The University of Edinburgh, The King's Buildings, Edinburgh, EH9 3JN, UK
5. *Pocket Guide to USAF Operational Risk Management*, can be downloaded from the internet by using the following, Pocket Guide to USAF Operational Risk Management, http://www.mitre.org/work/sepo/toolkits/risk/index.html, public domain
6. "Design defects of the Ford Pinto Gas Tank," *Engineering Disaster*, http://www.fordpinto.com/blowup.htm

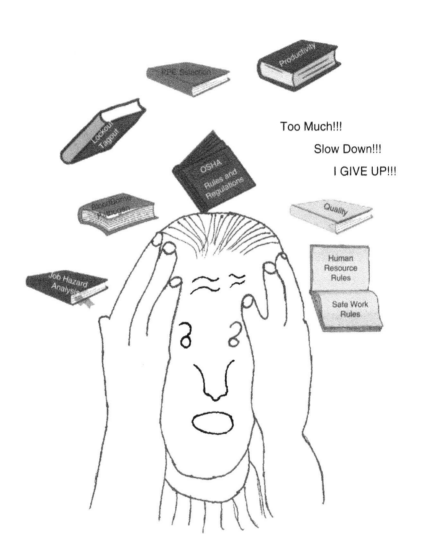

# 7

# Assessing Safety and Health Training Needs

Having a great JHA process that has the potential for improving operational safety is of no value without a workforce that understands the process and its importance. This chapter provides an overview of the concepts useful in training and educating employees to assure a better understanding of JHA and safety. At the end of the chapter you will be able to:

- Define a "good" training program
- Design a training program
- Discuss various training methods
- Develop learning objectives
- Cite the levels of training program evaluation.

"A picture in the head is worth more than a word in the ear."

—Richard Gandy

Just presenting training is not sufficient. Organizations have the responsibility to ensure that their employees are trained to achieve its objectives and that each employee understands the hazards of their work environment and how to protect themselves from specific hazards. If the training has been presented well and has been understood, each employee should be able to demonstrate knowledge of hazards and associated risk.

Training is more complicated than telling someone how to perform a job. Training is the transfer of knowledge from the trainer to the employees in such a way that everyone can accept and use their knowledge in the execution of their job. This knowledge should be specific and directed at the behavioral aspects of the employee [5, 6].

Safety training does not occur in a vacuum. For safety training to be successful, it must have the support of all levels of the organization. The organizational climate and behavioral norms are likely to be more powerful than the behavior taught in safety training sessions. The group can enforce its norms with continual rewards, encouragement, and production pressures. In the right culture supervisors should see themselves as coaches who continually reinforce safety training. The safety training is unlikely to have a long-term impact on the organization if it does not factor in the forces that will form obstacles to use instead of the desired skills [7].

The following examples are areas to think about as we discuss training in this chapter. We hope that you will get our message that in order to have an effective safety training program, repetition, in various forms, is the key to increasing use of training skills.

## Example 1

A radio talk show host had a caller that described his show this way: "I only listen to you on Mondays and Fridays because what I hear on Monday, I also hear on Friday." He further commented that "you repeat yourself over and over." Think about it: if you listen to anything over and over, you are bound to learn something.

## Example 2

A simple TV commercial for headache relief has hit the air waves. It does not talk about the product; instead it only talks about getting rid of a headache, but the message is repeated three times and becomes stuck in your memory. Now there are many other commercials echoing this commercial.

## Example 3

A children's show, "Blue's Clues," introduces a main character, a blue dog that discusses specific topics and uses repetition as a way for children to learn. While designed for the younger child, the use of repetition has shown increased learning of materials. This will be discussed in more detail later in the chapter.

## Example 4

You are in an airport and on the moving walkway. What do you hear repeatedly before getting to the end of the walkway? "Caution, the moving walkway is ending."

What do these examples have in common? In many respects, we learn the same way, no matter whether we are adults or children. We mostly learn through the use of repetition.

"Telling ain't training."

—Harold Stolovitch

# 7.1 HOW IS A GOOD TRAINER DEFINED?

According to Dan Petersen, learning theorists generally agree that individuals will learn most efficiently when they are motivated toward a goal that is attainable. It is necessary that this goal be desired by the person and the learning behavior directly related to achieving the goal [4].

We have an opportunity to train individuals on various safety subjects. So WHY, when we train employees in organizations and quiz them on the subject manner, do we find a percentage of the employees got the message while others seem clueless as to what was said? Does this mean that we are poor trainers? Not necessarily! Trainers need to ask several questions:

- Could it be the way that the training program was developed?
- Did I deliver training in a logical and consistent manner?
- Did the training atmosphere create a good learning environment?
- Are some employees smarter than others?
- Do we know our subject so well that we have a hard time conveying our message?
- Is it that we only have a limited time (in some settings as little as 15 minutes) to get the message across?
- Did the trainee have fun learning the subject matter?

"Training is as long as your butt can stand it."

—Unknown

There is a tendency to make safety training more difficult than it needs to be, throwing too many details and materials at the trainee in the time allowed. Therefore, safety training can become a "soap box" event and we only convey what we are told to do—

i.e., reading a prepared script that may or may not apply to our particular situation. We tend to do an "information dump" and expect employees to remember everything that was said without additional refreshers or other learning aides.

"Presenting features in search of a problem is called 'shot-gun selling,' 'spray and pray,' or in the vernacular of the trade, 'show up and throw up.'"

—Rick Page

Through the internet a wealth of safety resources is at our fingertips. Some sources design safety scripts of material to present to employees in an effort to promote safety. Unfortunately, off-the-shelf programs are not always applicable to existing conditions or work environments. We encountered one situation where an organization had purchased generic safety material so that they could provide a consistent message to multiple plants. At one location, the message for employees was how to drive in snow; however, the plant was in an area of the deep South that never gets snow.

Let's approach safety training from an objective point of view. We are all trainers in some respect, no matter what position we hold, whether at work or in our personal life. The safety professional trains other individuals to be "safe." To be effective in that effort, we must adhere to and apply principles of adult learning, as well as use some innovations from training research.

## 7.1.1 Basic Training Principles

Training of supervisors and employees need not be complex or lengthy. There are five basic training principles that can be used as guidelines:

- Communicate the purpose of the training: Everyone must understand the purpose of the training. The beginning of any training program should focus on *why* safety training is important and how everyone can benefit personally.
- Organize the presentation to maximize understanding: Refer to Table 7-1 for a sample training outline. If you are instructing someone, the sequence in which you present the material must match the steps that they must take to accomplish the task. This example will be used as a JHA in Part 3.
- Provide appropriate work practices: Arrange for employees to practice and apply new skills and knowledge as soon as possible. We learn best when we can immediately practice and apply newly acquired skills and knowledge. Instruction time should include information, demonstration, practice, and application within the initial session. The human learning curve drops quickly and the details of the training will be forgotten without immediate use.
- Provide immediate feedback (knowledge of results): As employees practice, you as an instructor need to know if training is effective. Practicing a task incorrectly creates the wrong behavior, which could result in an injury. Feedback for correct behavior enhances motivation and encourages formation of the desirable behavior.
- Account for individual differences: We are all unique individuals. We learn in different ways. A successful training program incorporates a variety of learning techniques, such as written instruction, audiovisuals, lectures, hands-on, coaching, some fun, etc. In addition, we learn at different speeds and the pace of the training should recognize these differences. The attention

## Table 7-1
## How to Change a Tire

The following sequence of events are example instructions on how to change a tire on a car.

- Pull off to the side of the road out of traffic.
- Park on level ground and provide safe distance from traffic.
- Apply the parking brake and place the transmission in park if it is an automatic. If the car has a standard transmission, place the shift in first gear or reverse.
- Turn off engine.
- Turn on your hazard warning lights.
- Watch out for oncoming traffic as you exit the car.
- Set safety flashers or reflectors five car lengths front and back.
- Pull spare tire and all tools out of the trunk.
- Remove wheel covers.
- Properly place jack under the car. Consult your owners' manual and find where the jack needs to be positioned.
- Jack the car up part way so tire is still touching ground. Raise the car just enough to take the weight off the tire.
- Using the lug wrench, loosen the lug nuts slightly but do not remove.
- Jack up the car so the tire is above ground.
- Remove the lug nuts from the bolts. Place on wheel covers where they will not be lost.
- Remove the flat tire.
- Replace the flat tire with the spare.
- Replace the lug nuts on the bolts and tighten, but not too tight, just enough to hold the tire in place while you lower the car.
- Lower the car until it touches the ground, and then firmly tightens all of the lug nuts.
- Gather everything and put in trunk.
- As soon as possible, have the flat tire repaired and reinstalled.

span and focus of employees must be considered. Refer to Figure 7-1 for an overview of retention rates. Brief sessions allow the trainee to incorporate the information. Long sessions tend to lose the attention of most of us.

- Keep the employees involved and active in training. One effective way to learn is by having employees personally instruct each other in a controlled environment. After the initial instruction and practice, divide the group into teacher/learner teams, pairing a rapid learner with a slower learner, giving them a chance to learn [1].

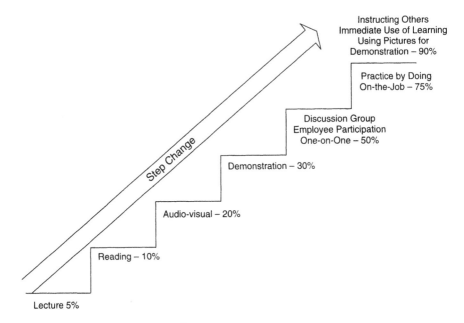

**Figure 7-1** Average Retention Rates

Adapted from Roughton, James, James Mercurio, *"Developing an Effective Safety Culture: A Leadership Approach,"* Butterworth-Heinemann, 2002, pp. 283

The following five principles of teaching and learning should be followed to maximize program effectiveness:

- Employees should understand the purpose of the training.
- Information should be organized to maximize effectiveness.
- People learn best when they can immediately practice and apply newly acquired knowledge and skills.
- As employees practice, they should receive immediate and detailed feedback.
- People learn in different ways, so an effective program will incorporate a variety of training methods.

OSHA web site, http://www.osha-slc.gov/SLTC/safetyhealth_ecat/comp4.htm, public domain

The many ingredients that make up a good training program make the task of blending the required elements challenging. We consider changing a tire to be a relatively simple job; however, there are a number of steps and tasks required as well as basic knowledge of driving, traffic, tool use, and physical skills!

For most presenters, the objective of training is to change or improve performance or behavior and increase the trainee's knowledge. Hazards are intangible concepts and we must make the concepts as clear as possible to trainees that may or may not understand the language of safety.

We tend to do an "information dump": how much I can get out of my mouth to show everyone what I know. As "content" experts, we have a tendency to focus on delivering detailed materials assuming it will be understood by everyone [5, 6].

> "Sometimes technical salespeople feel that if you just show them how smart you are, they'll buy from you, so tell 'em everything you know, whether they need it or not."
>
> —Rick Page

Instead of doing an ego-dump, we must assess employee knowledge of the subject matter before training is designed. We must validate employee retention at various times after training – this will aid in keeping our assumptions about any learning in check [5, 6].

## 7.1.2 Types of Safety Education

> "We are forced to rely on people, which is why we put so much emphasis on training them."
>
> —Henry Block, H&R Block

### 7.1.2.1 General Safety Instruction

Employees generally receive specific safety education through various work experiences (new employee orientation, safety meetings, one-on-one, etc.) [2, 3]. Refer to Table 7-2 for a definition of education and training.

The following list is considered the usual outline for safety "instruction." It may or may not be effective, as instructional objectives are not required and Knowledge, Skills, and Attitudes/Abilities (KSAs) are not evaluated.

- Conveys required and "nice-to-know" information.
- Develops training and learning goals as appropriate [2, 3].
- Knowledge and skills are not measured at the end of training session.
- May include learner goals, but does not require objectives.
- Attendance is the only requirement in order to get a certificate.
- Learners evaluate the quality of content and presentation.
- Learners are not evaluated by presenter [9].

**Table 7-2**
**Education and Training Defined**

| Education |
| --- |
| • That which leads one out of ignorance.<br>• Anything that affects our Knowledge, Skills, and Attitudes/Abilities (KSAs).<br>• The "WHY" in safety educates the employee about natural occurrences, hazard recognition, awareness and prevention, injury causation, consequences of exposures, and system consequences (coaching and/or reward) of behavior.<br>• Primarily increases knowledge and attitudes. |
| **Training** |
| • One method of education.<br>• The "How to do" and "What to do" in safety.<br>• Primarily increases knowledge and skills.<br>• A specialized form of education that focuses on developing or improving skills, with a focus on performance. |
| OR-OSHA 109 *Train-The-Trainer I: Managing the Safety Training Program 46*, Presented by the Public Education Section, Oregon OSHA, Department of Consumer and Business Services, public domain |

### 7.1.2.2 Technical Safety Training

To ensure that employees use the correct behavior, they not only need to know why using safe procedures and practices is important, but they also need to know exactly how to actually perform those procedures and practices [2, 3]. Technical training should be used over and above just instruction. As trainers we must understand that the most important thing is to remember that technical safety training is "hands-on" and "how to" [9]. The JHA will help to accomplish such technical training. The following list is a key to technical training:

- Convey "must know" information.
- Instruct employees on how to perform procedures and practices properly.
- Use operational learning objectives.
- Evaluate knowledge and skills in the learning environment.
- Provide oral/written exams and skill demonstration. Employees must pass a test in class to get a certificate.
- Describe general/specific policies, procedures, practices.
- Measure learner's knowledge and skills immediately after training.
- Evaluate learner in the actual work environment and before exposure to hazards.

An effective safety training program is only one part of an overall safety management system. A training program is itself a "subsystem" of a process with its own structure of inputs, processes and outputs. If the program is well-designed and the safety culture supports training, then the program performance will be effective [2, 3]. Refer to Table 7-3 for an overview of the ANSI Z490.1-2001 Standard.

An effective training program outcome requires design, input, and process, just as any safety or quality system does. Refer to Figure 7-2 for an effective training program outcome.

### Table 7-3
### Overview of the ANSI Z490.1-2001 Standard

ANSI Z490.1-2001, Criteria for Accepted Practices in Safety, Health, and Environmental Training is a broad-based voluntary consensus standard covering all aspects of safety training, which includes training development, delivery, evaluation, and management of the training function. The criteria in the standard were established based on accepted best practices in the training industry and the safety, health and environmental industries.

*What Does the Standard Say about Training Program Elements?*

ANSI 490.1-2001, Section 3.2, states that a safety training program should include written plans that tell how training development, delivery, documentation, recordkeeping, and evaluation will be accomplished. The following elements should be included in a training program:

1. **Training Development**: This section outlines procedures for developing a needs assessment, learning objectives, course content and format, resource materials, and criteria for course completion.
2. **Training Delivery**: This section provides ways to ensure that the quality of training is delivered by a competent trainer in a suitable training environment.
3. **Training Documentation and Recordkeeping**: This section provides procedures, forms, and reports that ensure the quality and maintenance of training delivery and program evaluations, and trainer and trainee certifications.
4. **Training Evaluation Plan**: The standard provides procedures that describe how evaluation of training program design and performance will be accomplished with a continuous improvement approach [2, 3].

American National Standards Institute (ANSI), Z490.1-2001, Criteria for Accepted Practices in Safety, Health, and Environmental Training

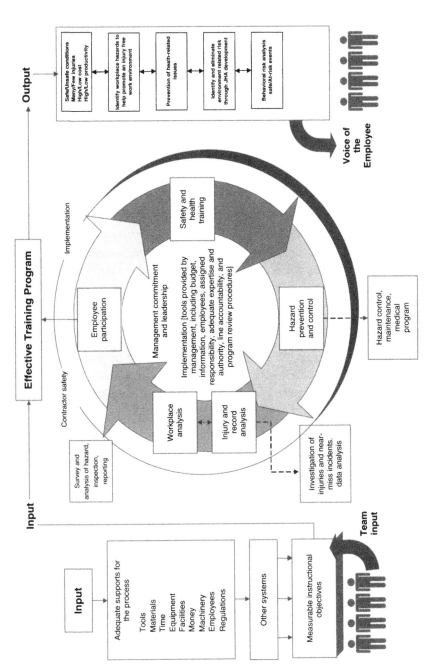

**Figure 7-2** Safety Management System, Effective Training Program Outcome

## 7.1.3 Training Plan Linked to Consequences

An important factor is that a training plan should be effectively linked to consequences.

"A lesson plan is an instructional prescription, a blueprint describing the activities the instructor and student may engage in to reach the objectives of the course. The main purpose is to prescribe the key events that should occur during the module."

—Robert Mager

There are two type of consequences: *natural* and *system*. The following paragraphs will describe each consequence.

*Natural consequences* occur automatically in response to our behaviors/actions. We are punished or rewarded by something for what we do. If we slip and fall down, there are two consequences: we either get hurt or we do not. In the context of safety, the natural consequences outcome is either "safe or we have an at-risk event." Whether we are injured or not injured involves the assessment of risks within the job task for which the training is being designed. If we do not identify and control the hazards and associated risk, the safety training will not be properly designed [9].

*System consequences* are organizational responses to our behaviors/actions. We are rewarded or punished based on what we do. Various consequences may occur; for example, someone else may administer discipline, apply pressure, apologize, recognize and reward, or ignore performance [9]. We may have an excellent safety training program in place, but if we do not identify and understand the system that generated the consequences, we may find the training programs less than effective.

A training program must be results-oriented to be effective. It must fully identify both categories of consequences and the subconsequences within each. The objective is to provide two-way communication that helps the employee learn required safety techniques and provides the trainer with feedback on the various consequences that need to be further considered [5, 6].

Training will vary based on the established goals and objectives [1]. Part of developing a training program is determining who needs training. This is the reason for completing a "needs assessment." Refer to Table 7-4 for an overview of who needs training.

Refer to Appendix L for a sample lesson plan.

**Table 7-4**
**Who Needs Training**

- Training on hazard and associated concepts should target new hires, contract employees, and employees in high-risk areas.
- Managers and supervisors should be included in training plan development. Supervisors should receive training in hazard recognition, hierarchy of controls, risk assessments, incident investigations, and how to provide and reinforce training.
- Training for managers should emphasize the importance of their role in visibly supporting safety and ensuring that the process is used.
- The long-term employee who changes a job as a result of new process or material changes should be provided training prior to start-up.
- The entire workforce needs periodic refresher training in general safety awareness.
- Evaluate the safety system when initially designing the training program. If the evaluation is done properly, it can identify the safety system strengths and opportunities for improvement, and provide a basis for future changes.
- Keeping training records will ensure that everyone gets timely training.

Adapted from, OSHA's web site, http://www.osha-slc.gov/SLTC/safetyhealth_ecat/comp4.htm, public domain

## 7.2 STEPS IN THE COURSE DEVELOPMENT PROCESS

"Though it seems hard to believe, instructors are frequently asked to develop courses intended to teach people what they already know, or to use instruction to solve problems that can't be solved by instruction."
                                                                            —Robert Mager

Refer to Table 7-5 for OSHA training development guidelines and Table 7-6 for ANSI/ASSE Z490.1-2001 guidelines on training development.

### 7.2.1 Conducting a Training Needs Assessment

Conducting a needs assessment is the most important objective when developing a safety training program. Too often, training programs are pulled "off the shelf" with limited or no effort made to adapt to the specific application. This generic training approach may be easy to implement but fails to ensure that employees truly know why they must follow a set of rules. There are areas

## Table 7-5
## OSHA Training Development Guideline

The OSHA model has seven elements:

1. Determine if training is needed
2. Identify training needs
3. Identify goals and objectives
4. Develop content and activities
5. Develop evaluation methods
6. Develop training documents
7. Develop improvement strategies

The model is designed to be used to enhance your training program. Using this model, you can develop and administer safety training programs that address site-specific issues, fulfill the learning needs of employees, and strengthen the overall safety program.

Adapted from OSHA's Publication 2254 (Revised 1992), "Training Requirements in OSHA Standards and Training Guidelines," public domain

## Table 7-6
## ANSI/ASSE Z490.1-2001 Guidelines on Training Development

Training should be developed to improve the occupational safety, health, or environmental knowledge, skills, or abilities used by the employees in the performance of their jobs.

Training development should include a systematic process, including needs assessment, learning objectives, course design, evaluation strategy, and criteria for completion.

ANSI/ASSE Z490.1-2001 Guidelines on Training Development Training Development

of site-specific assessment that must be conducted. Refer to Table 7-7 for types of safety training needs

A training needs assessment is a thorough analysis of an organization to determine the climate of the organization. Many trainers who do not perform a training needs analysis find that their program success is spotty: sometimes successful, but other times, even when delivered in the same way by the same trainer, wildly unsuccessful. We have provided the same type of training programs to groups where one group was highly involved and actively participating, while the other group could just as well have been rocks. The reason is that

**Table 7-7**
**Types of Safety Training Needs**

| |
|---|
| • Orientation training for new employees and contractors. |
| • SOPs. |
| • Hazard recognition training. |
| • Training required by regulatory agencies. |
| • Training for emergency response and emergency drills. |
| • Injury prevention and loss-producing event training. |
| Adapted from OSHA's web site, http://www.osha-slc.gov/SLTC/safetyhealth_ecat/comp4.htm, public domain |

no two training groups are exactly alike. Training needs, level of motivation, educational background, and many other factors can affect the training environment. The trainer must be able to assess training needs and adapt the training accordingly for each group [7].

Training requires a three-part organizational analysis that involves the study of the entire organization. It determines how to target where the organization training emphasis can and should be placed.

- The first step is to get a clear understanding of short-and long-term goals. What is the organization trying to achieve in its overall safety efforts in general and specifically by department?
- The second step is an inventory of the company's human and physical resources.
- The final step is an analysis of the climate of the organization. What is the level of energy that the organization has to develop and implement the training process? [4]

You want the training program to focus directly on safety concerns that can be most appropriately addressed by training. Specific hazards that employees need to know about should be identified through the JHA based on a site assessment, change analysis, injury records, near-miss reports, maintenance requests, employee suggestions and other means.

A safety problem may be considered a skill deficiency issue. Often you will find a safety problem is a precursor to general performance issues and is impacting productivity or other organizational drivers. The employee may not understand or have an understanding of using a specific sequence of actions. By changing the way a job is completed, the employee may fail to clearly see or comprehend the nature of hazards. The employee may be experienced and highly educated but for a variety of reasons did not find the training of benefit [5, 6].

### 7.2.1.1 Developing Learning Activities

An understanding of learning theory and adult learning is necessary to complete the needs analysis. We are faced with a workforce that is being raised on entertainment delivered by very creative media sources ranging from computer gaming, internet resources, reality television, movies, and other resources. As a result, many of the individuals entering the workforce will have years of experience with interactive and dynamic learning methods indirectly developed. They have experiences not found in the past generations and may be extremely well-versed in many topics through use of these media. Training must be creative and take into account the skill level of the audience. Be imaginative in your choice of methods and materials, and make sure that you use all internal and available external resources. Get ideas by looking at the training programs of other organizations in your industry, the internal training personnel and local training groups. Organizations such as the American Society of Training and Development (www.astd.org), while not directly related to safety, can provide methods and information on human performance improvement and education of adults.

## 7.2.2 Establishing Learning Objectives

"Now it's time to describe the instructional outcomes (the 'need-to-do's); it's time to construct a verbal word picture that will help guide you in developing the instruction and help guide students in focusing their efforts."

—Robert Mager

A *learning objective* is different from a learning goal. A learning objective defines the specific level of quality performance. A learning goal is defined as a general skill or behavioral change that will occur as the results of training [5, 6]. In this section we will focus on the learning objective.

A learning objective is a brief, clear statement of what the employee will be able to do as a result of training. The groundwork for the learning objective has already been laid once a thorough job analysis has been completed. A JHA should describe all of the steps and associated tasks involved in a job skill. The learning objective should focus only on the tasks to be included in the training session. Sometimes an entire job needs to be learned and sometimes only a portion of a task within a step needs to be learned. The training assessment lists the behavior that needs to be learned. A learning objective defines how well and under what conditions the task must be performed to verify that the task has been learned. Learning objectives are important because instructional strategies and evaluation techniques are an outgrowth of the learning objectives [7].

A learning objective is a statement describing an outcome. It describes results, rather than the means of achieving stated results. It defines the expectation for the trainee. Objectives must be a clear, competency-based statement that describes what the employee will be able to do at the end of the training [5].

Learning objectives help the instructor to:

- Design and select instructional content and procedures.
- Organize the employee's own efforts and activities.
- Evaluate or assess the success of instruction.

It is important to express training objectives in specific terms and to make your objectives measurable. This will focus the content of your course on the objectives.

### 7.2.2.1 Guidelines for Writing Learning Objectives

Learning objectives are always written from the viewpoint of what the trainee will do, not what the trainer will do:

Right: Employee will be able to change a flat tire.

Wrong: Instructor will cover the method on how to change a flat tire.

Verbs or action words are used to describe the behavior and must be as specific as possible. Words that must be avoided include vague words such as "know," "learn," "comprehend," "study," "cover," and "understand" [7].

Right: The employees will be able to complete the steps for changing a flat tire.

Wrong: The employees will learn how to change a flat tire.

The desired behavior must be observable and measurable so that the trainer can determine if the task has been learned [7].

Right: The employees will demonstrate their ability to change a flat tire.

Wrong: The employee will know how to change a flat tire.

Objectives should be given orally and in writing to the employees, so that they understand the purpose of the training session.

### 7.2.2.2 Components of Learning Objectives

There are four components that need to be considered each time a learning objective is developed: Target audience, behavior, conditions, and standards [7].

**Target Audience**

*Audience Analysis*: A critical step in a training assessment is to analyze the target audience. The trainer should determine the general educational background of the audience, their job duties, their previous training history, the length of employment, the general emotional climate of the organization, behavioral norms, and attitudes toward training. It is vital to determine if employees have mastered prerequisite skills and knowledge in order to target training appropriately [7].

It is important to take time "up front" to pinpoint the expectations of the employees that you are going to train. Determine how much support there is from the management team. Determine management's desired training objectives. Talk with employees from the target audience that you will be training. This discussion will help to determine the objectives and expectations of training. You should survey employees from the subordinate group. This part of the needs analysis does not have to be formal. Often a tour of a facility provides an opportunity to ask simple and direct questions of the employee. The key is to listen to employees to assess expectations. The ability to listen is very important because employees will often volunteer information to a sincere and skilled listener [7].

The target audience must be considered because the same topic may be approached differently based on the background of the groups to be trained. One thing that is important is to ensure that courses and activities are designed for the adult learner. The following examples must be considered based on the learning objectives for adult learning:

- Always include goals or objectives that let employees know what is coming.
- Adult learners must be given time to reflect or think about each point of learning. Focus on one thing at a time. They should not have to take a lot of notes when you want them to listen.
- Design in time to reflect or think about each point of learning into the process.
- Include samples, stories, and scenarios that apply the learning to something they can relate to. This could include a funny video clip that is related to the subject that is being taught.
- Include lists and acronyms. We use a lot of safety acronyms and assume that everyone understands them. Make sure that acronyms are explained.
- Flag important information. Use a "parking lot" list to keep ideas fresh in the trainee's mind.
- Adults do not effectively learn by simply being told. They must have a chance to digest the material and, whenever possible, apply the learning to something they can relate to. Ask questions and involve the trainees.

Include open-ended questions and exercises. Design active audience participation in the learning process whenever possible.

- Information more easily enters the long-term memory when it is linked to old memories or can be related to something the learner has experienced.
- The short-term memory is linear, works best through lists, and is the only conscious part of the brain.
- Giving adult learners an advance organizer, such as workshop goals or objectives, helps them to retain information.
- Let them know what is important: what to focus on every time there is a change in points or a new topic to discuss.
- The mind is attracted to what appears different or novel than what is expected or routine [4, 9].

---

Summary of Sequencing Safety Training

- Content: What subjects will the training cover?
- Connecting: How will each topic be related to the workplace? Why is it important?
- Loading: To what depth will each subject be covered?
- Sequencing: In what order will the topics be covered?

Known to Unknown – Common chemicals at home.
Simple to Complex – Simple lockout/tagout procedure.
General to Specific – fall protection systems.
Theory to Practice – Accident investigation.
Step by Step – Any procedure.

---

OR-OSHA 109 Train-The-Trainer I: Managing the Safety Training Program 46, Presented by the Public Education Section, Oregon OSHA, Department of Consumer and Business Services, public domain

## Behavior

The behavioral component of the training objective is the *action* component. It is the critical component of the objective, as it pinpoints the way employees will demonstrate that they have gained the necessary knowledge. Learning is measured by a change in employee behavior. How will employees prove what they have learned?

- Will they **explain** ... ?
- Will they **calculate** ... ?
- Will they **operate** ... ?

- Will they **repair** . . . ?
- Will they **troubleshoot** . . . ? [7]

The highlighted verbs in the above examples indicate the behavioral aspect required. The behavior component should be easy to determine based on the task analysis, when written in behavioral terms [7].

### *Types of Behavior in Learning Objectives*
The next step is to identify the domains of learning, the types of behavior that can be described within an objective. Behaviors are categorized in one of the following domains of learning: cognitive, psychomotor, or affective.

*Cognitive behaviors*: "Cognitive behaviors" describe observable and measurable ways that the employee can demonstrate that they have gained the knowledge and/or skill necessary to perform a task safely. Most learning objectives describe cognitive behaviors, as these behaviors are easy to master while others are much more difficult. In designing safety instruction, trainers move from the simple to the complex to verify that employees have the basic foundation they need before moving to the next level of skills [7]. This does not mean that the basic skills have to be re-taught if the trainer cannot verify through observations, pretests, training records, etc., that prerequisite skills have been mastered. However, training sessions have turned into a disaster because the trainer made the assumption that the employees had mastered basic skills and began the training at too high a level. In contrast, training sessions have bored the employees by being too basic. It is important for trainers to be able to label learning objectives and design training sessions appropriate to the level of cognitive behavior required to perform a task. The following paragraphs will describe examples of types of cognitive behaviors [7]:

*Knowledge-level cognitive behaviors* are the easiest to teach, learn, and evaluate. They often refer to rote memorization or identification. Employees often "parrot" information or memorize lists or name objects. Common action words such as: identify, name, list, repeat, recognize, state, match, and define [7].

*Comprehension-level cognitive behaviors* have a higher level of difficulty than knowledge-level cognitive behaviors because they require the employee to process and interpret information. The employee is not required to apply or demonstrate the behavior. Commonly used action words include verbs such as: explain, discuss, interpret, classify, categorize, cite evidence for, compare, contrast, illustrate, give examples of, differentiate, and distinguish between [7].

*Application-level cognitive behaviors* move beyond the realm of explaining concepts orally or in writing. They deal with putting ideas into practice and involve a routine process. Employees apply the knowledge they have learned. Examples of action words commonly used include the following: demonstrate, calculate, do, operate, implement, compute, construct, measure, prepare, and produce [7].

*Problem-solving cognitive behaviors* involve a higher level of skills than application-level cognitive behaviors. The easiest way to differentiate between the application-level and problem-solving level is to apply application-level to a routine activity and problem-solving level to nonroutine activities. Problem-solving skills require analysis (breaking a problem into parts), synthesis (looking at parts of a problem and formulating a generalization or conclusion), and evaluating (judging the appropriateness, effectiveness, and/or efficiency of a decision or process and choosing among alternatives). Examples of action words commonly used include: troubleshoot, analyze, create, develop, devise, evaluate, formulate, generalize, infer, integrate, invent, plan, predict, reorganize, solve, and synthesize [7]. This is the behavior that will be used in development of JHAs.

*Psychomotor Behaviors*: Learning new behaviors requires developing new cognitive skills (knowledge, comprehension, application and/or problem solving) as well as new psychomotor skills that may be required in the application phase of learning. Psychomotor behaviors pertain to the proper and skillful use of body mechanics and may involve both gross and fine motor skills. Safety training sessions for psychomotor skills should involve as many of the senses as possible. The trainer must adapt the format involving such physical movements to match the skill level of the learner and the difficulty of the task [7]. The key to teaching psychomotor skills is that the more the learner observes the task, explains the task, and practices the task correctly, the better the trainee performs the task.

*Affective Behaviors*: These behaviors include the attitudes, feelings, beliefs, values, and emotions. Trainers must recognize that affective behaviors influence how efficiently and effectively learners acquire cognitive and psychomotor behaviors. Learning can be influenced by positive factors (success, rewards, reinforcement, perceived value, etc.) and by negative factors (failure, lack of interest, punishments, fears, etc.) [8–10]. Other affective behaviors (attitudes and emotions) that must be considered go beyond positive or negative motivations toward learning [7]. Training objectives that state affective behaviors are usually much more difficult to observe and measure than cognitive behaviors. Nevertheless, they are critical to the ultimate success of the safety-training program. A critical factor to remember is that, while training can stress the importance of affective behaviors, employees are mostly influenced by the behavioral norms of an organization. Behavioral norms refer to the peer pressures that result from the attitudes and actions of the employees/management as a group. Behavioral norms are the behaviors a group expects its members to display. Before attempting to make changes in an organization, it is important to identify existing norms and their effects on employees [7].

## Conditions

The "condition" component of the objective describes special constraints, limitations, environment, or resources under which the behavior must be demonstrated. Note that the condition component indicates the conditions under which the behavior will be tested, not the conditions under which the behavior was learned [7]. The JHA should define the environment in which the job is completed.

## Standards

Finally, it is important that all safety training is conducted as effective as possible, where employees have a good learning environment where they can learn. Therefore, to determine if your training is effective an acceptable standards must be developed along with a measurement system to measure the effectiveness of the content presented. The objective specifies the acceptable standard of performance. It should outlines a quality criteria for acceptable achievement. Safety training usually requires full mastery of a skill and, therefore implied with phrases like:

- Correctly
- Accurately
- Without error

For example: "After training the employee must be able to perform all steps of the lockout tagout procedure without error"

### 7.2.2.3 Learning Styles

One of the pitfalls of instruction is that trainers tend to develop safety-training programs that accommodate the way the trainer learns best, not the way the employees learn best. For example, if the trainer learns best by reading, a manual may be provided to the employees and the trainee will be expected to master the procedure by reading the manual. If the trainer learns best through experimentation, they tend to throw employees into a situation workshop with little guidance. If a visual learner, they use videos, graphs, pictures.

It is important to emphasize individual growth and remember that employees have different learning styles with which they are most comfortable. Every employee is different and must be treated as an individual [7].

Passive Learners

- Reading manuals/books
- Watching audiovisual presentations
- Hearing a lecture
- Observation demonstrations

OR-OSHA 109 Train-The-Trainer I: Managing the Safety Training Program 46, Presented by the Public Education Section, Oregon OSHA, Department of Consumer and Business Services, public domain

Active Learners

- Participating in discussions
- Role-playing
- Performing a experiment
- Taking a field trip
- Hands-on learning
- Responding to a scenario
- Making a presentation

OR-OSHA 109 Train-The-Trainer I: Managing the Safety Training Program 46, Presented by the Public Education Section, Oregon OSHA, Department of Consumer and Business Services, public domain

Some learners prefer to learn by themselves while others prefer to work in groups. Some employees need a lot of organization and learn in small steps sequentially while others assimilate whole concepts with a flash of insight or intuition. Some employees are visual and learn best through drawings, pictorial transparencies, slides, demonstrations, etc., while others learn best through hearing words and enjoy lectures and reading transparencies and slides with words. Increased retention results from what we know of split hemisphere learning. Just as different areas of the brain control the body, so does the brain absorb and record different types of information [7].

In each case the trainer needs to understand that these differences in learning styles exist. The key is to try and combine as many types of activities and media as possible so that employees can have access to the way they learn best and also learn to adapt to other learning styles as applicable. This means that a training session might include a handout for readers, a lecture for listeners, and an experiment for doers, depending on the objective [7].

The key to accommodating learning styles is that all instructional strategies and media are selected as a means to help the employee and not as a convenience for the instructor. The safety trainer should constantly look for alternate strategies and media so that if one strategy or type of media is ineffective, the safety trainer has multiple strategies from which to select [7].

## 7.2.3 Course Content Development

Once specific objectives are identified, the training methods can be selected that ensure that training objectives can be achieved.

As a tool box of techniques, various training methods can be applied as appropriate to a specific group or audience. Refer to Table 7-8 for examples of common types of training methods.

### Table 7-8
### Common types of training methods

| | |
|---|---|
| Case study | Actual or hypothetical situation. |
| Lecture | Oral presentation of material, usually from prepared notes and visual aids. |
| Role play | Employees improvise behavior of assigned fictitious roles. |
| Demonstration | Live illustration of desired performance. |
| Games | Simulations of real-life situations. |
| Stories | Actual or mythical examples of course content in action. |
| Discussion | Facilitated opportunity for employees to comment. |
| Question | Employees question the trainer and receive answers to questions. |
| Small group | Employees divide into sub-groups for discussion or exercise. |
| Exercises | Various tasks related to specific course content. |
| Instruments/Job Aids | Tools, equipment and materials used on the job. |
| Reading | Employees read material prior to, during, and/or after the session. |
| Manuals | Handbooks or workbooks distributed to employees. |
| Handouts | Selected materials, usually not part of a manual. |

OR-OSHA 109 Train-The-Trainer I: Managing the Safety Training Program 46, Presented by the Public Education Section, Oregon OSHA, Department of Consumer and Business Services, public domain

Each method must be evaluated in relation to the stated objective and the level of expertise in the group. The objective is to make the training interesting and in an environment where the trainee can learn.

After you have completed the training program design, you should conduct a "pilot run," using a test group. Conduct the program just as you would for a group of employees. This pilot run will help to work out the kinks and allow you to make the necessary changes to make it flow properly. The pilot run will help identify problem areas and will help to evaluate the effectiveness of the overall goals and objectives [5, 6].

### 7.2.3.1 Delivering Effective Safety Training

A good learning environment maximizes learning by offering the employees a reason to learn, preparing them for the stated objectives of the organization, and developing trust with the instructor.

Unrealistic expectations are usually a result of a failure to understand what constitutes an effective training program. An example is a request to train 200 people with a wide variation in knowledge of background information and need-to-know. Another example of an unrealistic expectation might be a request to train employees who have just worked a shift from 11:00 p.m. to 7:00 a.m. at 7:00 a.m. on Saturday with no additional pay. It is easy to anticipate a problem in motivating the group. Be sure to set appropriate times and dates.

One fatal mistake for a trainer is to approach the trainer-learner relationship as a teacher-child relationship. Most of the trainer role models we have observed have acted like adults teaching children, pointing fingers at the trainees, hand on the side like a grade school teacher does, etc. It is essential for trainers to view themselves as facilitators of the adult learning process. Refer to Figure 7-3, Mager's decision tree for improving safety performance, and Figure 7-4, showing a comparative analysis of training [4]. Although no generalizations apply to every adult learner, it is helpful in planning training sessions to keep the following characteristics of adult learners in mind [7]:

- Rote memorization may take more time than in childhood, but purposeful learning can be assimilated as fast or faster by an older adult as by high school students. Despite the cliché that "old dogs can't learn new tricks," healthy adults are capable of lifelong learning.
- Adults want to know the "big picture." Most adults want satisfactory answers to their questions: "Why is it training important?" and "How can I apply the training?"

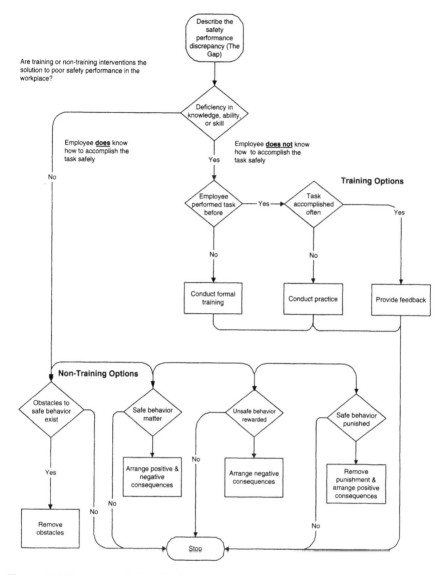

**Figure 7-3** Improving Safety Performance*
Adapted from Oregon State OSHA Consultation Program, OR-OSHA 105, Introduction to Training, 2006, pp. 9, public domain
Adapted from Mayer, Robert, Peter Pipe, Analyzing Performance Problems, pp. 3

- Adults function in roles which mean that they are capable of and desirous of participating in decision making about learning.
- Adults do not like to be treated "like children" and especially do not appreciate being reprimanded in front of others. Adults have specific objectives for learning and generally know how they learn best.

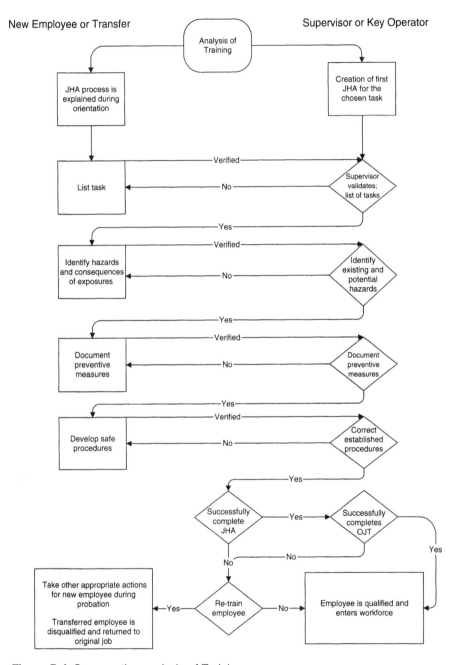

**Figure 7-4** Comparative analysis of Training

Adapted, OR OSHA 103, Comparative Analysis, pp, 14, http://www.cbs.state.or.us/external/osha/educate/training/pages/materials.html, public domain

- Adults have experienced learning situations before and have positive and/or negative impressions about learning.
- Adults have had a wealth of unique experiences to invest in learning and can transfer knowledge when new learning is related to old learning. Adults recognize good training and bad training when they see it [8, 10].

The trainer should try to determine the causes of the performance deficiencies and tailor the training to the needs of the organization. For example, when individuals are motivated to perform well but lack skills or knowledge, an ideal training opportunity exists [7].

## 7.2.4 Safety Program Evaluation

Training evaluation helps to determine if the training has achieved its goal of improving the employee's skills and awareness. When carefully developed and carried out, the evaluation will highlight a training program's strengths and identify opportunities for improvements [5, 6].

Evaluation analyzes the quality of the training program design, trainer, and training materials. The evaluation should offer an opportunity for the trainee to tell the trainer how to revise and improve the program [9]. Refer to Appendix L for a sample safety training program audit and for ANSI guidelines for evaluating a training program.

For an effective discussion of training evaluation using excerpted information from the public domain source, see Oregon-OSHA 108, Train-the-Trainer: Managing the Safety Training Program based on the ANSI Z490.1-2001 standard.

ANSI Z490.1-2001, Section 6.2, Evaluation, approaches detailed strategies for evaluating training and training programs. The standard outlines four levels of evaluation, as developed by Donald Kirkpatrick, author of *The Four Levels for Evaluation*. However, more recently research by educators and discussions with Donald Kirkpatrick, Level 4 Evaluation has been divided into two separate levels, creating a fifth level that discusses impact on profit. [11]

### 7.2.4.1 Level 1: Measuring Employees' Reactions

This first level of evaluation solicits feedback from the employee about what their perceptions are concerning various aspects of the training. Was the employee satisfied with the presentation and content of the training?

### 7.2.4.2 Level 2: Measuring KSAs in the Learning Environment

This level involves measuring the learning that took place during the training session. The evaluation occurs immediately after the training is presented. Quantifying the learning that took place is made by measuring increased knowledge and improved skills. Proficiency should be evaluated and documented by the use of a written assessment and skill demonstration. The following guidelines can be used when developing testing methods for safety training:

- Evaluate employee's knowledge and skills.
- The level of minimum achievement should be specified in writing. If a written test is used, a minimum of 25 questions should be used for the more complex subjects.
- If a skills demonstration is used, the tasks chosen and the means to rate successful completion should be fully documented.
- The content of the written test and the skill demonstration should be relevant to the objectives of the course.
- The written test and skill demonstration should be updated as necessary to reflect changes in the curriculum.

The proficiency assessment methods, regardless of the approach or combination of approaches used, should be justified, documented, and approved by management.

### 7.2.4.3 Level 3: Evaluating the Application of KSAs in the Work Environment

This level of evaluation measures both the learner and the safety culture. It gauges how well the learner applied the training in the actual work environment. Evaluation at this level may indicate the degree that the safety culture supports the training. This level of evaluation is required by ANSI Z490.1-2001. ANSI or compliances aside, it is just smart business policy to ensure that training is effective.

According to Donald Kirkpatrick, there are five supervisor behaviors that affect learner attitudes about safety training:

- Preventing: Supervision does not allow the employee to use the procedures or practices that have been taught.
- Discouraging: Supervision does not encourage behavioral changes by sending implicit (non-verbal) messages that they want behavior to remain the same.

- Neutral: Supervision does not acknowledge the training received. There is no objection to behavioral change as long as the job gets done on time.
- Encouraging: Supervision acknowledges the training and encourages the employee to use what they learned.
- Requiring: Supervision knows what training was received and insists that the learning is transferred to the job.

### 7.2.4.4 Level 4: Evaluating How Training Has Impacted Productivity

Determining how the organization has improved because of the hazard and risk control training program is level 4. The question is: can improvement be clearly determined?

*Safety improves process quality*: Evaluate how the training has impacted the quality (efficiency, effectiveness) of a job.

- In Six Sigma, "safety" is not usually considered a variation in the process. However, when safety training is defined the same way every time, training would be considered effective if the variations in the process have been reduced and consequently fewer injuries have occurred.

### 7.2.4.5 Level 5: Evaluates How Training Has Impacted Profits

Training affects the bottom-line results. If training has improved the bottom line profitability, a return on the investment (ROI) of the company can be determined.

Note that you must exercise caution when applying ROI to safety, as you may set up a negligent liability potential. "Safety vs. profit" is not good business.

- ROI is calculated by converting productivity and quality improvements to monetary values. This is the most difficult level of evaluation but converts the safety language into the language that management speaks. What value have you added to production, quality, and improvement of the organization?
- Have injury rates decreased? How has that improved direct and indirect costs?
- Have identified hazards and associated risk been reduced and how is it documented?
- This level of evaluation allows the training to show compliance has been attained with the added bonus of a more productive operation [9].

## 7.2.5 Recordkeeping

It should be obvious that it is to your advantage to maintain extensive training records. A simple form can identify the trainees and trainer, the topic or job, materials provided and the training date, with space for a brief evaluation of the employee's participation and success. These records will help ensure that everyone who needs training receives it, that refresher courses are provided at regular intervals, and that documentation is available, when needed, to show that appropriate training was provided. We live in a litigious world and on occasion must show the efforts of our work. See Appendix 7.4 for a Sample Training Certification [5, 6].

## 7.2.6 Develop Improvement Strategies

The question that should be asked after training is: "How can we continually improve the training?" Therefore, in any system there is always room for improvement. If it isn't the course material itself, it may be the culture, management system, that supports the training. Ultimately, improving training is all about changing the behavior of the organization including management and employees. Effective change in the management culture is crucial to any long term success.

Traditionally, the use of one-on-one (face-to-face) meetings, with daily, weekly, and monthly safety meetings has been stressed by safety professionals. In construction and other high-hazard industries where the work situation changes rapidly, "tool-box talks," "2-minute drills," and weekly and monthly meetings are recommended. These meetings provide a means of communicating the upcoming tasks and hazards that may be present, current injuries that have occurred, and any environmental changes and procedures needed [5, 6]. As with any meeting, these can be beneficial if not allowed to become a meeting just for the sake of having a meeting.

## 7.3 "BLUE'S CLUES" TRAINING TECHNIQUES

This chapter has provided a discussion on different ways to train employees. These are all great techniques, but now let's turn our attention to a somewhat different approach. We offer a challenge to everyone that reads this section to ask: "What if . . . " We encourage you to keep your mind open and just think about what will be presented from this point on. What if we make safety training so simple that you see the results in action as employee do their jobs?

We are introducing "Blue's Clues." "Blue's Clues" is a cartoon for children aged 3 to 5 years. The website: "Building Your Business, Advancing Your Career, Getting Ahead, A resource provided by the Business Journal of Youngstown, Ohio," 11/13/02, published an article titled "Follow Blue's Clues To Find Business Success" written by Diane Tracy that provides some insight on training. This article starts as follows: "Do you feel that going to the office requires that you leave your personal values at the door and don the mantle of corporate toughness? Do you assume that work is work and fun is fun and never the twain shall meet? Do you believe that embracing "softer" qualities like compassion, joy, curiosity, and playfulness would negatively impact profits? If your answer is yes, you're not alone. Too many of us have allowed our 'childlike' values to be programmed right out of us—and as a result, too many companies are failing to live up to their full potential." Reference http://www.business-journal.com/survival/articles/startrun/BluesClues.html

To continue the concepts, we looked at Diane Tracy's book, *Blue's Clues for Success: The Eight Secrets behind a Phenomenal Business*, where she explores the extraordinary success of the Nickelodeon children's show that's become a highly merchandized international phenomenon [12].

> "If you use your mind and take a step at a time, you can do anything you want to do."
>
> —Diane Tracy

In this book, she provides an overview on how the creators of "Blue's Clues" took their fundamental core values, their inherent love for children, and a basic premise that work should be enjoyable and parlayed it into a business that has generated over $3 billion in merchandising within a few years, airs in more than 60 countries and has become a household name. Note the key comments: Core Values, Love of Children, Employee Participation, and Work as Enjoyable [12].

This summary will provide a flavor of what we are trying to accomplish with this chapter.

On the website "CHCM Research" it was indicated that children watching the same episode of "Blue's Clues" repeatedly show increased material comprehension, especially in using problem-solving strategies. They further indicate that regular viewers of "Blue's Clues" tend to interact with TV programming more frequently than other children. CHCM studies suggest that a longitudinal study indicated that watching "Blue's Clues" increased information-acquisition skills such as sequencing, patterning, relational concepts, and transformations. The website also notes that the program improved children's problem solving and flexible thinking such as solving riddles, exhibiting creative thinking, and nonverbal and verbal expression skills.

Refer to website http://www.cmch.tv/research/fullrecord.asp?id=1773

## 7.3.1 Improved Self-Esteem?

The *Nick jr* website asks the question: "Why do kids love "Blue's Clues?" According to Dr. Alice Wilder, director of research for "Blue's Clues," it is because the show "empowers, challenges, and builds the self-esteem of preschoolers while making them laugh."

Dr. Wilder states that, "We use a multi-layered approach. In every episode, the learning concepts become more difficult. We start them off with concepts that preschoolers can easily grasp. Once they get the hang of it, they can try something more difficult. They also achieve mastery through repetition. *We find that the kids might not understand everything on Monday, but by Friday they will.*" (Italics added for emphasis.)

The website made one more important statement: "We take bigger themes and break them down to something they can grab on to. Everything starts from this foundation." Reference: http://nickjr.co.uk/shows/blues/more.aspx

The punch line here is simple: how can this concept be applied to adult training? Adults have different behaviors than children, but the process of learning appears basically the same. Of importance is the understanding that we cannot provide one orientation or one session and expect long-term retention of information. Training programs have long stressed the learning curve of employees and how significant knowledge is lost in a very short period of time. When viewed from the perspective of risk, without feedback and with the lack of injury and loss experience of many employees and management, the importance of your control programs can be lost. An ongoing effort of simple, straightforward refreshers must be part of the ongoing JHA process.

## 7.4 SUMMARY

A well-designed training program provides an integrating of KSAs, all blended together to mold specific programs and desired learning under specified conditions.

The content of a training program and the methods of how it is presented should reflect specific training needs and the particular characteristics of the organization. Identification of training needs is an important and essential early step in training design. Involving management and employees in this process and in the subsequent teaching provides the vision and guidance of leadership for all levels of the organization.

Experienced training professionals often encounter a misconception that all problems are training problems and that developing a training program can solve them. Training is not the answer to all operational problems. What other obstacles or behavioral consequences are driving behavior towards at-risk actions? Can all employees clearly explain existing and potential hazards to which they are exposed? Do they know how to protect themselves and their coworkers from these hazards?

In Chapter 8, we will discuss what we call "plug-N-play programs." JHAs, hazard recognition, PPE assessment, and other similar programs are usually built around a "traditional" safety model which stems from regulatory compliance. We will provide some insight into the many programs that are typically floating around an organization, utilizing many resources, in many cases in a non-value-added manner and not always getting the "biggest bang for the buck," and often not achieving the results that the organization desires.

## CHAPTER REVIEW QUESTIONS

1. What defines a good trainer?
2. When is safety education most effective?
3. What is the purpose of communicating training?
4. What are the two types of safety education?
5. How would you define the terms education and training?
6. What is technical safety training?
7. How would you define a training program?
8. What are natural consequences?
9. How would you describe who needs training in the organization?
10. What is training as defined by the OSHA Training Development Guideline?

## REFERENCES

1. Oklahoma Department of Labor, *Safety and Health Management: Safety Pays*, 2000, http://www.state.ok.us/~okdol/, Chapter 13, pp. 81–90, public domain
2. OSHA Publication 2254 (Revised 1992), "Training Requirements in OSHA Standards and Training Guidelines," public domain
3. U.S. Department of Labor, Office of Cooperative Programs, Occupational Safety and Administration (OSHA), *Managing Worker Safety and Health*, November 1994, public domain

4. Peterson, Dan, *The Challenge of Change, Creating a New Safety Culture*, Implementation Guide, CoreMedia, Development, Inc., 1993, Resource Manual, Category 16, Supervisor Training, pp. 73–75

5. Roughton, James, Whiting, Nancy, *Safety Training Basics, A Handbook for Safety Training Program Development*, Government Institute, 2000

6. Roughton, J., J. Lyons, April 1999, "Training Program Design, Delivery, and Evaluating Effectiveness: An Overview," American Industrial Association, The Synergist, pp. 31–33

7. *FAA System Safety Handbook*, Chapter 14: System Safety Training, pp. 14-1–14–15, December 30, 2000, public domain

8. OR-OSHA 105, *An introduction to effective safety training*, Introduction to Safety Training, Presented by the Public Education Section Department of Business and Consumer Business Oregon OSHA, public domain

9. OR-OSHA 109 *Train The Trainer I: Managing the Safety Training Program* 46, Presented by the Public Education Section, Oregon OSHA, Department of Consumer and Business Services, public domain

10. OR-OSHA 113, *Train The Trainer II: Conducting Classroom Safety Training*, Presented by the Public Education Section Oregon OSHA, Department of Consumer and Business Services, public domain

11. OR-OSHA 113, *Train The Trainer III: Developing Training Courses*, Presented by the Public Education Section Oregon OSHA, Department of Consumer and Business Services, public domain

12. Tracy, Diane, *Blue's Clues for Success: The Eight Secrets behind a Phenomenal Business*, Kaplan Business; 1st edition, 2002

# Appendix L

This appendix was adapted from the following public domain sources:

OR-OSHA 105, An introduction to effective safety training, Introduction to Safety Training, Presented by the Public Education Section Department of Business and Consumer Business Oregon OSHA, public domain

OR-OSHA 109 Train-The-Trainer I: Managing the Safety Training Program 46, Presented by the Public Education Section, Oregon OSHA, Department of Consumer and Business Services, public domain

OR-OSHA 113, Train-The-Trainer II: Conducting Classroom Safety Training, Presented by the Public Education Section Oregon OSHA, Department of Consumer and Business Services, public domain

OR-OSHA 114 Train the Trainer III: Developing Training Courses, Presented by the Public Education Section, Oregon OSHA, Department of Consumer and Business Services, public domain

OR-OSHA 115 Train the Trainer IV: Conducting On-the-Job Training (OJT), Presented by the Public Education Section, Oregon OSHA, Department of Consumer and Business Services, public domain

The website is located at http://www.cbs.state.or.us/external/osha/educate/training/pages/materials.html

## L.1 SAMPLE SAFETY AND HEALTH TRAINING POLICY

### L.1.1 Purpose

Training is one of the most important elements in our company's safety program. It provides employees an opportunity to learn their jobs properly, brings new ideas to supervision, reinforces existing ideas and practices, and puts the Safety Program into action.

Everyone in our company will benefit from safety training through fewer workplace injuries, reduced stress, and higher morale. Productivity, profits, and competitiveness will increase as production costs per unit, turnover, and workers' compensation rates lower.

## L.1.2 Management commitment

_____ will provide the necessary funds and scheduling time to ensure effective safety training is provided. This commitment will include paid work time for training and training in the language that the employee understands. Both management and employees will be involved in developing the program.

To most effectively carry out their safety responsibilities, all employees must understand (1) their role in the training program, (2) hazards, potential hazards, and consequences of exposure that need to be prevented or controlled, and (3) the ways to protect themselves and others. We will achieve these goals by:

- Educating all managers, supervisors and employees on their safety management system responsibilities;
- Educating all employees about the specific hazards and control measures in their workplace;
- Training all employees on hazard identification, analysis, reporting and control procedures;
- Training all employees on safe work procedures.

Our training program will focus on safety concerns that determine the best way to deal with a particular hazard. When a hazard is identified, we will first try to remove it entirely. If that is not feasible, we will then train employees to protect themselves, if necessary, against the remaining hazard. Once we have decided that a safety problem can best be addressed by training (or by another method combined with training), we will follow up by developing area-specific training goals based on those particular needs.

_Employees._ At a minimum, all employees must know the general safety rules of the organization, site-specific hazards and the safe work practices needed to help control exposure, and the their role in all types of emergency situations. We will ensure all employees understand the hazards to which they may be exposed and how to prevent harm to themselves and others from exposure to specific hazards.

We will commit available resources to ensure that all employees receive safety training during the following:

- Whenever a new employee is hired, a general safety orientation will be provided that includes an overview of company safety rules, and why those rules must be followed.
- Whenever an employee is given a new job assignment, formal classroom training will be reviewed and the supervisor will provide initial specific task training. It is extremely important that supervisors emphasize safety during initial task assignment.
- Whenever new work procedures are begun, formal classroom training will be updated and supervisors will provide additional on-the-job training.
- Whenever new equipment is installed if new hazards are introduced.
- Whenever new substances are used, the hazard communication program will be reviewed.
- The bottom line: train for safety whenever an employee is exposed to a new hazard.

Employees must know they are responsible for complying with all company safety rules, and that most injuries will be prevented if they follow the specific safe work practices. They must be very familiar with any PPE required for their jobs. Further, they must know what to do in case of emergencies.

Each employee needs to understand that they are not expected to start a new work assignment until they have been properly trained. If a job places an employee at-risk, then they must report the situation to their supervisor.

*Supervisors.* Supervisors will be given special training to help them in their leadership role. They need to be instructed on how to look for hidden hazards in areas under their supervision, to insist on maintenance of the physical protection in their areas, and to reinforce employee hazard training through performance feedback and, when necessary, fair and consistent enforcement.

We will commit necessary resources to ensure supervisors understand the following responsibilities and the reasons for:

- Detecting and correcting hazards in their work areas before they result in injuries;
- Providing physical resources and personal support that promote safe environment.
- Providing performance feedback with effective recognition and discipline techniques.
- Conducting on-the-job training.

In our organization, supervisors are considered the primary safety trainers. All supervisors will complete train-the-trainer classes to learn training techniques and how to test employee knowledge and skills. They will also receive training on how to apply fair and consistent recognition and discipline. Supervisor training may be provided by the supervisor's immediate manager, by the Safety Department, the Training Department or by outside resources.

*Managers*. All line managers must understand their responsibilities in the Safety Program. This may require classroom training and other forms of communication that will ensure that managers fully comprehend their duties and safety responsibilities. The safety program elements will be covered periodically as a part of regular management meetings. Managers will be trained in the following subject areas:

- The elements of a safety management system and how the impact of the various processes in the system can have a positive impact on corporate objectives,
- Their responsibility to communicate the Safety goals and objectives to their employees,
- Their role also includes making clear assignments of Safety responsibilities, providing authority and resources to carry out assigned tasks, and holding subordinate managers and supervisors accountable.
- Actively requiring compliance with mandatory Safety policies and rules and encouraging employee participation in discretionary safety activities such as making suggestions and participation in the safety committee.

Training will emphasize the importance of managers visibly showing their commitment to the safety program. They will be expected to set a good example by scrupulously following all the safety rules themselves.

*Recognition and Reward*: The purpose of an effective system of recognition is to motivate employee participation and build ownership in the safety system. When employees make suggestions that will improve safety training, we will recognize them. When employees make a significant contribution to the success of the company we will recognize and reward their performance. Employees will submit all suggestions directly to immediate supervisors. Supervisors are authorized to reward employee's on-the-spot when the suggestion substantially improves the training process or content.

## L.1.3 Training and Accountability

To help make sure our efforts in safety are effective, we have developed methods to measure performance and administer consequences. Managers must understand that they have a responsibility to first meet their obligations to our employees prior to administering any discipline for violating safety policies and rules.

Managers and safety staff will be educated on the elements (processes) of the safety accountability system. They will be trained on the procedures to evaluate and improve these elements. Training will focus on improving the Safety Program whenever hazardous conditions and unsafe or at-risk events are detected. At the same time, we will use effective education and training to establish a strong "culture of accountability."

Safety orientation will emphasize that compliance with safety policies, procedures, and rules as outlined in the safety plan is a condition of employment. Discipline will be administered to help the employee increase desired behaviors, not to in any way punish. Safety accountability will be addressed at every training session.

## L.1.4 Types of Training

We will also make sure that additional training is conducted as deemed appropriate.

_____ (Responsible individual) will ensure Safety training is in full compliance with any OSHA standards.

*New Employee Orientation.* The format and extent of orientation training will depend on the complexity of hazards and the work practices needed to control them. Orientation will include a combination of initial classroom and follow-up on-the-job training.

- For some jobs, orientation may consist of a quick review of site safety rules; hazard communication training for the toxic substances present at the site; training required by relevant OSHA standards, e.g., fire protection, lockout/tagout, etc; and a run-through of the job tasks. This training be presented by the new employee's supervisor or delegated employee.
- For larger tasks with more complex hazards and work practices to control them, orientation will be structured carefully. We will make sure that new employees start the job with a clear understanding of the hazards and how to protect themselves and others.

We will follow up on supervisory training with a mentor system, where an employee with proper experience is assigned to watch over and coach a new employee, either for a set period of time or until it is determined that training is complete.

Whether the orientation is brief or lengthy, the supervisor will make sure that before new employees begin the job, they receive instruction in responding to emergencies. All orientation training received will be properly documented.

*Temporary Employees* will receive training (coordinated with their management if from an outside organization) to recognize our specific workplace's hazards or potential hazards.

*Experienced Workers* will be trained if the installation of new equipment changes their job in any way or if process changes create new hazards or increase previously existing hazards.

*All Employees* will receive refresher training as necessary to keep them prepared for emergencies and alert to ongoing housekeeping problems.

*Personal Protective Equipment (PPE).* Employees needing to wear PPE and persons working in high risk situations will need special training. Supervisors and employees must be instructed in the proper selection, use, and maintenance of PPE. Since PPE sometimes can be cumbersome, employees may need to be motivated to wear it in every situation where protection is necessary. Therefore, training will begin with a clear explanation of why the equipment is necessary, how its use will benefit the wearer, and what its limitations are. Remind your employees of your desire to protect them and of your efforts, not only to eliminate and reduce the hazards, but also to provide suitable PPE where needed.

Individual employees will become familiar with the PPE they are being asked to wear. This is done by handling it and putting it on. Training will consist of showing employees how to put the equipment on, how to wear it properly, and how to test for proper fit and how to maintain it. Proper fit is essential if the equipment is to provide the intended protection. We will conduct periodic exercises in finding, donning, and properly using emergency personal protective equipment and devices.

*Vehicular Safety.* All employees operating a motor vehicle on the job (on or off premises) will be trained in its safe vehicle operation, safe loading and unloading practices, safe speed in relation to varying conditions, and proper vehicle maintenance. We will emphasize in the strongest possible terms the benefits of safe driving and the potentially fatal consequences of unsafe practices.

*Emergency Response.* We will train our employees to respond to emergency situations. Every employee in every area will understand:

- Emergency telephone numbers and who may use them,
- Emergency exits and how they are marked,
- Evacuation routes, safe havens and shelter-in-place and
- Signals that alert employees to the need to take action.

We will practice emergency drills at least semiannually, so that every employee has a chance to recognize the signal and evacuate in a safe and orderly fashion. Supervisors or their alternates will practice counting personnel at evacuation gathering points to ensure that every worker is accounted for.

We will include procedures to account for visitors, contract employees, and service employees such as cafeteria employees. At sites where weather or earthquake emergencies are reasonable possibilities, additional special instruction and drilling will be given.

*Periodic Safety Training* In some areas, complex work practices are necessary to control hazards. Elsewhere, occupational injuries are common. At such sites, we will ensure that employees receive periodic safety training to refresh their memories and to teach new methods of control. New training also will also be conducted as necessary when OSHA standards change or new standards are issued.

Where the work situation changes rapidly, weekly meetings will be conducted needed. These meetings will remind employees of the upcoming week's tasks, the environmental changes that may affect them, and the procedures they may need to protect themselves and others.

*Identifying types of training.* Specific hazards that employees need to know about should be identified through total site health and safety surveys, job hazard analysis, and change analysis. Company accident and injury records may reveal additional hazards and needs for training. Near-miss reports, maintenance requests, and employee suggestions may uncover still other hazards requiring employee training.

## L.1.5 Safety Training Program Evaluation

We will determine if the training provided has achieved its goal of improving employee safety performance. Evaluation will highlight training program strengths and identify areas of weakness that need change or improvement.

_____(The training department and safety committee/coordinator) will evaluate training through the following methods:

- Observation of employee skills.
- Surveys and interviews to determine employee knowledge and attitudes about training.
- Review of the training plan and lesson plans.
- Comparing training conducted with hazards in the workplace.
- Review of training documents.
- Compare pre- and post-training injury and accident rates.

If evaluation determines program improvement is necessary, the safety committee/coordinator will development recommendations.

## 6.0 Certification

All training will be reviewed by _____ (Signature) Date
All training will be approved by _____ (Signature) Date

## L.2 ANSI GUIDELINES FOR EVALUATING TRAINING PROGRAMS

ANSI Z490.1-2001, Section 3.4, Program Evaluation, recommends evaluating three important elements of a safety training program.

### L.2.1 Evaluate training program management

Training works best when it's designed and implemented as an integrated system, rather than a series of unrelated training sessions. Elements to evaluate include:

- Responsibility and accountability
- Staffing and budgets
- Facilities and equipment
- Development and delivery
- Documentation and records
- Evaluation processes

### L.2.2 Evaluate the training process

Training should be conducted using a systematic process that includes a needs assessment, objectives, course materials, lesson plans, evaluation strategies, and criteria for successful completion. Areas of emphasis include:

- Quality (clarity, appropriateness, relevance) of training goals
- Adequacy of the learning environment
- Quality (operational, clarity, relevance) of learning objectives
- Effectiveness of the training process

## L.2.3 Evaluate the training results

By evaluating the results of training, it's possible to make improvements to existing plans and gain awareness of the need for new training. Items to evaluate include:

- Quality of the strategic training plan of action
- Support for long-term (life-long learning)
- Quality of program management and manager competency
- Application of a systems approach that links training program elements
- Identification of completing demand and setting priorities
- Adequate support and funding

## L.3 SAMPLE SAFETY TRAINING PROGRAM AUDIT

## L.3.1 Introduction

This audit evaluates criteria for safety, health, and environmental training programs, including development, delivery, evaluation, and program management detailed in ANSI/ASSE Z490.1-2001.

## L.3.2 Purpose

The purpose of this audit is to measure the degree to which the employer is utilizing accepted practices for safety, health, and environmental training.

## L.3.3 Instructions

Completing this audit is primarily a two-step process. First read each question below and the five categories below to conduct an analysis. Next, evaluate each question using a 0-5 point rating system described below to justify your ratings.

### Analysis
Analyze each of the following five categories to develop a justification for the rating that more accurately determines the rating.

- **Standards**. Analyze policies, plans, programs, budgets, processes, procedures, appraisals, job descriptions, rules. Are they informative and directive? Are they clearly and concisely communicated?
- **Conditions**. Inspect the workplace for hazards that might indicate the effectiveness of training. The absence of hazards indicates effectiveness.
- **Behaviors, actions**. Observe both employee and manager behaviors and activities. Are they consistent and appropriate? Do they reflect effective safety education and training?
- **Knowledge, attitudes**. Analyze what employees are thinking by conducting surveys and interviews. Do employees have full knowledge, positive attitudes? High trust and low fear indicate effectiveness.
- **Results**. Analyze training records that validate knowledge, skills and abilities are effectively applied in the workplace. Continually improving results indicate effectiveness.

### Evaluate

Enter your rating to the left of each statement. Use the following guidelines for your rating.

**5-Fully Met**: Analysis indicates that the condition, behavior, or action described is fully met and effectively applied. There is room for continuous improvement, but workplace conditions and behaviors indicate effective application. (Employees have full knowledge and express positive attitudes. Employees and managers not only comply, but exceed expectations. Effective leadership is emphasized and exercised. Safety policies and standards are clear, concise, fair, informative and directive, communicate commitment to everyone. Results in this area reflect continual improvement is occurring.)

This area is fully integrated into line management. First line management reflects safe attitude and behavior. Safety is considered a "value and not a priority."

**3-Mostly Met**: Analysis indicates the condition, behavior, or action described is adequate, but there is still room for improvement. Workplace conditions, if applicable, indicate compliance in this area. Employees have adequate knowledge, express generally positive attitudes. Some degree of trust between management and labor exists.

Employees and managers comply with standards. Leadership is adequate in this area. Safety policies and standards are in place and are generally clear, concise, fair, informative and directive. Results in this area are consistently positive, but may not reflect continual improvement.

**1-Partially Met**: Analysis indicates that the condition, behavior, or action is partially met. Application is most likely too inadequate to be effective. Workplace conditions, if applicable, indicate improvement is needed in this area. Employees lack adequate knowledge, express generally negative attitudes. Mistrust may exist between management and labor. Employees and managers fail to adequately comply or fulfill their accountabilities. Lack of adequate management and leadership in this area. Safety policies and standards are in place and are generally clear, concise, fair, informative and directive. Results in this area are inconsistent, negative, and does not reflect continual improvement.

**0-Not Present**: Analysis indicates the condition, behavior, or action described in this statement does not exist or occur.

| Section 3.0 Training Program Administration and Management | |
|---|---|
| Rating (circle one) | Criteria |
| 0 1 3 5 | Safety training is integrated into an overall safety, health and environmental management system. |
| 0 1 3 5 | The training program addresses responsibility and accountability for the training program. |
| 0 1 3 5 | The training program identifies and allocates resources available to the trainer and trainee. |
| 0 1 3 5 | The training program includes an effective course development process. |
| 0 1 3 5 | The training program includes an effective course design process. |
| 0 1 3 5 | The training program describes effective course presentation using appropriate techniques. |
| 0 1 3 5 | The training program describes appropriate and effective delivery strategies. |

## L.4 SAMPLE TRAINING CERTIFICATION

| Training Subject | | Date | Location |
|---|---|---|---|
| **Trainee certification.** I have received on-the-job training on those subjects listed (see other side of this sheet):<br><br>This training has provided me adequate opportunity to ask questions and practice procedures to determine and correct skill deficiencies. I understand that performing these procedures/practices safely is a condition of employment. I fully intend to comply with all safety and operational requirements discussed. I understand that failure to comply with these requirements may result in progressive discipline (or corrective actions) up to and including termination. | | | |
| Employee Name | | Signature | Date |
| | | | |
| | | | |
| | | | |
| **Trainer certification.** I have conducted orientation/on-the-job training to the employees(s) listed above. I have explained related procedures, practices and policies. Employees were each given opportunity to ask questions and practice procedures taught under my supervision. Based on each student's performance, I have determined that each employee trained has adequate knowledge and skills to safely perform these procedures/practices. | | | |
| Trainer | | Signature | Date |
| | | | |
| **Training Validation.** On _____ (date) I have observed the above employee(s) successfully applying the knowledge and skills learned during the training. | | | |
| Supervisor | | Signature | Date |
| | | | |

**(Page 2 of certification) Sample Hazard Communication Training Outline**
The following information was discussed with students:

* Overview of the hazard communication program – purpose of the program
* Primary, secondary, portable, and stationary process container labeling requirements
* Discussion of the various sections of the MSDS and their location
* Emergency and Spill procedures
* Discussion of the hazards of the following chemicals to which students will be exposed
* Symptoms of overexposure
* Use/care of required personal protective equipment used with the above chemicals
* Employee accountability

The following procedures were practiced:

* Chemical application procedure
* Chemical spill procedures
* Personal protective equipment use
* Emergency first aid procedure

The following (oral/written) test was administered.
(You may want to keep these tests as attachments to the safety training plan and merely reference it here to keep this document on one sheet of paper. OSHA recommends at least 25 questions for technically complex training.)

1. What are the labeling requirements of a secondary container? (Name of chemical and hazard warning)
2. When does a container change from a portable to secondary container? (when employee loses control)
3. What are the symptoms of over-exposure to _____? (stinging eyes)
4. Where is the "Right to Know" station (or MSDS station) located? (in the production plant)
5. What PPE is required when exposed to_____? (short answer)

# Part 3

---

## Developing an Effective
## Job Hazard Analysis

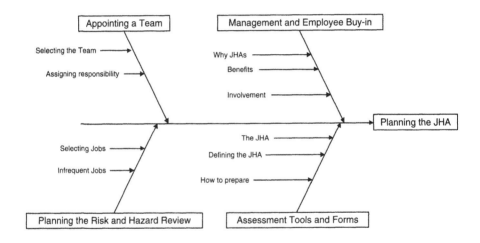

# 8

# Planning for the Job
# Hazard Analysis

The objective of the JHA process is to assure that a self-sustaining process is implemented and remains effective. An essential element is to establish a JHA team that can lead the effort. The importance of the JHA is to use a structured team approach when developing methods for assuring that the team understands the JHA as discussed in this chapter. At the end of the chapter you will be able to:

- Identify the importance and benefits of the JHA.
- Understand how to select a JHA team.
- Define methods on how to select and prioritize jobs.

"We do not want production and a safety program, or production and safety, or production with safety. But, rather, we want safe production."
—Dan Petersen

A wide array of safety material is available from many media vendors specializing in compliance-related safety programs. This compliance-related safety material has become the basis for developing safety programs at many organizations. Generic materials can be purchased that allow you to put your company name on the item, place it on a bookshelf, and—there you go—a written program. When needed, you can just pick it off the bookshelf. We refer to this type of program as "plug-N-play."

These plug-N-play programs are usually built around a "traditional" safety model that stems from a need for regulatory compliance. Safety programs are usually considered to be just another required program and may not be considered a part of the overall operational process. Safety is an "add-on" and not in the mainline production process.

Companies like to compare or benchmark themselves against similar organizations. On the surface, these companies seem to have all of the correct answers. Their safety record is good (the injury rate is below the national average). They tend to adopt programs identified in the benchmark as a model to use for their organization. However, the assumption here is that all of the elements in the model organization are the same as those in the company searching for improvement. Subtle changes in leadership, budget, and tasks may mandate a new set of solutions.

Why do organizations who implement all of these programs still have loss-producing events? Organizations are highly complex environments. To simplify the complexity, the approach is often to repeat what appears successful in other organizations. Many organizations work on the program and do not regard safety efforts as a process. Organizations have a tendency to operate in a reactive mode: if something does not work immediately, try something else quickly. This typically becomes the "program of the month," sending mixed messages to employees. In the rush to achieve goals, we tend to pick things that are easy to solve, put a bandage on the problem, and go on to the next concern, without really fixing anything.

## 8.1 WHERE DO I BEGIN?

Implementing hazard recognition systems, behavioral safety method, risk assessments, or incentives creates the view that safety is separate from general operations. While these approaches may cover a high percentage of the safety needs of an organization, they have not adequately addressed specific risk elements.

You begin by viewing the various safety programs as a "treatment" or "intervention" for specifically defined hazards and risks. This shifts your perception regarding the controls or actions you need to control hazards. If you are providing a "treatment" or "intervention," the boundary created by a "program" dissolves. Other options may appear that were not noticed before when you were focused only on program development.

## 8.1.1 Regaining the "Feel" of the Workplace

If not constantly alert to change, we may lose the "feel" of the workplace. Turnover, merger, and technology bring individuals to work areas who have never worked in a production or service environment. The challenge is to help management understand the benefits of developing JHAs. Management must know: "How does this effort bring value to the organization?" To implement JHAs, managers must understand the connection with the safety management systems, and understand that the JHA process is a long-term investment and commitment.

## 8.1.2 Conducting the JHA

The goal of the JHA is to help management understand that the correct process provides excellent insights and improvements into the safety system if implemented properly and maintained. The following elements are required to begin a JHA process:

- Develop a list of *jobs* at a facility. Use the risk assessment to rank the jobs by hazard severity.
- Develop a list of *hazardous jobs*. Establish a method to rank and set priorities for jobs. List those jobs with hazards that present unacceptable risks, based on those most likely to occur and with the most severe consequences of exposures. These jobs should be your first priority for analysis. Refer to Chapter 6 for a simple risk matrix to accomplish this task.
- Involve employees in developing the JHA. Involving employees in the process will help minimize oversights of specific hazards, ensuring a quality analysis. This also allows employees to "buy in" to the solutions identified. They will share ownership in the safety process.
- Review injuries and other loss-producing events, including history of injuries, losses that required repair or replacement, and any near misses— any events where an injury or loss did not occur, but could have. Review the data collected with all employees. These events are "warning signs" (indicators) that the existing preventative measures (if any) may not be adequate.
- Conduct a preliminary task review. Enlist employees to help break down the job into its basic steps and then into specific tasks. Discuss hazards with employees that they know exist in the current work environment. Discuss possible changes with employees so that they can help to eliminate hazards.
  - Correct safety issues that are identified as soon as possible.
  - Outline each step and its related tasks. All jobs can be broken down into basic job steps and then into specific tasks. When preparing to

develop a JHA, watch the employee perform the job, listing each step and then each related task as they are being performed. Record enough information to describe each action without getting too detailed. Avoid detailing the breakdown of tasks so that it becomes a long list of items. You may find it valuable to get input from other employees who have performed the same job. We will expose you to the "cause and effect" diagram in Chapter 9, Figures 9-3 and 9-4, which will help you to brainstorm the steps and their related tasks.

Next, review tasks with the employee to make sure you have not overlooked important elements. Include the employee in all phases of the analysis, from reviewing the job description and related tasks and procedures to discussing uncontrolled hazards, associated at-risk events, and preventative measures. For those hazards determined to present unacceptable risks, evaluate types of hazard controls.

## 8.1.3 Why Is a JHA Important?

The JHA is the foundation of a safety process. The JHA is used to help close the gaps between what is being done and a successful safety management system.

As you conduct a JHA you find that you can and will develop an in-depth understanding of task-specific training requirements. The JHA helps to define the step-by-step and task-by-task analysis for performing each job.

Obvious hazards are quickly identified by the JHA, while less obvious hazards can only be identified through a systematic analysis of each job by defining each step and their related tasks to the JHA. The risk analysis (Chapter 6) combined with identifying at-risk events can lead to better defined preventative measures.

The JHA provides a consistent method for developing an effective training program that will help to provide a safer and more efficient work method. A properly designed JHA provides the training aid that helps employees make safe choices when performing specific tasks. Refer to Table 8-1 to review a list of why JHAs are important.

## Table 8-1
## Why Are JHAs Important?

- Helps to detect existing and potential hazards and consequences of exposure.
- Helps to assess and develop specific training requirements. Refer to Chapter 7 for a discussion on training needs.
- Helps supervision understand what each employee should know and how they are to perform their job.
- Helps to recognize changes in procedures or equipment that may have occurred.
- Improves employee participation in the work design process.
- Helps to define specific at-risk events that might be occurring.
- Helps to outline preventive measures needed to modify or control associated at-risk events.

Adapted and modified, OR OSHA 103o, Why are JHA's Important, pp. 5, http://www.cbs.state.or.us/external/osha/educate/training/pages/materials.html, public domain

---

The JHA process provides a better method to identify opportunities for improvement in the safety management system and can help provide the foundation for a risk-based analysis. As the structured process is followed, issues may be uncovered that require management decisions to include: Is this a task we should avoid? Can the risk be modified or reduced?

The JHA brings to light the complexity of the job.

Once you understand the complexity and current design of a job, you can develop and/or update the job procedures to reflect the needed safety revisions and modifications.

---

If used as a tool to identify at-risk events and associated risk of hazards, the JHA can establish the performance measurement system needed to determine how effectively a task is being safely performed.

Figure 8-1 provides an overview of the JHA development process and Figure 8-2 provides an overview of the JHA implementation process.

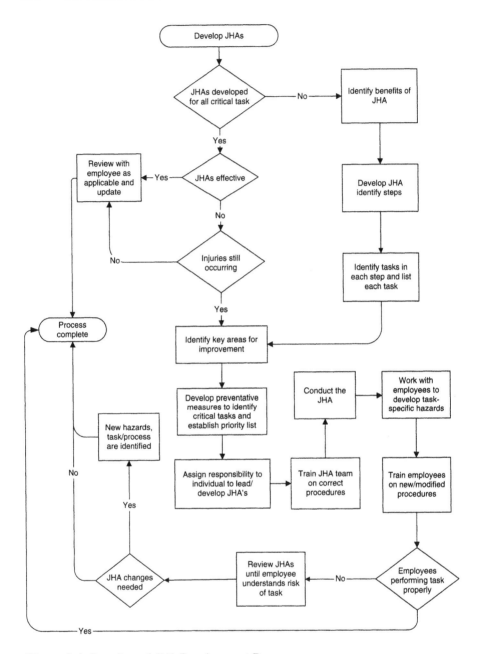

**Figure 8-1** Overview of JHA Development Process

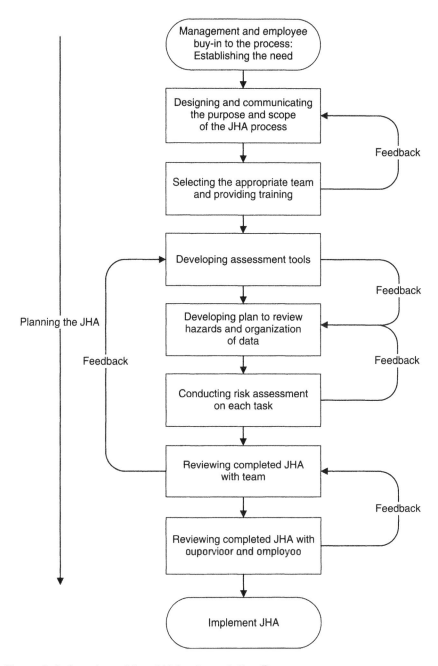

**Figure 8-2** Overview of the JHA Implementation Process

### 8.1.3.1 Benefits of Developing JHAs

The benefits of a JHA go beyond compliance with OSHA. The JHA process helps to bridge gaps in the management system and provides a method to bring reality to discussions about the nature of the work being performed. Both supervision and employees develop an understanding and appreciation for various tasks.

> Supervisors can use the findings of a JHA to eliminate and/or prevent hazards, helping to reduce injuries; providing safer, more effective work methods; reduced workers' compensation costs; and increased productivity. The analysis can be a valuable tool for training new employees in the tasks required to perform their jobs safely.

> Benefits of a JHA
>
> - Injury-causing hazards are eliminated.
> - It provides instruction on how to safely perform the tasks within each step for use in job orientation/training. It allows for refresher instructions on infrequent/periodic jobs.
> - Employees, teams, and supervisors know better how the total job is done.
> - Job methods improve, efficiency increases, quality is enhanced and operation costs are reduced.
> - The employee is kept closely involved in safety.

The JHA provides a consistent message for all new, relocated, and/or seasoned employees. It can be applied to specific job steps or tasks that have been modified. It can be used as an awareness tool for those who need updated training and/or specific review of nonroutine tasks [3, 8].

> JHAs help managers, supervisors and employees to identify risks by:
>
> - Working together to record each step and its tasks as currently performed.

- Consulting with all involved individuals on identifying existing and potential hazards and consequences of exposure involved in each situation.
- Enlisting help in developing a list of potential at-risk events and preventative measures and developing an action plan to correct noted hazards.

### 8.1.3.2 Drawbacks of a JHA

A JHA program takes time and effort, both to document and implement effectively. Although a JHA is an effective tool it must be considered part of a continuous improvement process, a "living document" that will change from time to time as new opportunities for improvements are discovered. This potential for improvement will outweigh any drawback [1].

Because each analysis takes considerable time and attention, direct management commitment and support is necessary to make a JHA process possible. Resources must provide access to the needed equipment, materials, tools, and individuals for effective observation and evaluation of the task. Open communication and effective two-way dialog will help to resolve any issues with this process [2, 8].

## 8.2 WHY IS IT IMPORTANT TO GET EMPLOYEES INVOLVED IN THE PROCESS?

As discussed in Chapter 5, employee participation is one of the key component in any successful safety management system. Employees become more receptive to changes when they are provided an opportunity to help in the decision to develop any change. Getting employees involved in the decision-making process helps the JHA developer to better understand the job. A job's high number of "simple" tasks originally thought to be "low hanging fruit," an easy point for beginning, can turn out to be surprisingly complex as the data is collected [8].

When employees are involved in the JHA development process they will be in a better position to understand the concepts and why it is important to develop the JHA. Both safety and task performance can be improved.

It is important that employees understand the intended purpose of the JHA review. You are studying the related steps and tasks of the job from a safety viewpoint and not evaluating their productivity. However, the JHA will direct attention to areas where the overall job quality may have potential for improvement.

## 8.3 SELECTING A TEAM

Developing and administering the JHA requires much time and effort, no matter the size of an organization. To help minimize this effort, a cross-functional team should be developed to help administer the process, which includes: risk assessment, identifying at-risk events, and clearly defining the job, its steps and tasks. This team can promote a more efficient process and ensure that a variety of perspectives and opinions are considered. It is important for supervision to have a role in the JHA development process and be part of this team. An individual with authority and the ability to facilitate the process and overcome any barriers must be identified.

The team must provide a point person who will serve as a clearinghouse for development of all JHAs. The individual selected should have a combination of the following traits:

- Be a good communicator and show teaching skills.
- Be respected by their peers as knowledgeable about operations.
- Be able to apply problem-solving techniques to uncover hidden hazards.
- Grasp the basic concepts of the process and risks [6].
- Understand the relationship of injuries and how they apply to the process.
- Be able to recognize hazards and develop preventative measures to eliminate the hazards of the task [7, 9].

This individual should be in a position to overcome barriers that may occur during implementation and able to reinforce management's commitment to the process and development of the JHA.

The team can administer the JHA process using the following techniques:

- Providing input and advice to the developer of the JHA.
- Evaluating the effectiveness of the JHAs during site reviews, investigations, and employee discussions.
- Periodically evaluating all JHA development ensuring that they use the current operational process.

- Reviewing the JHAs when operational conditions change, such as when new employees start work, when employees are relocated, when new equipment, tools, materials or chemicals are introduced into the process, etc.
- Having a follow-up process in place to ensure that any new hazards have not been created.
- Working with employees to understand what is expected of them and determine if they are satisfied with the new procedures [4].

This team should:

- Develop a charter approved by management.
- Hold regular meetings and provide written documentation of meeting minutes to management.
- Encourage employee participation.
- Review employee complaints, suggestions, and/or concerns and bring to the attention of management.
- Provide feedback without fear of reprisal.
- Analyze data concerning risk assessment and hazard recognition, and make recommendations for preventative measures.

Before training the team, clear and concise expectations must be defined. A specific charter (Chapter 5, Appendix K) must be developed to show both the team's and management's commitment. It is important that this charter clearly define and outline the team's roles and responsibilities as to exactly what they are to do and the scope of their responsibilities and authorities. The charter will address such issues as:

- The overall objective of the team.
- The time allowed on developing JHAs.
- Expectations in terms of document quality.
- Resources that will be provided for training and skill development.
- Long-term objectives and time requirements for completing selected JHAs [2, 8].

All of these expectations must be clearly communicated to all managers, the JHA developers, and the general employee population.

Once the team's charter has been established and team members appointed, they can be educated on the concepts of the JHA, which will include the methods for the following: implementing the process, conducting risk assessment, identifying at-risk events and hazard recognition, and controlling the process. The team should be provided with training on how to be a team, team concepts, and general methods used to ensure effective meetings are held.

## 8.4 HOW DO I KNOW THAT A JHA WILL WORK FOR ME?

Employees in their first year of employment account for a significant percentage of injuries. Possible explanations could be:

- Lack of knowledge of the associated risk, related hazards, and proper procedures.
- Lack of physical ability to complete specific tasks.
- Improper use of "on-the-job training" where a new employee is trained by a "seasoned" (long time) employee.
- Cultural perception of what are acceptable at-risk events and what are not, peer pressure, a "macho" environment or belief that injuries are one's fate [4, 5].

Does an employee's perception of existing and potential hazards and consequences of exposure differ from that of the employer? The employee sees a hazard and wants it addressed immediately. The employer may respond and address the hazard quickly but is often slowed down by internal structures, budget process, correction criteria, priorities, etc. Answers to critical questions must be clearly defined: Is the safety issue real? How big is the risk? What are the options? What is the best way to correct the identified risk? Who is going to correct the hazard? How long will it take to develop and implement preventative measures? How much will it cost? Is there need for additional training? One must remember that risk is based on probability even with a blatant hazard; no loss-producing events may have been developed.

The JHA can be used as a problem-solving tool that can define the root causes of injuries. A job can be viewed as a series of causes and effects of various elements that work together to complete an action or service. Each of the elements brings inherent hazards that have a level of risk (severity and probability). The JHA provides a structure that allows these elements to be analyzed as well as the proper controls for each step and task required by the job. This approach improves the probability that the JHA will be feasible and can be implemented as problem constraints and potential opportunities are made visible.

Refer to Figure 8-3 for a basic JHA model.

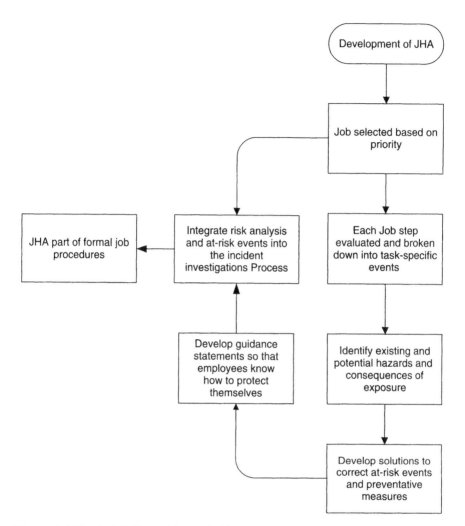

**Figure 8-3** Basic Job Hazard Analysis Model

Adapted from Five-Step JHA Model To Aim for Safety and Efficient Operation, Alexander C. S. Chan, The Hong Kong Polytechnic, Hong Kong, 1995

JHA is a technique that can help an organization focus on specific tasks as a way to identify hazards before they occur. It focuses on the relationship between the employee, the task, risk of the hazard, at-risk events, specific tools, material and equipment, and the work environment. Ideally, after uncontrolled hazards are identified, steps can be taken to eliminate or reduce the risk of hazards to an acceptable level.

The JHA can be considered an element of a proactive management system which, applied properly, will ensure that employee safety is fully considered during the planning stages of any activity [5].

## 8.5 DEFINING THE JHA

Many employees do not view the JHA as a process but think of it as just another form to be completed [4]. The main reason: if a JHA is not properly implemented or is not well defined, it is often regarded as a meaningless bureaucratic "paper" program that was only developed to satisfy some requirements. We have seen excellent JHAs developed and then not used as an active part of the operational process. The best JHAs are those that are focused, defining task-specific activities, clearly documented, communicated to employees, implemented, reviewed, used consistently, and updated regularly.

---

Developing a hazard assessment before a specific job is begun has been utilized in some form or another for many years. The Corps of Engineers uses a form of the JHA process called Task Hazard Analysis (THA) on construction sites. The THA is developed prior to performing any new task. No task will proceed until a THA has been conducted and all affected site employees are trained.

---

Table 8-2 provides an overview of basic terms that are used in developing the JHA.

### Table 8-2
### JHA Basic Terms

| Term | Discussion |
|------|------------|
| Job | A job can be defined as any activity (mental or physical or both) that has been assigned to an employee as a responsibility and has both positive and/or negative consequences based on its performance [4, 5]. A job can be further defined as a sequence of steps with specific tasks that are designed to accomplish a desired goal. |

## Table 8-2
### *Continued*

| Term | Discussion |
|------|-----------|
| Steps | Steps are specific elements in a job. A step can be further defined as a segment of an operation necessary to advance "one of a series of actions, processes, or measure taken to achieve a goal." A job is completed by following a sequence of steps. |
| Task | A task can be defined as "a piece of work assigned or done as part of one's duty. A function to be performed; an objective." A list of specific tasks is defined once the job steps have been observed. Tasks are detailed actions taken to complete a step. |
| Hazard | A hazard is the potential for harm and is often associated with a condition or activity that, if left uncontrolled, can result in an injury. Hazards can be divided into specific categories. Refer to Appendix M for a list describing the most common hazards [4]. Hazards increase the probability that injury or harm will occur. |
| Risk | A risk is the chance of a hazard to cause harm. Risk is the combination of the probability and potential severity of an injury or harm. Risk is based on an array of factors: human activity or design, material, construction, i.e., working at a height frequently is a higher risk than at ground level. Risk is defined by a combination of probability (how frequently an item is exposed to a potential loss-producing event) and the potential severity of the event. Risks are either avoided, reduced or controlled. Refer to Chapter 6 for a discussion on Risk Assessment. |
| Analysis | Analysis is the art of breaking down a job into its basic steps and then into the many tasks that make up each step and then evaluating each task to identify specific inherent hazards and risk. Each hazard is corrected or a method of protection (engineering controls, administrative practices, PPE, etc.) is identified and implemented as part of the operating procedures [9, 10]. |

## Table 8-2
### *Continued*

| Term | Discussion |
|---|---|
| Consequences of Exposure | Consequences are defined as "Something that logically or naturally follows from an action or condition. The condition relation of a result to its cause." American Heritage, p296, 1 and 2. An exposure occurs when anyone (employees, objects, etc.) enters a "danger zone" by virtue of the proximity to the hazard [9, 10]. |
| At-risk events | This is the number of events observed where someone is putting themselves at risk. For example, you count the number of times a person turns a lever, if the lever has an inherent hazard. |
| Residual Risk | The risk remaining after response or mitigation (existing measures and incremental strategies). Residual risk must be less than the initial risk analysis. |

# 8.6 SELECTING THE JOBS FOR ANALYSIS

To evaluate specific jobs effectively, you should have knowledge or experience of the work process, understand how to conduct the JHA, and have an understanding of preventative measures that are applicable [4].

When selecting the jobs for analysis, you are looking at combinations of actual physical hazards, the actions of the employee, and/or gaps in the management system. You may find that a job has been broadly described, resulting in too much flexibility allowed in the completion of its tasks, or it may be too narrowly defined, allowing the employee limited ability to self correct [4]. The ability to estimate reasonable outcomes and develop a course of action using problem-solving tools, risk assessment models, and hierarchy of controls is essential to the process [4].

Refer to Figure 8-4 for an overview of the JHA process, selecting the job for analysis.

You should begin with jobs that have the highest rates of disabling injuries, based on the result of your data collected. These jobs can provide the potential for immediate improvement in safety and showcase the workers' compensation claims savings and other cost reductions. New jobs with specific tasks where changes have been made in a process and/or procedure should be the next priority. Jobs with many tasks are good candidates for analysis, as their process complexity leads

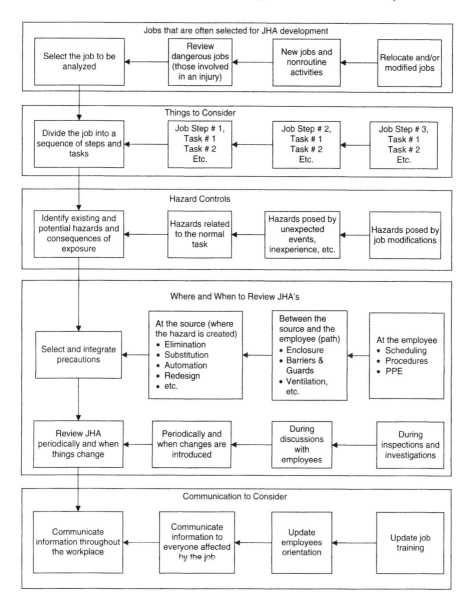

**Figure 8-4** Overview of JHA Process Selecting the Job For Analysis

Adapted from Saskatchewan Labour Website, Occupational Health & Safety, How to Identify Job Hazards, Partners in Safety, How to identify job hazards 1, Occupational Health and Safety Division, http://www.lab.gov.sk.ca/, WorkSafe Saskatchewan http://www. worksafesask.ca/, September, 2002, pp. 7

to potential procedural issues. The jobs in which you "push one button," while generally a lower risk and appearing simple in nature, can offer real surprises. The selection of jobs for review is driven by immediate needs (high frequency/ high severity), high risk, or new and unfamiliar areas [10]. Refer to Table 8-3 for some typical types of jobs that should be considered high priority.

**Table 8-3**
**High Priority Jobs**

- High frequency of injuries, illnesses, or damage.
- High degree of risk as found in industry history or from risk assessment.
- Long duration of task.
- High physical forces.
- Posture required of the person; i.e., ergonomics.
- Point of operation requiring employee vs. machine interface or exposure.
- High pressure, mechanical, pneumatic, fluid, etc.
- Excessive vibration.
- Environmental exposures.
- Nonroutine tasks involving high risk.
- Temporary employees in the operation.
- High turnover or rotation of employees.
- Near misses or close calls; "almost" at-risk events.
- Recent process or operational changes or relocation of equipment.
- New job and/or tasks with little or no injury statistical data.
- New equipment or process.
- Look for jobs requiring industrial hygiene, ergonomic or environmental special surveys or monitoring.

Adapted and modified from OR OSHA 103o, What Jobs Need JHA's? pp. 6, http://www.cbs.state.or.us/external/osha/educate/training/pages/materials.html, public domain.

Jobs Appropriate For JHA

A JHA can be conducted on many jobs in the workplace. A priority must be established for the following types of jobs:

- The highest injury rates (injuries per/100 employees.)
- The potential to cause severe or disabling injuries, even if no history of previous injuries, illnesses, or harm exists.
- One simple human error that could lead to a severe injury or events.
- New employees to the operation.
- New operations and/processes.
- Changes in processes and procedures.
- Job complex enough to require written instructions.
- Nonroutine tasks.

To further help identify jobs that need review, refer to Table 8-4 for other considerations for the JHA process.

Ideally, all jobs should have a JHA developed. However, constraints imposed by time, resources, and effort require the setting priorities [4].

### Table 8-4
### Examples of the Types of Situations to Consider

| Situation | Discussion |
|---|---|
| Known dangerous jobs | These are jobs based on the number of injuries or near misses documented over a given time. For example, jobs where an injury has occurred, any hazardous condition or exposure to harmful substances has been caused, or hazards that could have caused a serious injury [4]. |
| Severity | Tasks that have been involved in serious injuries. There may be a basic problem in the work environment, the management system, or in the job performance itself [8]. Refer to Figure 8-5 for an overview of incident investigations. |
| Frequency | The higher the frequency rate, the greater the reason for developing and implementing a JHA. |
| Task-specific | Those tasks where there is a potential for injuries, even if no injury has occurred [4]. |
| New tasks | Tasks where there is no history or statistical information about potential risk. Injury potential is high in a task where the employee is not accustomed to the job [9][10]. For example, hazards in new jobs may not be obvious or anticipated [4]. |
| Infrequently performed jobs (nonroutine activities) | Employees who undertake nonroutine jobs or who work alone may be in more danger than under normal conditions [4]. |
| Relocated and/or modified jobs | Changes in job procedures or conditions are sometimes frequent and may create new hazards [4]. |

Adapted and Modified Missouri OSHA Website, On-Site Safety and Health Consultation, Managing Worker Safety and Health, http://www.dolir.state.mo.us/ls/onsite/ccp/index.html, public domain

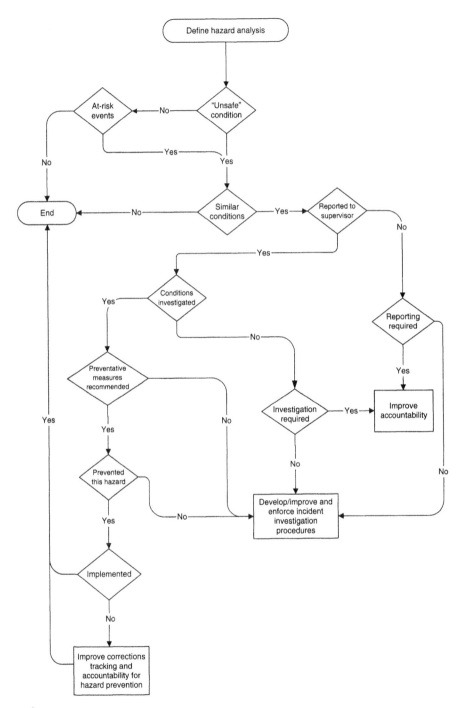

**Figure 8-5** Overview of Incident Investigation Methods

Adapted from Missouri Department of Labor, http://www.dolir.mo.gov/ls/safety consultation/ccp/appendix_9_4.html, public domain

When conducting a JHA, make sure that you take into consideration applicable safety standards for your industry. Compliance with these standards is mandatory, and by incorporating their requirements in the JHA, you can ensure that your safety process meets known requirements.

## 8.6.1 Nonroutine Tasks

Nonroutine jobs can be difficult to identify or locate. One way of preparing nonroutine JHAs is to have a group of experienced employees and supervisors develop the job procedure through discussion with employees who have performed the job [1]. The schedule for completing the job is established and the JHA verified for correctness at the next nonroutine activity.

To be effective, the JHA must be reviewed before any nonroutine activity is performed. This review will ensure that individuals understand the existing and potential hazard and associated risk and consequences of exposure.

## 8.7 SUMMARY

This chapter has defined the planning requirements needed to develop a successful JHA. It is important for both management and employees to understand the importance of the JHA, the benefits and value of the process. The JHA provides the tools to "think outside of the box" and to improve the work environment.

The JHA is one of the most misunderstood and misused tools in safety. It has many names, such as: Job Safety Analysis (JSA), Activity Hazard Analysis (AHA), or Task-Specific Hazards Analysis (THA). These are all basically similar in nature, with only slight variations. No matter what you call the process, the JHA is a tool used to develop safe work practices and/or procedures.

JHA is only one part of an overall good safety management system. It must be used along with other techniques and methods to drive the organization to the safety excellence level that you want to achieve. The JHA is a continuous improvement tool and is by definition forever changing. It allows opportunities for improvements in the management system to be identified quickly.

Once you have identified the opportunities for improvement, you can develop preventive measures.

It is important to note that every time an aspect of the job or task changes, the JHA must be modified to ensure continued safe and effective performance. It is important to note that each change may develop new and undefined risks with new and uncontrolled hazards. Management and employees must work together to record each task as it is being performed.

Employees who do the job usually know the most about the related risks and should participate in the JHA process. Sometimes, we fail to simply ask employees about the risk of the task and bring them into the review process. Even worse, we do not always take their concerns seriously, as we do not know the seriousness of the hazards [4]. By consulting with the employee to help identify the risk of hazards and at-risk events, preventative measures can be better determined. By developing a JHA collectively, a sense of ownership is created that encourages teamwork. The systematic gathering of information through teamwork is essential to determine true job conditions and the avoiding of snap judgments, bad disciplinary actions and lack of understanding of risk and hazards [1, 4, 5].

In Chapter 9 we will continue to focus on breaking the job down into individual components to identify specific tasks. A checklist will be introduced to help in preparing the JHA. Checklists for hazard analysis usually move gradually from general condition questions to questions that focus on individual procedures. Once the hazards have been identified and risk review completed, the solutions can be developed to protect the employee from physical harm.

## CHAPTER REVIEW QUESTIONS

1. What is the best approach when developing a JHA?
2. Why is it important to educate management on JHAs?
3. Why is a JHA important?
4. What are the benefits of developing JHAs?
5. What are the major drawbacks of a JHA?
6. Why is it important to get employees involved in the process?
7. Why is it important to select a JHA team?
8. What is the role of the JHA team?
9. What are the possible causes of injuries for new employees?
10. Why is it important to define the JHA?

# REFERENCES

1. U.S. Department of Labor, Occupational Safety and Health Administration OSHA 3071 Job Hazard Analysis, 1998, 2001 (Revised), public domain
2. Oregon OSHA Website, *Job Hazard Analysis*, http://www.cbs.state.or.us/external/osha/pdf/workshops/103w.pdf, public domain
3. Saskatchewan Labour website, Occupational Health & Safety, *How to Identify Job Hazards, Partners in Safety, How to identify job hazards* 1, Occupational Health and Safety Division, http://www.lab.gov.sk.ca/, WorkSafe Saskatchewan http://www.worksafesask.ca/, September, 2002, public domain
4. Missouri Department of Labor, *Establishing Complete Hazard Inventories*, Chapter 7, http://www.dolir.mo.gov/ls/safetyconsultation/ccp/, public domain
5. Montana Department of Labor and Industry, Occupational Safety and Health Bureau, Department of Labor and Industry, *Job Safety Analysis Identification of Hazards*, no date, public domain
6. Roughton, J., Nancy Whiting, "Safety Training Basics: A Handbook for Safety Training Development," Government Institute, Rockville Maryland, 2000
7. Roughton, J., October 1995, "Job Hazard Analysis: An Essential Safety Tool," J. J. Keller's OSHA Safety Training Newsletter, pp. 2–3
8. Roughton, J., April 1995, "How to Develop a Written Job Hazard Analysis," Presentation, National Environmental Training Association Conference, Professional Development Course (PDC) Orlando, Florida
9. Roughton, J., C. Florczak, January 1999, "Job Safety Analysis: A Better Method," Safety and Health, National Safety Council, pp. 72–75
10. Roughton, James, James Mercurio, *Developing an Effective Safety Culture: A Leadership Approach*, Butterworth-Heinemann, 2002

# Appendix M

## M.1 DESCRIPTION OF COMMON HAZARDS

| Hazards | Hazard Description |
|---|---|
| Acceleration/ Deacceleration | When something speeds up or slows down too quickly |
| Biological | Airborne and bloodborne viruses, animals, insects, vegetation |
| Chemical (Corrosive) | A chemical that, when it comes into contact with skin, metal, or other materials, damages the materials. Acids and bases are examples of corrosives. |
| Chemical (Flammable) | A chemical that, when exposed to a heat ignition source, results in combustion. Typically, the lower a chemical's flash point and boiling point, the more flammable the chemical. Check MSDS for flammability information. |
| Chemical (Toxic) | A chemical that exposes a person by absorption through the skin, inhalation, or through the blood stream that causes illness, disease, or more severe consequences. The amount of chemical exposure is critical in determining hazardous effects. Check Material Safety Data Sheets (MSDS). |
| Chemical Reactions (Potential Explosion) | Chemical reactions can be violent; can cause explosions, dispersion of materials and emission of heat. |

| Hazards | Hazard Description |
|---------|--------------------|
| Electrical (Fire) | Use of electrical power that results in electrical overheating or arcing to the point of combustion or ignition of flammables, or electrical component damage. |
| Electrical (Loss of Power) | Safety-critical equipment failure as a result of loss of power. |
| Electrical (Shock/Short Circuit) | Contact with exposed conductors or a device that is incorrectly or inadvertently grounded, such as when a metal ladder comes into contact with power lines. As little as 60 Hz alternating current (common house current) is very dangerous because it can stop the heart |
| Electrical (Static/ESD) | The moving or rubbing of wool, nylon, other synthetic fibers, and even flowing liquids can generate static electricity. This creates an excess or deficiency of electrons on the surface of material that discharges (spark) to the ground resulting in the ignition of flammables or damage to electronics or the body's nervous system. |
| Electrical (Contact) | Inadequate insulation, broken electrical lines or equipment, lightning strike, static discharge etc. |
| Ergonomics (Human Error) | A system design, procedure, or equipment that is error-provocative, i.e. a switch turns something on when pushed up. |
| Ergonomics (Strain) | Damage of tissue due to overexertion (strains and sprains) or repetitive motion. |
| Ergonomics, Eight risk factors | High Frequency: There are a lot of repetitions of the same movement in a task. High Duration: The employee must repeat the same movement over an extended period of time. High Force: The employee must exert force to complete the task. This may include lifting, pushing, pulling, reaching, etc. Posture: Stress from over-extending body parts, or improper body position is part of the task. Point of Operation: The location of the employee or tool in relation to the material or product, increases the stress impact of other risk factors |

| Hazards | Hazard Description |
|---|---|
| | Mechanical Pressure: Hand-held tools have hard, sharp edges or short handles. |
| | Vibration: Impact tools, power tools, bench-mounted buffers and grinders (for example) produce excessive vibration. |
| | Environmental Exposure: The employee works in or temperature extreme environments. |
| Excavation (Collapse) | Soil collapses in a trench or excavation as a result of improper or inadequate shoring. Soil type is critical in determining the likelihood of a hazard. |
| Explosives (Chemical Reaction) | Explosions result in large amounts of gas, heat, noise, light, and over-pressure |
| Explosion (Over Pressurization) | Sudden and violent release of a large amount of gas/energy due to a significant pressure difference such as rupture in a boiler or compressed gas cylinder |
| Fall (Slip, Trip) | Conditions that result in falls (impacts) from height or traditional walking surfaces (such as slippery floors, poor housekeeping, uneven walking surfaces, exposed ledges, etc.) |
| Fire/Heat | Temperatures that can cause burns to the skin or damage to other organs. Fires require a heat source, fuel, and oxygen |
| Flammability/Fire | For combustion to take place, the fuel and oxidizer must be present in gaseous form |
| Mechanical | Pinch points, sharp points and edges, weight, rotating parts, stability, ejected parts and materials, impact. |
| | Skin, muscle, or body part exposed to crushing, caught-between, cutting, tearing, shearing items, or equipment. |
| Mechanical Failure | Typically occurs when devices exceed designed capacity or are inadequately maintained. |
| Mechanical/ Vibration (Chafing/failure) | Vibration that can cause damage to nerve endings, or material fatigue, i.e., examples, abraded slings and ropes, weakened hoses and belts |

| Hazards | Hazard Description |
|---|---|
| Noise | Noise levels (> 85 dBA 8 hr TWA) that result in hearing damage or inability to communicate critical information |
| Pressure | Increased pressure in hydraulic and pneumatic systems. |
| Radiation | Non-ionizing(burns), Ionizing (destroys tissue). |
| Radiation (Ionizing) | Alpha, Beta, Gamma, neutral particles, and X-rays that cause injury (tissue damage) by ionization of cellular components. |
| Radiation (Non-Ionizing) | Ultraviolet, visible light, infrared, and microwaves that cause injury to tissue by thermal or photochemical means. |
| Struck Against | Injury to a body part as a result of coming into contact of a surface in which action was initiated by the person. Example, a screwdriver slips. |
| Struck By (Mass Acceleration) | Accelerated mass that strikes the body causing injury or death. (Example, falling objects and projectiles.) |
| Temperature Extreme (Heat/Cold) | Temperatures that result in heat stress, exhaustion, or metabolic slow down such as hypothermia. |
| Violence in the Workplace | Any violent act that occurs in the workplace and creates a hostile work environment that affects employees' physical or psychological well-being |
| Visibility | Lack of lighting or obstructed vision that results in an error or other hazard. |
| Weather Phenomena (Snow/Rain/ Wind/Ice) | Naturally occurring events. |

Adapted and combined, Oregon Website, Job Hazard Analysis, http://www.cbs.state.or.us/external/osha/pdf/workshops/103w.pdf, public domain and U.S. and Department of Labor, Occupational Safety and Health Administration OSHA 3071 Job Hazard Analysis, 2001, Appendix 2 (Revised), public domain

# 9

# Breaking the Job Down into Individual Components

As discussed in a previous chapter, the tendency is to leap from problem identification to solution without a full understanding of the problem or solution. To reduce the potential for leaping into action before both the problems and possible impacts of change created by the solution are understood, a method should be followed that brings all the elements of a job into focus. This chapter outlines how to bring the job components into such focus using the JHA. At the end of the chapter you will be able to:

- Define the basic steps in JHA development.
- Identify methods for breaking the job down into steps and tasks.
- Construct a cause-and-effect diagram.

"Seek first to understand, then to be understood."

—Stephen Covey

In Chapter 8 we discussed how to plan for the JHA and why it is important that every time aspects of the job change, the JHA must be modified to ensure that it continues to be a safe and effective procedure.

## 9.1 BASIC STEPS IN THE JHA DEVELOPMENT PROCESS

When trying to evaluate the basic steps of a job, every action, including movements, is critical and must be considered. Care must be taken not to make the JHA too detailed and spin into a time and motion study. Too much detail will make the JHA ineffective and unenforceable [1].

When breaking a job down into its specific steps and related tasks, begin by talking to as many employees as possible: new, experienced (seasoned), transferred, temporary employees, and other individuals who are familiar with the job. This will include management, maintenance, employees who have performed the tasks before, safety professionals, and other individuals who may have information about the job. In your discussions with these individuals, hazards and consequences of potential exposures that may not have been visible will become more apparent.

The JHA process begins with selection of the job and ends with the development of a written standard operation procedure (SOP). Completion of the JHA provides a baseline for the development and refinement of job procedures and work instructions. The JHA is not intended to replace work instructions or SOPs, as these provide the formal details of the job. It can however provide data for the refinement of the SOP. The JHA is an approach for detailing exposures to hazards and developing preventative measures [5].

We may have developed a management system built around traditional hazard recognition programs and have communicated specific hazards to employees as being a necessary part of doing the job. We must be aware of how management and employees perceive hazardous tasks. Table 9-1 provides a summary of the management perception vs. employee perception.

### Table 9-1
### Management's Perception vs. Employee's Perception

| Management's Perception | Employee's Perception | |
|---|---|---|
| | **Know about hazard** | **Do not know about the hazard** |
| Know about hazard | Ideal; both management and employees know about hazard. | Management recognizes the hazard but the employee does not recognize the hazard. |
| Do not know about the hazard | Employees know the hazard of the task and management does not! | Neither management nor employees know the hazard of the task. Worst case! |

In this example, we want to be in the upper left quadrant where knowledge is open and there are no hidden perceptions, hazards are communicated, and the appropriate preventative measures can be taken to correct the hazard [6].

When breaking a job down into its individual components, we want you to think about the job as a whole, not just the sequence of steps. We discussed the job of changing a tire in Chapter 7, Table 7-1. With this job in mind, picture a job as a funnel from which you must filter out the hazards and associated risks. Many elements may be listed in a job description, intended to instruct the employee on how to perform the job. At the top of this funnel, you have a large amount of information to decipher. As the job is analyzed, the job description is filtered into a manageable subset of "steps." These steps are analyzed for finer elements, with specific tasks identified as necessary to complete each step [7]. Refer to Figure 9-1 for a demonstration of this funnel concept.

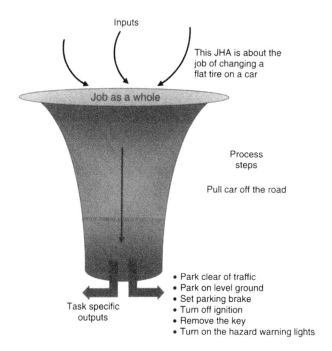

**Figure 9-1** Most variations are caused by a few critical X's

As we proceed to break the job into the steps and then into the tasks, basic elements of the JHA process provide the essential information for safety improvement:

- Select the appropriate job for analysis, as discussed in Chapter 8. Assign the risk based on severity, probability, and frequency of the exposure to the hazard, as discussed in Chapter 6.
- Break the job down into its individual steps (sequence of actions required to complete the job).
- Break the steps down into individual tasks. Define the risk level by reviewing tools, equipment, materials, and environment associated with each task. Use the example risk matrix as discussed in Chapter 6.
- Determine if any task is nonroutine.
- Define the existing and potential hazards and the consequences of exposure for each task.
- Document any at-risk events. What are potential actions and behaviors that may be present? What are the factors that may be driving a specific behavior? [2, 4]. Identify preventative measures that can be used to eliminate and/or minimize the hazard. Use the "hierarchy of controls" as listed in Table 9-2. The key is to ensure that all preventative measures identified are in the employee's control.
- Assess residual risk for each task using the risk assessment tools as discussed in Chapter 6. One important point to remember is that you want to ensure that the residual risk is less than the initial risk. If there is not a substantial decrease in risk, additional engineering controls must be implemented.
- The final stage of a good JHA is to develop an SOP. The SOPs must be developed so that risk control methods are integrated into the related tasks.

All of this can be summarized as a basic illustration of the JHA process. Refer to Figure 9-2.

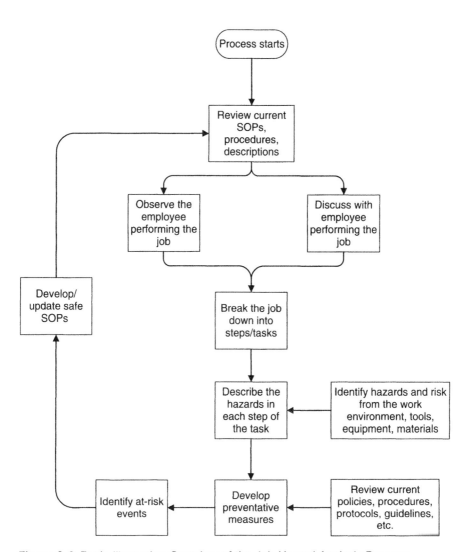

**Figure 9-2** Basic Illustration Overview of the Job Hazard Analysis Process

Adapted from Chan, Alexander C.S. An Overview Of Job Hazard Analysis Technique, Illustration of the 5-Step Job Hazard Analysis Process, The Hong Kong Polytechnic University, Hong Kong, www.ic.polyu.edu.hk/oess/papers/JHA_paper_issst.doc

## Table 9-2
## Hierarchy of Controls

| |
|---|
| Measures vary, with certain actions being more effective than others at reducing risk levels. Information obtained from a JHA is useless unless preventative measures recommended in the analysis are incorporated into each task and are under the employee's control. Managers should recognize that not all preventative measures are equal. |
| The order of precedence and effectiveness of preventative measures include:<br><br>• Avoid or reduce the hazard strength.<br>• Engineering controls.<br>• Administrative controls.<br>• PPE. |
| Engineering controls include:<br><br>• Elimination and/or minimization of the hazards or risks.<br>• Designing/redesigning the facility, work station, equipment, or process to remove a hazard, or substituting processes, equipment, materials, or other factors to lessen the hazard.<br>• Enclosing the hazard, i.e., using enclosures for areas, such as noisy equipment.<br>• Isolating the hazard, i.e., using devices such as interlocks, machine guards, welding curtains, etc.<br>• Removing or redirecting the hazard, such as with local and exhaust ventilation. |
| Administrative controls include:<br><br>• Written operating procedures, work permits, and safe work practices.<br>• Exposure time limitations (used most commonly to control temperature extremes and ergonomic hazards with rotations.)<br>• Monitoring the use of highly hazardous materials.<br>• Alarms, signs, and warnings (lights, sound, etc.)<br>• Buddy systems or team approach using two or more employees.<br>• Training, refresher training. |

**Table 9-2**
*Continued*

---

PPE

PPE may include such equipment as respirators, hearing protection, protective clothing/footwear, safety glasses, and hardhats. The PPE is only acceptable as a preventative measure in the following circumstances:

- When engineering controls are not feasible or do not eliminate the hazard.
- While engineering controls are being developed.
- When safe work practices do not provide sufficient additional protection.
- During emergencies when engineering controls may not be feasible.

Use of one preventative measure over another measure higher in the control precedence may be appropriate for providing interim protection until the hazard is corrected permanently. If the hazard cannot be eliminated, the adopted preventative measures will likely be a combination of all three items instituted simultaneously.

---

## 9.2 TASKS DEFINED

Using the example of changing a tire, as presented in Chapter 7, we now will provide a method to break down the task of changing a tire into manageable pieces. The method provides a visual picture of what we are trying to say. This method is called a "cause and effect" diagram, or Ishikawa diagram, known as a "fishbone." We have used the fishbone and found that it allows the JHA developer to list all of the elements of the job as well as all of the tasks as they relate to specific steps. We offer several versions of the fishbone that we have used to collect the necessary data, a macro level (high level) and a micro view (low level). Refer to Figure 9-3 for the macro fishbone outline of all of the elements defining the structure of the existing and potential hazards of changing a tire. Figure 9-4 provides a more defined breakdown of the specific tasks and provides a micro view of the steps and associated tasks. What we are trying to show here is that there are several ways of looking at a job; one will be at a macro level with all elements made visible, and one at a micro level that allows the tasks within each of the job steps to be defined [8].

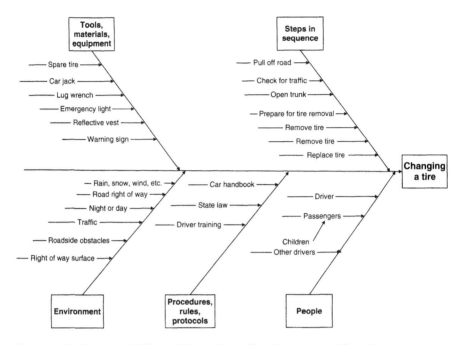

**Figure 9-3** Cause and Effect of Changing a Tire, Macro Level Part 1

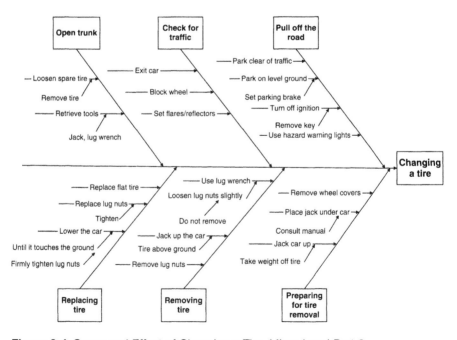

**Figure 9-4** Cause and Effect of Changing a Tire, Micro Level Part 2

The Cause and Effect Diagram

The Ishikawa diagram, also known as the "cause and effect" or "fishbone" diagram, is a deceptively simple yet powerful tool to use in developing the JHA. It provides a graphic presentation of all the "causes" required to create an "effect." It does look like a fishbone, hence its nickname.

It can be used with the JHA to bring together all the elements (causes) of a job (effect).

The diagram consists of a box (the "fish head") placed on the right-hand side. A brief description of the job, step, task, or issue is placed in the box. A horizontal "backbone" is drawn from this box to the left. "Ribs" that represent the selected causes are placed along the backbone.

At the "macro" level we divide the job into five areas: the steps required, the tools/equipment, material needed to complete the task, the current "policies/procedures" in use and, finally, the people who will do the job.

The "micro" level drills down further, with each element studied in more detail to determine hazards that may be present within the task and other details.

For our example, we have used five ribs: Steps/Tasks, Tools/Equipment/ Materials, Environment, Procedures, People. The horizontal arrows next to the ribs are where the various subelements are placed. Along "People," place all the job titles that might be associated with the job. Next to "Environment," list all the things that are in the area of the job and create the work environment (air quality, conditions, building structure, pathways, lighting, temperature, etc.) and so forth. Each rib provides a place to put all the items, conditions, procedures, etc. that go towards creating the job.

Once developed, the diagram allows the JHA developer to step back and look at how the various elements interact. Questions can be asked: "How do\the steps followed match to the environment?" "Are the procedures adequate for the identified tools/equipment/materials?" "Can the people identified handle the tools/equipment/materials?" The risk and hazards associated with the combination of elements can be further assessed against the risk matrix.

The cause and effect diagram brings into view the many components needed to complete a job, cause an incident or create a risk, providing an additional method to search for root cause.

You will see the cause and effect diagram through this book with several variations.

## 9.3 USING A CHECKLIST

The safety management system over a period of time can drift from established procedures due to equipment wear or malfunctions, jam-ups, inadequate tools, wrong type tool for task, damage (mushroomed hammers, bent tools, etc.), incorrect materials, and lack of training or employee behavior changes [12]. Each workplace has its own unique hazards. A clearly defined checklist can further define the requirements and conditions needed to keep this "scope drift" from occurring.

The JHA may find that the various elements of the job do not match. Physical hazards may not match the proper preventative measures. The tasks design may not consider the current environment where performed. Different people may now be doing the job and not receiving the level of training needed. To provide guidance with this analysis, the checklist provides a consistent method to identify common hazards. Refer to Appendix M for description of hazards. Appendix N provides a sample checklist that you can use when you are conducting a JHA.

## 9.4 METHODS FOR BREAKING DOWN THE JOB INTO STEPS AND TASKS

Two methods can be used in combination with the checklist to identify the best way to evaluate the hazards and risks of job task: the *discussion method* and the *observation method* [9].

### 9.4.1 Discussion Method

The discussion method is the simplest and least expensive. The JHA developer and the employees discuss the specific tasks noting experienced hazards and at-risk events. On the initial review, only obvious hazards are identified based on recollections and observations of employees who have performed the job. The discussion method collects the range of employee experience linked to the tasks, allowing a comparison of experiences with what is considered the acceptable risk [4, 10].

In the discussion method, try to determine what "stress" level the employee is experiencing. Employee stress is very important because, as production pressures and demands change, the employee's behavior will change based on the given conditions. You must take stress into account when discussing the job tasks. If employees feel pressured to meet various production goals in order to meet performance goals, they may feel compelled to take risks. This perceived pressure may take focus off of the tasks at hand, unconsciously placing the employees in harm's way. Stress and job pressure must be factored into the JHA.

---

When we discuss stress, a serious injury comes to mind: an employee stuck his hand into a running nip point while trying to make a minor machine adjustment. When the injury was investigated, the review team could not understand the condition that was present and why the employee would stick his hand into the moving piece of equipment. The message for many years had been that all equipment must be shut down and locked out before doing any type of adjustment. We asked several witnesses to describe what happened and received a very clear lesson. The witnesses stated, "What do you (management) do when a piece of equipment stops? You always ask the question, 'Why is the equipment down?' and will constantly ask the same question over and over until the equipment is up and running. We do not like that type of pressure. So we (employees) have decided that if a minor adjustment is to be made, we will not shut the equipment down but take a chance and make the adjustment while it is running." The moral of the story: Management has unknowingly driven the wrong behavior by pushing employees to keep the equipment running without understanding the consequences of exposures. Since few injuries had occurred, the risk of being caught in a piece of equipment was perceived as less than the consequences of dealing with management! When the equipment now goes down, the first question asked is: "Is everyone okay?" and then "Why is the equipment down?"

---

## 9.4.2 Observation Method

This method involves going to where the job is and observing it being performed. While observing the job, you can discuss with the employee their perception of existing or potential hazards, their understanding of the consequences of exposures, as well as how they perceive the inherent risk within each task [12]. This method provides the opportunity to ask questions at each action and in "real time" allowing the team to see first-hand actions, movements, the environment, etc. directly [11].

After you have recorded job observations and reviewed the data collected, all employees should be involved in reviewing the analysis to ensure that all of the basic elements have been noted in the right sequence [12]. When tasks are listed out of order, a greater opportunity exists to miss hazards or create a new hazard. Your job hazard analysis must focus on what is *actually* being done rather than the theory of how the task should be done.

---

The observation method involves watching experienced employees perform the job step-by-step and task-by-task. The major advantages of job observation, group discussion and verification include:

- Tapping the knowledge of experienced employees and supervision who are directly involved in the job task.
- Not relying only on SOPs or memory and perception of how the task is to be performed.
- Reducing the probability that existing and potential hazards and consequences of exposure created by unexpected events are not overlooked or missed.
- Prompting recognition of hazards and the development of suitable preventative measures for each defined task.
- Involving and encouraging supervision and employees to "own" the process and follow the JHA.
- Encouraging employees and supervisors to coach or train inexperienced employees in safe job completions [4].
- Defining a risk management protocol to assess risk in order to reduce the probability and/or severity of an injury or loss-producing events.
- Defining measures that should be used to reduce the risks of the identified hazards [4].
- Identifying other at-risk events or behaviors that may be present.

---

Once either of these methods is completed, they allow the JHA developer to verify the tasks through the following applications:

- "Try outs" of the draft JHA through field observations.
- Group discussions to review and provide feedback about the job.
- Establish routine structured reviews of the job on a regular basis.
- Establish measurement and monitor implementation of JHA in department or areas usually found to be "weak" [4].

## 9.5 WHAT TOOLS CAN BE USED TO ENHANCE THE JHA PROCESS?

A suggested JHA kit should include a camera and/or video equipment, graph paper, tape measure, flashlight or other aids to document and measure what is occurring. The JHA team should have access to software that allows photos and drawings to be assembled into a completed JHA product. The combination of electronic spreadsheets, word documentation, and graphics software, such as Microsoft Word, Excel, and Visio are essential to developing a quality product.

### 9.5.1 Cameras and Video Equipment

With the major changes in digital equipment, digital photos and videos can be easily produced for use in further analyzing and studying the task. Digital cameras provide a quick and economical way to customize a JHA. Photos can be used in discussions with employees so that they know what you are trying to accomplish.

A video of an employee performing assigned duties can be used to analyze the job tasks under the existing work conditions. A video is an excellent way to evaluate body position, stress points, repetitiveness, etc., as movements, and use of tools, equipment, and materials can be captured more effectively and used by the JHA team. Videotaping allows the assessment and evaluation of the duration of hazardous-exposure-defined at-risk events. As tasks are performed in rapid motion, direct visual observation is difficult. Slow motion and stop action can provide additional insights for evaluation. Video cameras that provide internal stopwatches and character generators can be used for timing and labeling an analysis [4, 5].

As the video is reviewed, you may find subtle and different elements that you may not have directly observed during the review. Employees may not be aware of certain movements and actions, as they are following the habitual way (normal behavior) of doing the job. Reviewing the video in a group setting involving employees and supervisors can be especially useful as a training and awareness communication tool.

The video can be used as a gauge to make sure that everyone sees and understands the same issues, concerns, and areas for improvement.

### 9.5.2 Drawings and Sketches

A drawing or sketch of the work area provides the details on the placement of materials, tools and equipment as well as the general patterns of movement, travel or flow of the process. Drawings provide a way to show the various camera and video angles and perspectives. A simple drawing of the work process and workstation might be all that is needed to begin the evaluation of the workflow. The drawing can be shared with employees who provide indications of the dominant features of the work environment [3, 5].

## 9.6 "CAN'T SEE THE FOREST FOR THE TREES"

The observer can be the employee's supervisor, but the process should involve other employees.

The job must be observed as performed under normal conditions, movements, equipment, etc. and the environmental conditions (noise, lighting, dust, etc.) that might be present. If the job is performed only on the second or third shift, the JHA should be conducted at those times. Similarly, only regular tools, materials and equipment should be used, not items specifically requested for the observation [12]. However, be open to comments made by employees that imply the job steps or tasks have on occasion been completed under extreme conditions or hazardous situations. These nonroutine situations will require further review and consideration for special controls or administrative policies to be developed.

Reality Check! If the process outlined to this point is followed, the organization should be supportive of JHA development.

---

Questions for a Quality Review

- What resources (tools, equipment, materials, supplies, etc.) are required to do the job?
- What basic step starts and completes the job? Establish an agreed-upon beginning and ending of the job. Approximately 15 steps is the maximum number that should be analyzed.
- Determine the sequence of what occurs, task-by-task, as listed under each step. Note departures from the normal procedures, as these can potentially cause injuries and may be an indication of process problems that need immediate attention.
- Describe each task in sequence. Begin each task with an easy-to-understand action word, such as remove, lift, pry, etc. [12].

The JHA provides a proven methodology that can be used for the review and development of safe, consistent job practices, and instruction. The objective is to make the JHA a user-friendly document that all affected employees can use to understand the stated task.

# 9.7 SUMMARY

Before you start the JHA development, decide who is going to be involved. Ideally, all employees involved with the job should work on the development of the JHA so that comprehensive input and buy-in is developed.

With each job step, there are usually a number of tasks with inherent risks and hazards. Once the basic tasks have been identified, the existing and potential risks and hazards and the consequences from exposure can be defined for each listed task needed to complete each step.

The JHA puts a spotlight on specific job conditions or issues. By developing a JHA, hazards and risks not seen or previously experienced can be brought into the light of analysis.

---

Overview for developing a JHA.

Step 1: Select the job. Don't make it too broad (i.e., change a tire) or too narrow (i.e., pushing a button). Priority must be given to the jobs with the worst injury rates, those which tend to produce disabling injuries, jobs with a high severity potential, and new jobs.

Step 2: Break the job down into steps listing all tasks that are associated with each step using the "cause and effect diagram." Describe concisely what is being done. Use employees who are experienced in the job. Ensure that you discuss with the employee that the objective of the JHA is to study the job to make it safer. Work through the process, asking the employee what they would do, what's next and why? Record each observation using action words, such as lift, pull, close, open, etc. and outline exactly what object is receiving the action, such as a lever, a cover, an arm, a tool, etc. Ensure that all tools, materials, equipment, and the environment are identified. Finally, verify with the employees you observe to ensure that the tasks are correct and shown in the correct sequence.

Step 3: Identify hazards in each step and task of the process. Can anyone be caught in, on, or by objects? Can they slip, trip, or fall? Do employees strain or overexert themselves? What is the work environment?

Are environmental hazards present? Is layout or placement of materials, equipment, and tools an issue? Are tools and equipment correct for the job and in good condition? Does the equipment need repair? Will a change in one task create a hazard in another task? Once the hazards are identified, check again with the employee and anyone else familiar with the job.

Step 4: Eliminate the hazards. Find creative ways to eliminate the hazards by preventing existing and potential hazard and consequences of exposure. Find a better way to do the task. Start with the goal of the job to identify tasks and work along several routes to the goal, finding the one element which is not only the safest, but the most economical and practical. Change the physical conditions that may have created the problem, such as pick up or move materials, change a work height, replace a guard, modify a guard, etc. Change the procedure. Can tasks be conducted less frequently if exposure is a problem? Check solutions with the employee. Watch the job steps and carefully evaluate if the tasks and actions are aligned with the completed JHA. At this stage, it would be helpful to record the steps on video for use in future training.

In Chapter 10 we will walk through the process of putting together an actual JHA and will explain each section in detail.

## CHAPTER REVIEW QUESTIONS

1. In Table 9-1, Management's Perception vs. Employee's Perception, where would the appropriate quadrant be located for optimal knowledge to be open, hazards communicated, and the appropriate actions taken?
2. Why is it important to use a checklist?
3. What are the two methods for breaking down the job?
4. How is the observation method verified?
5. What is a great reality check to see whether your system is working?
6. What are some questions that can be asked to see if you are getting a quality review?
7. What is the major objective of the JHA?
8. Each job task has inherent risks; How do you define the potential risk?
9. What is the method of selecting the jobs for developing a JHA?
10. What are some of the method used to break the job down into individual steps and tasks?

# REFERENCES

1. U.S. Department of Labor, Occupational Safety and Health Administration OSHA 3071 *Job Hazard Analysis*, 1998, 2001 (Revised), public domain
2. Oregon OSHA Website, *Job Hazard Analysis*, http://www.cbs.state.or.us/external/osha/pdf/workshops/103w.pdf, public domain
3. Saskatchewan Labour website, Occupational Health & Safety, *How to Identify Job Hazards, Partners in Safety, How to identify job hazards* 1, Occupational Health and Safety Division, WorkSafe Saskatchewan http://www.worksafesask.ca/
4. Missouri Department of Labor, *Establishing Complete Hazard Inventories*, Chapter 7, http://www.dolir.mo.gov/ls/safetyconsultation/ccp/, public domain
5. Missouri Department of Labor, *Catching the Hazard That Escapes Control*, Chapter 9, http://www.dolir.mo.gov/ls/safetyconsultation/ccp/, public domain
6. Roughton, J., Nancy Whiting, "Safety Training Basics: A Handbook for Safety Training Development," Government Institute, Rockville Maryland, 2000
7. Roughton, J., October 1995, "Job Hazard Analysis: An Essential Safety Tool," J. J. Keller's OSHA Safety Training Newsletter, pp. 2–3
8. Roughton, J., April 1995, "How to Develop a Written Job Hazard Analysis," Presentation, National Environmental Training Association Conference, Professional Development Course (PDC) Orlando, Florida
9. Roughton, J., C. Florczak, January 1999, "Job Safety Analysis: A Better Method," Safety and Health, National Safety Council, pp. 72–75
10. Roughton, James, James Mercurio, *Developing an Effective Safety Culture: A Leadership Approach*, Butterworth-Heinemann, 2002
11. Chan, Alexander C.S., *An Overview of Job Hazard Analysis Technique*, The Hong Kong Polytechnic University, Hong Kong, www.ic.polyu.edu.hk/oess/papers/JHA_paper_issst.doc
12. Roughton, J., January/February 1996, "Job Hazard Analysis: A Critical Part of your Job as Supervisor is Evaluating and Controlling Workplace Hazards," Occupational Health and Safety, Canada, pp. 41–44

# Appendix N

SAMPLE FACILITY CHECKLIST
YOU CAN ALSO USE THE SELF-INSPECTION CHECKLIST
AS DETAILED IN CHAPTER 2, APPENDIX E

| Yes | No | NA | Checklist Item |
|---|---|---|---|
| | | | Are there materials on the walking surface or floor that could cause a tripping hazard? |
| | | | Is lighting required? Is lighting adequate? |
| | | | Are electrical, chemical, thermal, mechanical or kinetic, biological, acoustic (noise), or radiation hazards associated with the task? Are any of these hazards likely to develop? [1][3][4][5] |
| | | | Do environmental hazards such as: chemicals, radiation, welding rays, heat, or excessive noise, result from the performance of the task? [1][3][4][5] |
| | | | Are dusts, fumes, mists, or vapors in the air? [1][3][4][5] |
| | | | Can contact be made with hot, toxic, infectious, or caustic substances? Could the employee be exposed to radiation sources? |
| | | | Are tools, machines, and equipment in need of repair? |
| | | | Does excessive noise exist that could affect communication? Is there likelihood that the noise can cause hearing loss? |

| Yes | No | NA | Checklist Item |
|-----|-----|-----|----------------|
| | | | Are job procedures in place? Are these procedures understood by all employees and followed and/or modified as applicable? |
| | | | Are emergency procedures defined for the job? Are employees trained in emergency procedures? |
| | | | Are employees who operate vehicles and equipment authorized and properly trained? |
| | | | Are industrial trucks or motorized vehicles properly equipped with brakes? Are there overhead guards, backup signals, horns, steering gear, seat belts, etc.? Are they properly maintained? |
| | | | Are employees wearing the appropriate PPE? How was it specified and selected? Was a PPE assessment conducted? Refer to Appendix F and G for an example of PPE assessment. |
| | | | Have any employees complained of headaches, breathing problems, dizziness, strong odors, or other indicators of health related issues? |
| | | | Are there confined or enclosed spaces? Have they been tested for oxygen deficient atmospheres, toxic vapors, or flammable materials? Is ventilation adequate? Who has the reports and when were they completed? Have they been reviewed recently? |
| | | | Are workstations and tools designed to account for ergonomically related movement or issues, bending, twisting, stretching, motions etc.? Is the employee required to make movements that are rapid, stressful, a stretch, bend, etc.? Is the employee required to make movements that could lead to or cause hand or foot injuries, strain from lifting, and the hazards of repetitive motions? Can employees strain themselves by pushing, pulling, lifting, bending, or twisting? Can they overextend or strain themselves while doing a task? Is the flow of work properly organized? Is the employee at any time in an off-balance position? |

| Yes | No | NA | Checklist Item |
|---|---|---|---|
| | | | Are body positions, machinery, pits or holes, and/or hazardous operations adequately guarded? Are fixed objects in the area whose location and design may cause injury, from sharp edges, location, or shape? |
| | | | Are lockout procedures used for machinery deactivation during maintenance and/or unjamming equipment? |
| | | | Is the employee positioned at a machine in a way that is potentially dangerous? |
| | | | Can reaching over moving machinery parts or materials cause an injury to the employee? Is the danger of striking against, being struck by, or contacting a harmful object present? Could employees be injured if they are forcefully struck by an object or contact a harmful material? |
| | | | Can employees be caught in, on, by, or between objects? Can they be injured if their body, their clothing or equipment is caught on an object that is either stationary or moving? |
| | | | Can employee be pinched, crushed, or caught between either a moving object and a stationary object, or two moving objects? |
| | | | Is equipment difficult to operate and does it have the potential to be used incorrectly? |
| | | | Is there a potential for a slip, trip, or fall? Can employees fall from the same level or a different level? |
| | | | Can suspended or residual energy cause harm to employee? |
| | | | Can weather conditions, i.e., ice, snow, water, etc., affect employee safety? Is the employee exposed to extreme heat or cold? |
| | | | What other hazards not discussed may be present? Have the potential causes of injuries been traced to their source (root cause)? |

Adapted and Combined U.S. Department of Labor, Occupational Safety and Health Administration OSHA 3071 Job Hazard Analysis, 1998 (Revised), public domain, and Saskatchewan Labour website, Occupational Health & Safety, How to Identify Job Hazards, Partners in Safety, How to identify job hazards, Occupational Health and Safety Division, http://www.lab.gov.sk.ca/, WorkSafe Saskatchewan http://www.worksafesask.ca/, September, 2002

# 10

## Putting Together
## the Puzzle Pieces

As we continue the process to effectively implement the JHA, this chapter will use the information previously presented to actually develop a JHA. At the end of the chapter, you will be able to:

- Identify the key considerations for conducting a JHA
- Complete the JHA form
- Assess and integrate into the JHA
- Identify the importance of considering residual risk in the JHA
- Cite the basics for listing controls and preventive measures
- Discuss the importance of reviewing the JHA with employees.

"It isn't the changes that do you in, it's the transitions. Change is not the same as transition. Change is situational: the new site, the new supervisor and/or manager, the new team roles, the new policy. Transition is the psychological process people go through to come to terms with the new situation. Change is external, transition is internal. Unless transition occurs, change will not work,"

—William Bridges

To continue the process development, this chapter will use the information previously presented to actually develop a JHA. We will continue to use our "changing a tire" example for the remainder of the book to show the proper method of developing a JHA.

As we discussed in Chapter 4, we make many choices each day, acting or making decisions either safely or at-risk. For example, if you are driving a car on a major highway, you are making decisions all of the time. You may have made a decision to move from one lane of traffic to another while using your cell phone, eating, or talking to a passenger. The fact that you changed lanes successfully was a factor of timing, other drivers, and/or nothing being in striking distance! We assume high risk and are reinforced when nothing bad happens. However, the consequences of all these exposures can catch up with us as a function of probability, when all of the elements align to create a loss-producing event and the controls are not adequate or present [1].

"Begin with the end in mind."

—Steven Covey

In a perfect world, we would not have hazards or behaviors leading to at-risk events. But in the real world, we have to deal with the hand that we are dealt. This is where the JHA comes into play. It provides a structure and method to ensure that a job and its required steps and tasks are consistently performed, reducing the potential consequences of exposure to hazards. This potential is reduced because the job has been assessed for related hazards [6].

The cause and effect diagram discussed in Chapter 9 pulled together all of the elements necessary to do the job. We can now begin the analysis of the job and structure the data for decision making by reviewing the following documents:

- JHA pre-hazard assessment worksheet. This worksheet is used to organize the data. It allows the user to take the steps and tasks data and combine them with the hazard and risk criteria. Refer to Appendix O. This worksheet is useful in collecting your thoughts in a more logical manner.
- Set of instructions on how to change a tire on a car and a complete JHA on Changing a Tire. Refer to Chapter 7.
- Annotated JHA on Changing a Tire. The sample tire changing JHA has been annotated to provide a model that highlights the key points. Refer to Appendix O.
- The traditional method of conducting the JHA on changing a tire vs. the new revised method. Refer to Appendix O. We will using as an example from the Canadian Centre for Occupational Health and Safety Website, http://www.ccohs.ca/oshanswers/hsprograms/job-haz.html: "What is a Job Hazard Analysis," reprinted with permission.

"A work of art is never finished, only abandoned."

—Leonardo da Vinci

# 10.1 COMPLETING THE JHA FORM

The key considerations when conducting a JHA is to paint a story (take a picture) of the potential and existing hazards and consequences of exposure that exist and then provide clear feedback to the employee on methods for protecting themselves.

Think about it this way: when a journalist writes a story, the information collected should provide the reader with a clear visual picture of what they are trying to communicate. They collect this information in a similar manner to a safety professional seeking answers to questions on how an employee was or might be injured. We in the safety profession know this as getting to the root of the cause. This is commonly known as the "Five Ws and one H": Who, What, Where, When, Why, and How. If you think about it, we are writing a story— taking a snapshot of the job, breaking this job down into steps and then into individual tasks—when we develop a JHA. Depending on the complexity of the JHA (the story), the safety professional (the journalist) might ask questions in several different ways. Refer to Table 10-1 for a description of the Five Ws and one H concept that will help in developing an effective JHA.

Many of us use mental checklists, as outlined in Table 10-1, to ensure that we have covered all of the important elements of an incident investigation. Therefore, it is important for the developer to have the basic knowledge, skills, and ability in conducting the following activities:

- Be able to review jobs, looking first at the steps and then the tasks, and ask specific questions to bring into focus existing and potential hazards and the consequences of exposure.
- Recognize a nonroutine task. What "hidden" tasks are being done whose timing keeps them invisible to daily review?
- Identify at-risk events that are associated with the steps and tasks.
- Develop preventative measures to control, reduce, and/or prevent the hazard.
- Train supervision and employees on the modified or new control to be completed [3, 4].

The following sections will guide you through a step-by-step example on how to develop the JHA form information.

## 10.1.1 The Header

The first section of the JHA form to be completed is the header. This header is designed to allow anyone reviewing the JHA to know exactly how the entire job is to be performed.

## Table 10-1
## The Five Ws and One H

| WHO: | WHAT: | WHERE: |
|---|---|---|
| Is involved?<br>Is affected by the JHA?<br>Is the best person to develop the JHA?<br>Is missing in the review of the JHA?<br>Has more information about the hazards in the related task?<br>Else should I talk to about developing the JHA? | Happened?<br>Is the end result that I am trying achieve?<br>Surprised me?<br>Is the most important fact I learned?<br>Is the history here?<br>Happens next?<br>Can people do about it? | Did this happen?<br>Else should I go to get the information to develop the JHA?<br>Is this JHA going be used?<br>Are employees involved in the process? |
| WHEN: | WHY: | HOW: |
| Did an injury occur?<br>Should I report hazards noted with the JHA? | Is this happening? (Is it an isolated case or part of a trend?)<br>Are people behaving the way they are?<br>Should everyone be able to read and understand the JHA?<br>Am I sure that I have the entire hazard documented? | Did an injury occur?<br>Will things be different after the JHA has been developed?<br>Will this JHA help the reader or reviewer?<br>Clear is the information presented?<br>Would someone else be able to describe to other employees the elements of the JHA?<br>How long will this JHA be effective? |

Adapted, *Handbook of Independent Journalism*, "Getting the Story," USINFO.Gov, http://usinfo.state.gov/products/pubs/journalism/getting.htm, public domain and The Difference That Continual Improvement Makes, Principle #4, David Strubler, Mgt 631, http://www.kettering.edu/~dstruble/ContImp.htm

The following elements make up the header:

- Job description
- Department
- Date developed
- Page numbering
- Performed by
- Employee signature
- PPE required

### 10.1.1.1 Job Description

This is a very basic statement about the job. Try to describe the job in as few words as possible, i.e., changing a tire, lawn mower operation, casing a die, etc. This description should be written and described in terms familiar to everyone performing the job. The detailed description of the steps and task is presented in the body of the JHA under the "Job Steps" and "Task-Specific" description, where you break the job down into finer components.

The job description is the starting point and outlines each task in the job. A complete job description defines all of the task-specific work elements in the order they are being performed. Once the job step and task sequence is detailed in the body of the JHA, everyone will have a better understanding of how the job and related tasks are being performed. To get the maximum benefit, the job description and breakout of steps and tasks should be written in everyday, plain language (typically an $8^{th}$ grade reading level) so that anyone can read it and get the same message. The use of pictures to enhance the message aids in communicating the steps and tasks. Keep your message as simple as possible by limiting technical jargon and use bullets to convey actions. This method will provide the most effective way of ensuring that all employees have an opportunity to read the same description and come to the same conclusions.

For a reality check, ask yourself: "If I gave this JHA with its details to anyone with limited or no knowledge of the job, would they understand what is intended to be accomplished?"

### 10.1.1.2 Department

This section documents the correct department or area where the job is to be performed. Use department names that are familiar to everyone, i.e., graphics, grinding, compactor, sand blasting, etc. Again, this should be a simple description.

### 10.1.1.3 Date Developed

The date that the JHA was completed and placed into operation. This date is used to document the initial review and is changed when there is a review, and/or update of the JHA. The old JHA should be kept for historical purposes to ensure that all changes can be tracked.

### 10.1.1.4 Page Numbering

Page numbers are used to ensure that there is a complete JHA on file. This will ensure that the user will have materials associated with the job.

### 10.1.1.5 Performed by

This section documents the name of the individual or individuals who conducted the review and/or developed the JHA. This provides a way to directly control the developers as well as have knowledge of what expertise was used.

### 10.1.1.6 Approvals

Levels of approvals should be required when the JHA is completed. This ensures that both supervision and all affected employees have been made aware of the JHA and its details. The approvals create awareness for those who have been involved in the analysis that their work will be reviewed to ensure that they understand what is documented in the JHA.

The required approvals include:

- Supervisor: The person responsible for enforcing all rules and requirements for the job, its steps, and tasks and who must understand the existing and potential hazards and consequences of exposures, at-risk events, and the preventative measures to ensure that the job is performed safely. The supervisor provides the leadership to ensure that the JHA is being used as intended and any changes as necessary are incorporated and updated. The supervisor reviews the JHA with the employees and signs it once satisfied that all employees understands the JHA.
- Employee: All employees who do the job must be consulted during the development of the JHAs. The employee signs the JHA training document once they have demonstrated to the supervisor an understanding of the existing and potential hazard and consequences of exposures, at-risk events, and the preventative measure.

### 10.1.1.7 PPE

PPE is the last section to complete after all of the hazards of each task and the specific PPE requirements are defined. This section is intended to be used only for operation-wide PPE usage, i.e.; safety glasses, hearing protection, safety shoes, etc. Refer to Chapter 3, Appendix G.

## 10.1.2 Body of JHA

This section captures the data collected from the cause and effect diagram and observation of the job task using the pre-hazard assessment worksheet in Appendix O.

Integration is a key factor of success. This is a strongly supported concept that integrates all of the necessary elements into the safety process. The JHA is one component of an overall strategy of a safety process. From the example, changing a tire, by integrating the non-routine activities, at-risk events and risk assessment of these events into the JHA you have a better tool defining your organizational goals and objectives to provide a safer workplace.

If kept separate, management may only concentrate on production and safety will become a priority and not a value to the organization. It is only through integration that employees will not forget or miss some elements because it has made safety completely tied to the management process.

This section of the JHA provides detailed instructions to the employees on hazards associated with specific task and provides a comment for each task on specific actions or control to be implemented.

The following section makes up the main body of JHA:

- Job steps and task-specific descriptions
- Non-routine (NR) task
- Risk assessment (RA)
- Existing and potential hazards and consequences of exposure
- At-risk events and preventative measures
- Residual risk (RR)

### 10.1.2.1 Job Steps and Task-Specific Description

This section details the job in a logical sequence by reviewing job steps and then defining the specific tasks that relate to a specific step.

The use of the cause and effect diagram allows the team to brainstorm through the sequence of events and make observations to verify the logic. Documenting the sequence of steps is a way of determining each element of the job as it is performed. This is one of the most critical parts of conducting a JHA.

Tracking the sequence of steps is important because, if the steps are out of sequence, you will not be able to properly capture the relationship between the steps and the existing and potential hazard and consequences of exposures, at-risk events, and the preventative measure that may exist. The cause and effect diagrams should have provided a detail analysis of the required steps and their related tasks that will be used to complete this section.

In our experience, it is interesting to watch experienced supervisors and employees struggling to identify the sequence of steps even through they may have done the job for years. They clearly know how to do the job but over time habits and individual changes, shortcuts and modification have been developed by each. The JHA exercise looks at each step to see if specific elements can

| Job Steps and Task-Specific Description | NR | RA | Existing and Potential Hazards and/or Consequences of Exposure | At-Risk Events and Preventive Measures | RR |
|---|---|---|---|---|---|
| **Pull car off the road**<br><br>• Park clear of traffic<br>• Park on level ground<br>• Set parking brake<br>• Turn off ignition<br>• Remove the key<br>• Turn on the hazard warning lights | | | | | |
| **Check for Traffic**<br><br>• Exit car<br>• Block wheel<br>• Set flares/reflectors | | | | | |

**Figure 10-1** Job Steps and Task-Specific Description

be combined and is why a visual observation is critical. For example, you want to group specific steps together such as: Pull car off the road, Check for traffic, Open trunk, Prepare for tire removal, Remove tire, Replace tire, etc. This grouping will allow each step to be identified and then the sequence of tasks specific to that step needed to complete this job. These tasks are listed below the appropriate step. This method is used when there are many steps in the JHA. Too many steps can be overwhelming and the beginning and ending parts of the job must be agreed upon. By grouping similar steps together, the JHA is simplified and placed in buckets of activities.

One important point to remember is that if there are fewer than three steps, then a JHA may not be needed. If there are more than 15 steps for a particular job, then more than one JHA may be required. The key is not to make the JHA too complicated or it will be useless [2]. For example, on the tire change JHA, if you are only going to change a tire, the entire JHA does not need to be reviewed. Instead you would only review the section that applies to that operation. Refer to Figure 10-1 for details on how to apply job steps and task-specific descriptions. To re-cap, the job is broken down into a group of steps with tasks detailed for each step.

## 10.1.2.2 Nonroutine Task

There are nonroutine jobs, nonroutine steps, and finally nonroutine tasks, depending on the situation; each job may have built-in flexibility. Nonroutine elements are critical tasks in every operation and can be identified by placing an **"X"** in this column listed as NR. When breaking the steps down into its individual tasks, look for those steps or tasks that are not performed on a routine basis. As the name non-routine implies, these tasks are done infrequently. They are critical to the process because this is where specific hazards are not experienced each day. For example, start up of a piece of equipment, performing preventative maintenance, piecework where specific equipment is used on a random basis, etc. There are actions that may only be required under certain conditions and controls may be forgotten. You may say, "It will only take a second or minute to do this job, so the controls or PPE are not needed." If the hazard is present, then only a split second is needed for it to show its effects.

The Job Steps and Task-Specific Description will take the guesswork out of trying to review the entire JHA when only a part of the tasks is being performed. By using this method, the supervisor, employee, or anyone about to perform a nonroutine tasks can quickly scan the JHA for these nonroutine tasks noted with an **X** and then review specific hazards associated with the task before staring work.

Let's look at an example. How many of you have password for websites that you rarely use? Most of us! You suppose that I ask you to tell me your passwords at this minute. You would have to stop and think about it for a while and maybe even forget the passwords for short period of time. But if you had to use this password in your normal job each day, you could recall it freely. How quickly did you remember? I think that you get the connection. You will have to determine the nonroutine tasks for your own environment.

If a JHA has nonroutine actions, then before each step or task begun, a formal review and refresher of the controls should be conducted.

Reality check: When defining the each job, step, or task, you should always ask the questions: "Is this a nonroutine item?" "How does it affect the task?" Refer to Figure 10-2 for details on how to apply nonroutine.

## 10.1.2.3 Risk Assessment

After conducting the risk assessments of the specific steps or tasks, the level of risk is included under the "RA section." As in the nonroutine section, a column is used to make it easier to be reviewed by the employees and the supervisors.

| Job Steps and Task-Specific Description | NR | RA | Existing and Potential Hazards and/or Consequences of Exposure | At-Risk Events and Preventive Measures | RR |
|---|---|---|---|---|---|
| **Pull car off the road**<br><br>• Park clear of traffic<br>• Park on level ground<br>• Set parking brake<br>• Turn off ignition<br>• Remove the key<br>• Turn on the hazard warning lights | X | | | | |
| **Check for Traffic**<br><br>• Exit car<br>• Block wheel<br>• Set flares/reflectors | | | | | |

**Figure 10-2** Documenting Nonroutine Tasks

Once the tasks have been defined you should now be able to establish the risk level based on the risk analysis as presented in Chapter 6 and the expanded version in Chapter 13. Refer to Figure 10-3 to review the Risk Assessment section.

This provides a quick review of the potential level of risk within each step. Tasks with high risk are given a priority and may require immediate actions that will impact on that particular step and the entire job. A step may include tasks with a marginal risk and have limited safety requirements. Another step may have catastrophic potential that requires strong mandates and training to be implemented. This column provides critical information that will determine the scope of the at-risk events and preventive measures.

| Job Steps and Task-Specific Description | NR | RA | Existing and Potential Hazards and/or Consequences of Exposure | At-Risk Events and Preventive Measures | RR |
|---|---|---|---|---|---|
| **Pull car off the road**<br><br>• Park clear of traffic<br>• Park on level ground<br>• Set parking brake<br>• Turn off ignition<br>• Remove the key<br>• Turn on the hazard warning lights | X | H | | | |
| **Check for Traffic**<br><br>• Exit car<br>• Block wheel<br>• Set flares/reflectors | | H | | | |

**Figure 10-3** Documenting Initial Risk Assessment to the Task

### 10.1.2.4 Existing and Potential Hazards and Consequences of Exposure

In this section you record the exposure to specific hazards. Anyone reviewing the JHA must understand the existing and potential hazards and consequences of exposure as they are related to each task. Refer to Chapters 1, 2, and 3 for details on identifying hazards. Do not determine the risk and hazards until all tasks are identified and mapped.

This section is used to detail hazards in order to ask: "What are the consequences of exposure to a specific hazard?" As defined by Webster, consequences are "Something that logically or naturally follows from an action or condition; the relation of a result to its cause." Exposure can be defined as: "An act of subjecting or an instance of being subjected to an action or influence." Based on these definitions, we are trying to identify what will happen if someone or thing is exposed to a hazard and there are no controls in place to remove or control the hazard. The JHA defines the nature of the hazards and the types of loss-producing events that may occur.

The reviewer must "think outside of the box"—i.e., think of all of the combinations of consequences of exposure (where someone could get hurt) and come back to reality. This is where the use of the cause and effect diagram can provide insight, by looking at and questioning how the five elements can combine to create risk and hazards.

> For example, if an employee places a hand into a running piece of equipment, what is the probability of getting the hand caught? What could be the severity of the event? The more times someone is exposed to an at-risk event, the greater chance that the hand will be caught. The exposure to this type of hazard increases the probability of getting caught in the piece of equipment and the severity could be serious.

The existing and potential hazards and consequences of exposure must be identified for each listed task. The linkage to the task is based on observations of the steps and tasks, evaluation of potential injuries and causes, and personal experience. The steps and tasks are reviewed against the tools, equipment, and materials to be used, (what are their hazards) and the environment and the hazards asking the question: how do these get controlled within the steps and tasks under analysis. Refer to Figure 10-4 for an example on how to apply existing and potential hazards and consequences of exposure.

| Job Steps and Task-Specific Description | NR | RA | Existing and Potential Hazards and/or Consequences of Exposure | At-Risk Events and Preventive Measures | RR |
|---|---|---|---|---|---|
| **Pull car off the road**<br>• Park clear of traffic<br>• Park on level ground<br>• Set parking brake<br>• Turn off ignition<br>• Remove the key<br>• Turn on the hazard warning lights | X | H | Traffic – speed and conditions<br>Getting hit by an oncoming vehicle<br><br>Environment – weather, ground surface, conditions<br><br>Vehicle condition | | |
| **Check for Traffic**<br>• Exit car<br>• Block wheel<br>• Set flares/reflectors | | H | Traffic speed<br><br>Getting hit by an oncoming vehicle | | |

**Figure 10-4** Documenting Existing and Potential Hazards and Consequences of Exposure

## 10.1.2.5 At-Risk Events and Preventative Measures

In this section, note that at-risk events are used to incorporate the behavioral component of the preventative measures. This section will define the specific controls for the at-events as they relate to the steps and tasks. Listed below each at at-risk event and preventative measure is a guidance statement used to convey the safe way of doing the tasks and methods on controlling the at-risk events. Refer to Chapter 4, Appendix J for a sample generic list of behaviors (at-risk events.) At-risk events and preventative measures are now matched to the existing and potential hazards and consequences of exposure. Refer to Figure 10-5 for an example on describing at-risk events and preventative measures.

As we learned in Chapter 4, we all make choices in life. In many cases our behavior controls our destiny. In the JHA example we are only identifying the at-risk events in the example of changing a tire, i.e., someone making the choice of not following procedures when changing a tire. The results of the consequences could be pinned under the car and under the worst case, an amputation due to improper placement of the hand in relations to the jack and vehicle.

Typically, a behavioral observational process uses a list of important events tied to the job and a general guidance statement about the at-risk event and the preventative measure. Consideration must be given to potential decisions and actions made by individuals as they complete each task. Also to be considered is the human element and physical and mental attributes required to complete each task or each step.

| Job Steps and Task-Specific Description | NR | RA | Existing and Potential Hazards and/or Consequences of Exposure | At-Risk Events and Preventive Measures | RR |
|---|---|---|---|---|---|
| **Pull car off the road** | | | | | |
| • Park clear of traffic<br>• Park on level ground<br>• Set parking brake<br>• Turn off ignition<br>• Remove the key<br>• Turn on the hazard warning lights | X | H | Traffic – speed and conditions<br>Getting hit by an oncoming vehicle<br><br>Environment – weather, ground surface, conditions<br><br>Vehicle condition | <u>Operating Vehicle Safety</u> | |
| **Check for Traffic** | | | | | |
| • Exit car<br>• Block wheel<br>• Set flares/reflectors | | H | Traffic speed<br><br>Getting hit by an oncoming vehicle | <u>Exit/Entry from Car</u> | |

**Figure 10-5** Documenting At-Risk Events

## Preventative Measures

The JHA now establishes methods and actions to eliminate or control the hazards identified using preventative measures. These are the actions taken to reduce the physical or personal hazards that can create injury or harm. These measures are developed followed the hierarchy of controls (refer to Chapter 9, Table 9-2). The selected controls are identified in this column and provided with a guidance statement outlining how an employee performing their job knows exactly how to protect themselves from the identified hazard.

To recap the hierarchy of controls:

*Avoid the hazard.*

This is the most effective measure. Avoid the risk and hazard. The following techniques are used to avoid the hazards:

- Cease doing the activity.
- Choose a different process.
- Modify an existing process.

*Engineering.*

Several control methods fall under the engineering heading:

- Eliminate or minimize the hazards or risks.
- Substitute or redesign with less hazardous substance, process, or equipment.

- Contain or isolate the hazard by using enclosures, machine guards, ventilation, welding curtains, etc.
- Modify or change equipment, tools, or materials to redirect the hazard.

*Administrative Controls.*

- Revise work procedures to modify tasks that are hazardous, changing the sequence of tasks, adding additional tasks such as locking out energy sources, training, giving warning, adding signage, etc.
- Monitor the use of very hazardous materials.
- Buddy systems or team approach.
- Training, refresher training.

*Personal Protective Equipment (PPE).*
The use of appropriate PPE may be required. To reduce the severity of an injury, emergency facilities, such as eyewash stations, may need to be provided.

- Reduce the exposure. These methods are the least effective and should only be used if no other solutions are possible. One way of minimizing exposure is to reduce the number of at-risk event (times) the hazard is encountered [7].

In listing the preventive measures, use of general statements such as "be safe," "be careful" or "use caution" should be avoided. Specific statements that describe both what action is to be taken and how it is to be performed are preferable. This is similar to what we discussed in Chapter 7 on learning objectives. The preventative measures are listed in the right-hand column of the worksheet to match the hazard in question. Refer to Figure 10-6 for details on documenting the preventative measures and general guidance statements.

## 10.1.2.6 Residual Risk

The final stage of a JHA is to conduct a final risk assessment to make sure that hazards have been eliminated or controlled. The objective is to determine what risk remains. Conducting a risk assessment after the controls are in place is necessary as the "residual risks" of each task are critical to every operation and are identified in this column by an "RR." The residual risk analysis is developed after completion of the at-risk events and preventative measures. As a key important issue that must be taken into account, the residual risk must always be less hazardous than the initial risk. If the residual risk is the same, then "what have you accomplished?" Refer to Chapter 6 for a detailed discussion on assessment and controlling risk. Refer to Figure 10-7 for how to document the residual risks of the task.

| Job Steps and Task-Specific Description | NR | RA | Existing and Potential Hazards and/or Consequences of Exposure Consideration | At-Risk Events and Preventive Measures | RR |
|---|---|---|---|---|---|
| **Pull car off the road**<br><br>• Park clear of traffic<br>• Park on level ground<br>• Set parking brake<br>• Turn off ignition<br>• Remove the key<br>• Turn on the hazard warning lights | X | H | Traffic – speed and conditions<br>Getting hit by an oncoming vehicle<br><br>Environment – weather, ground surface, conditions<br><br>Vehicle condition | Operating Vehicle Safety<br><br>• Ensure that brake is set so that vehicle will not move<br>• Make sure that vehicle is parked off of the roadway so that it is clear of traffic<br>• Ensure that vehicle is parked on a level surface to prevent it from moving when it is jacked up<br>• Vehicle maintenance | |
| **Check for Traffic**<br><br>• Exit car<br>• Block wheel<br>• Set flares/reflectors | | H | Traffic speed<br><br>Getting hit by an oncoming vehicle | Exit/Entry from Car<br><br>• Have clear view of oncoming traffic when exiting vehicle | |

**Figure 10-6** Documenting Preventative Measures and Guidance Statements

| Job Steps and Task-Specific Description | NR | RA | Existing and Potential Hazards and/or Consequences of Exposure | At-Risk Events and Preventive Measures | RR |
|---|---|---|---|---|---|
| **Pull car off the road**<br><br>• Park clear of traffic<br>• Park on level ground<br>• Set parking brake<br>• Turn off ignition<br>• Remove the key<br>• Turn on the hazard warning lights | X | H | Traffic – speed and conditions<br>Getting hit by an oncoming vehicle<br><br>Environment – weather, ground surface, conditions<br><br>Vehicle condition | Operating Vehicle Safety<br><br>• Ensure that brake is set so that vehicle will not move<br>• Make sure that vehicle is parked off of the roadway so that it is clear of traffic<br>• Ensure that vehicle is parked on a level surface to prevent it from moving when it is jacked up<br>• Vehicle maintenance | M |
| **Check for Traffic**<br><br>• Exit car<br>• Block wheel<br>• Set flares/reflectors | | H | Traffic speed<br><br>Getting hit by an oncoming vehicle | Exit/Entry from Car<br><br>• Have clear view of oncoming traffic when exiting vehicle | M |

**Figure 10-7** Documenting Residual risks

## 10.2 BENEFIT REVIEW: GETTING THE BIGGEST BANG FOR THE BUCK

To get the maximum benefits from the JHA, an effort should be made to determine the quality and performance of the reducing at-risk events and preventative measures.

When developing at-risk events and preventative measures, the use of general statements such as "watch out," "be careful," "use caution," "be safe," "think about," or "be mindful of" must not be used. What do these statements mean? Do they link at-risk events to safe performance of the job? No! The statements to be used must be specific preventative measures, actions and controls. Specific statements must be used that describe the actions, what actions will be taken, and how the steps/tasks are to be safely performed:

- Are the at-risk events and preventive measures specifically designed around what the employee is to actually do?
- Is the objective or the general guidance statement well defined?
- Are action words used describing how to prevent an employee from being injured?
- Are preventive measures practical?
- Is there a common sense approach, given the nature of the tasks performed?
- Do employees have control over at-risk events and preventative measures?
- Can employees control their actions or does it take management commitment and leadership to provide the tools and resources necessary to perform the job safely?

One question that must be asked: does the at-risk event and the preventive measures identified clearly protect the employee from the existing and potential hazards and consequence of exposure when performing the job steps and tasks in a safe manner?

Once hazards and operational concerns have been identified for each of the steps, then the at-risk events and preventative measures for each hazard are recorded. These control measures become the operational expectation and the established standard.

## 10.3 OKAY, I HAVE COMPLETED THE JHA: NOW WHAT?

Before the completed JHA is submitted for approval, a group of experienced employees and supervisors need to review the document one more time to ensure that nothing has been overlooked [5]. Ask several new or inexperienced employees to review the document to see if they have an understanding of the instructions and safety precautions. Rewrite sections that have vague or missing information or are confusing to the employees [5].

## 10.3.1 Review JHAs until Employee Understands Hazards of Job

When the JHA is completed, the employees may need to be retrained in the new procedures. Employees must demonstrate that they understand the hazards. The training may vary depending on the complexity of the steps and tasks. It may consist of an informal communication using a team approach or it could be formal on-the-job training in a class setting. The training depends on how different or unique the JHA procedure is from what the employee was currently doing before the change [5].

When training is completed, each employee should sign the JHA training record to acknowledge and demonstrate that they understand the identified hazards, and acknowledge the identified at-risk events and preventative measures. The JHA should be reviewed periodically (typically, two times per year), prior to performing any nonroutine task, new or modified task, or if an injury or damage occurs [5].

## 10.4 REVISING THE JHA

The JHA is only effective if reviewed and updated periodically. Even if the job itself has not changed, you may detect a hazard missed in an earlier analysis [5].

If a loss-producing event occurs, you should review the JHA immediately to determine if the JHA was followed and if changes in the job procedures are necessary. Close calls, near misses, or minor injuries are good examples of incidents that indicate the need for further examinations. If upon examination, a close call, near miss, minor injuries resulted from an employee's failure to follow job procedures, you should discuss the matter with all employees performing the job and take the appropriate action [5].

Any time the JHA is revised, employees that are affected by the change should be retrained in the new job methods, procedures, and at-risk events and preventative measures. A JHA is used to train new or transferred employees on the basic job steps and tasks [5].

Finally, once the JHA is complete, use it, and do not just file it away! If possible, post it at the job site where the employee has an opportunity to review the JHA. Make it an integral part of the SOP and checklist. Review and update the JHA periodically.

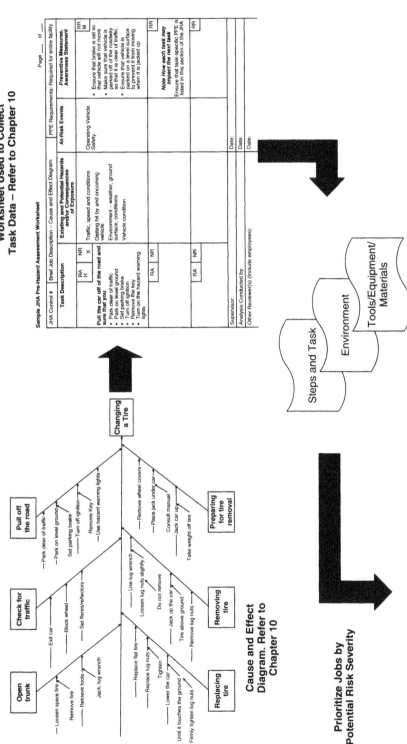

Pre-Hazard Assessment
Worksheet Used to Collect
Task Data – Refer to Chapter 10

Cause and Effect
Diagram. Refer to
Chapter 10

Steps and Task

Environment

Tools/Equipment/
Materials

Prioritize Jobs by
Potential Risk Severity

| Brief Description: Changing a Tire | Department: Highway | Date April 2007 | Last Revision: |
|---|---|---|---|
| Performed By: J. Roughton/Nathan Crutchfied | Employee: Joe Tire | Supervisor: Jim Production | |
| Personal Protective Equipment: None required | Note: Recommended PPE will be listed under each specific task based PPE Assessment | | |

Note: NR = Non-Routine Task – Place a check mark for these types of task. RA = Risk Analysis, RR = Residual Risk (After JHA Development)

| Job Steps and Task-Specific Description | NR | RA | Existing and Potential Hazards and/or Consequences of Exposure | Potential At-Risk Events and Preventive Measures | RR |
|---|---|---|---|---|---|
| **Pull car off the road**<br>• Park clear of traffic<br>• Park on level ground<br>• Set parking brake<br>• Turn off ignition<br>• Remove the key<br>• Turn on the hazard warning lights | | H | Traffic – speed and conditions<br>Getting hit by an oncoming vehicle<br>Environment – weather, ground surface, conditions<br>Vehicle condition | **Operating Vehicle Safety**<br>• Ensure that brake is set so that vehicle will not move<br>• Make sure that vehicle is parked off of the roadway so that it is clear of traffic.<br>• Ensure that vehicle is parked on a level surface to prevent it from moving when it is jacked up | M |
| **Check for Traffic**<br>• Exit the car<br>• Block the wheel<br>• Set flares/reflectors on the road | x | H | Traffic speed<br>Getting hit by an oncoming vehicle | **Exit /Entry from Car**<br>• Pay attention to on-coming traffic when existing vehicle | M |
| **Open Trunk**<br>• Loosen spare tire<br>• Remove tire<br>• Retrieve Jack and lug wrench | | M | Strain/sprain from removing tire and tools from trunk | **Eyes on Task**<br>• Keep eyes on task when removing tire<br>**Tools and Equipment**<br>• Keep back as straight possible when removing tire an tools from trunk<br>**Lifting and Lowering**<br>• Use proper lifting techniques when lifting tire from trunk | L |
| **Preparing for Tire Removal**<br>• Remove wheel covers<br>• Place jack under car<br>• Jack car up<br>• Take weight of tire | | M | Strain/sprain from removing wheel covers<br>Struck against vehicle when placing jack under car<br>Struck by vehicle when placing jack up car<br>Terrain/ground surface and condition, shift off of jack | **Tools and Equipment**<br>Select the correct tool for removing wheel covers<br>**Eyes on Task**<br>• Keep eyes on task when removing wheel covers<br>• Keep eyes on task when placing jack under car<br>• Keep eyes on task and highway when jacking up car | L |

| | Catastrophic | Critical | Marginal | Negligible |
|---|---|---|---|---|
| Frequent | HIGH | HIGH | SERIOUS | SERIOUS |
| Probable | HIGH | HIGH | SERIOUS | SERIOUS |
| Occasional | HIGH | SERIOUS | MEDIUM | LOW |
| Remote | SERIOUS | MEDIUM | MEDIUM | LOW |
| Improbable | MEDIUM | LOW | LOW | LOW |

**Defines Risk Assessment, Refer to Chapter 6**

**Determine Modification or Controls**

**Hierarchy of Controls, Refer to Table 9-2**
- □ Avoid or reduce the hazard strength
- □ Engineering controls
- □ Administrative controls
- □ PPE

**SOP, Refer to Chapter 11**

Standard or Safe Operating Procedures

**Completed JHA, Refer to Chapter 10**

**Figure 10-8** An overview of the entire JHA Process, soups to nuts

## 10.5 SUMMARY

The JHA is the end result of a process that develops a consistent approach to each job.

Your JHA files will now consist of the collected job data, the cause and effect diagram, the pre-hazard assessment, risk assessments, PPE assessment, the JHA, training records, and SOPs.

It is important to assign both authority and specific responsibility to implement at-risk events and preventative measures. A trainer or supervisor is needed to provide the training and scheduling; the manager provides resources and safe tools and equipment; and the employees follow the required guidelines and report back problems. The safety department and its resources are available for guidance and review.

There is a child's game that demonstrates what we have discussed. A group of people sit in a close circle. A designated person starts by whispering a brief story to the person sitting next to them. The story is then passed on by that person to the next person in the circle and on through the entire group. By the time the story gets back to the original storyteller, the message has drastically changed. If the first participant had a written script similar to a JHA, the story would have retained a consistent message. Regardless of the whether incident, training, observations, or other type of activity, the JHA provides the communication of a consistent message.

In Chapter 11 we discuss why developers of the SOP tend to forget that the JHA provides extensive documentation.

## CHAPTER REVIEW QUESTIONS

1. What are key considerations when conducting a JHA?
2. What is a *job hazard analysis*?
3. What is the *job description*?
4. How do you describe the job steps and task-specific description section of the JHA?
5. How would you define *nonroutine task*?
6. How would you describe the existing and potential hazards and consequences of the exposure section of the JHA?
7. How would you describe the at-risk events and preventative measures section of the JHA?
8. How would you define at-risk events?

9. What is residual risk?
10. How would you describe the preventative measures section of the JHA?

## REFERENCES

1. U.S. Department of Labor, Occupational Safety and Health Administration OSHA 3071 Job Hazard Analysis, 1992, 1998, 2002 (Revised)
2. Oregon OSHA Website, Job Hazard Analysis, http://www.cbs.state.or.us/external/osha/pdf/workshops/103w.pdf, public domain
3. Saskatchewan Labour website, Occupational Health & Safety, How to Identify Job Hazards, Partners in Safety, How to identify job hazards, Occupational Health and Safety Division, WorkSafe Saskatchewan http://www.worksafesask.ca/
4. Missouri Department of Labor, *Catching the Hazard That Escapes Control*, Chapter 9, http://www.dolir.mo.gov/ls/safetyconsultation/ccp/, public domain
5. Roughton, J., Nancy Whiting, *Safety Training Basics: A Handbook for Safety Training Development*, Government Institute, Rockville Maryland, 2000
6. Roughton, James, James Mercurio, *Developing an Effective Safety Culture: A Leadership Approach*, Butterworth-Heinemann, 2002
7. "What is a Job Hazard Analysis?" http://www.ccohs.ca/oshanswers/hsprograms/job-haz.html, OSH Answers, Canadian Centre for Occupational Health & Safety (CCOHS), 2006, reproduced with the permission of CCOHS, October 2006

# Appendix O

## O.1 SAMPLE INSTRUCTIONS ON HOW TO CHANGE A TIRE ON A CAR

- Pull off to the side of the road out of traffic. Make sure the car is parked off the road and clear of traffic.
- Park on level ground.
- Set the parking brake. Apply the parking brake and place the transmission in park if it is an automatic. If the car has a standard transmission, place the shift in first gear or reverse.
- Turn off the ignition and remove the key.
- Turn on the hazard warning lights.
- Check for traffic and exit the car. Watch out for oncoming traffic as you exit the car.
- Find something to block the front wheels if changeing a back tire or block the back wheels if changing the front tire (to keep the car from rolling).
- Set safety flares or reflectors five car lengths front and back.
- Open the trunk and remove the tire.
- Pull out spare tire.
- Retrieve others tools out of the trunk, such as jack and lug wrench.
- Remove wheel covers.
- Properly place jack under the car. Consult your owners' manual and find where the jack needs to be positioned.
- Jack the car up part of the way so that the tire is still touching the ground. Raise the car just enough to take the weight off the tire.
- Using the lug wrench, loosen the lug nuts slightly but do not remove.
- Jack up the car so the tire is above ground.
- Remove the lug nuts from the bolts.
- Remove the flat tire.
- Replace the flat tire with the spare.
- Replace the lug nuts on the bolts and tighten, but not too tight, just enough to hold the tire in place while you lower the car.

- Lower the car until it touches the ground, and then firmly tighten all of the lug nuts.
- Gather everything and put in trunk.
- As soon as possible, have the flat tire repaired and reinstalled.

# O.2 SAMPLE JHA PRE-HAZARD ASSESSMENT WORKSHEET

Page ____ of ____

| JHA Control # | Brief Job Description – Cause and Effect Diagram | | PPE Requirements: Required for entire facility | |
|---|---|---|---|---|
| **Task Description** | **Existing and Potential Hazards and/or Consequences of Exposure** | **At-Risk Events** | **Preventive Measures, Awareness Statement** | |
| RA / NR<br>H / X | | Operating Vehicle Safety | Ensure that brake is set so that vehicle will not move.<br>Make sure that vehicle is parked off of the roadway so that it is clear of traffic.<br>Ensure that vehicle is parked on a level surface to prevent it from moving when it is jacked up | RR<br>M |
| Pull the car off of the road and be sure that you | Traffic, speed and conditions | | | |
| | Getting hit by an oncoming vehicle | | | |
| Park clear of traffic<br>Park on level ground<br>Set parking brake<br>Turn off ignition<br>Remove the key<br>Turn on the hazard warning lights | Environment - weather, ground surface, conditions | | | |
| | Vehicle condition | | | |
| RA / NR | | | *Note how each task may impact the next task* | RR |
| RA / NR | | | Ensure that task-specific PPE is listed in this section of the JHA | RR |

| | |
|---|---|
| Supervisor: | Date: |
| Analysis Conducted by: | Date: |
| Other Reviewer(s) (Include employees): | Date: |

Adapted and modified, Oregon OSHA Website, Job Hazard Analysis 103 Workshop, http://www.cbs.state.or.us/external/osha/educate/training/pages/materials.html and U.S. Department of Labor, Occupational Safety and Health Administration OSHA 3071 Job Hazard Analysis, 1998, 1999, 2002 (Revised), www.OSHA.gov, public domain

# O.3 JHA - CHANGING A TIRE

| Brief Description: Changing a Tire | Department: Highway | Date April 2007 | Last Revision: |
|---|---|---|---|
| Performed By: J. Roughton/Nathan Crutchfield | Employee: Joe Tire | Supervisor: Jim Production | |
| Personal Protective Equipment: None required | Note: Recommended PPE will be listed under each specific task based PPE Assessment | | |

*Note: NR = Non-Routine Task – Place a check mark for these types of task. RA=Risk Analysis, RR=Residual Risk (After JHA Development)*

| Job Steps and Task-Specific Description | NR | RA | Existing and Potential Hazards and/or Consequences of Exposure | Potential At-Risk Events and Preventive Measures | RR |
|---|---|---|---|---|---|
| **Pull car off the road**<br><br>Park clear of traffic<br>Park on level ground<br>Set parking brake<br>Turn off ignition<br>Remove the key<br>Turn on the hazard warning lights | X | H | Traffic – speed and conditions<br>Getting hit by an oncoming vehicle<br><br>Environment - weather, ground surface, conditions<br><br>Vehicle condition | **Operating Vehicle Safety**<br>Ensure that brake is set so that vehicle will not move.<br>Make sure that vehicle is parked off of the roadway so that it is clear of traffic.<br>Ensure that vehicle is parked on a level surface to prevent it from moving when it is jacked up | M |
| **Check for Traffic**<br><br>Exit the car<br>Block the wheel<br>Set flares/reflectors on the road | X | H | Traffic speed<br><br>Getting hit by an oncoming vehicle | **Exit/Entry from Car**<br>Pay attention to on-coming traffic when existing vehicle | M |
| **Open Trunk**<br><br>Loosen spare tire<br>Remove tire<br>Retrieve jack and lug wrench | | M | Strain/sprain from removing tire and tools from trunk | **Eyes on Task**<br>Keep eyes on task when removing tire<br><br>**Tools and Equipment**<br>Keep back as straight possible when removing tire and tools from trunk<br><br>**Lifting and Lowering**<br>Use proper lifting techniques when lifting tire from trunk | L |

| Job Steps and Task-Specific Description | NR | RA | Existing and Potential Hazards and/or Consequences of Exposure | Potential At-Risk Events and Preventive Measures | RR |
|---|---|---|---|---|---|
| **Preparing for Tire Removal**<br><br>Remove wheel covers<br>Place jack under car<br>Jack car up<br>Take weight off tire | | M | Strain/sprain from removing wheel covers<br>Struck against vehicle when placing jack under car<br>Struck by vehicle when placing jack up car<br>Terrain/ground surface and condition, shift off of jack | **Tools and Equipment**<br>Select the correct tool for removing wheel covers<br><br>**Eyes on Task**<br>Keep eyes on task when removing wheel covers<br>Keep eyes on task when placing jack under car<br>Keep eye on task and highway when jacking up car | L |
| **Removing Tire**<br><br>Use lug wrench<br>Loosen lug nuts slightly<br>Jack up the car<br>Remove lug nuts | | H | Strain/sprain from using lug wrench<br>Struck against vehicle when loosen lug nuts<br>Struck by vehicle when jack car up | **Tools and Equipment**<br>Select the correct tool for removing lug nuts<br>Ensure that jack is positioned under car before continuing<br><br>**Eyes on Task**<br>Keep eyes on task when using lug wrench<br>Keep eyes on task when loosing lug nuts<br>Keep eye on task and highway when jacking up car | L |
| **Replacing Tire**<br><br>Replace flat tire<br>Replace lug nuts<br>Lower the car | | H | Strain/sprain from using lug wrench<br>Struck against vehicle when replacing lug nuts<br>Struck by vehicle when lowering car<br>Terrain/ground surface and condition, shift off of jack | **Eyes on Task**<br>Keep eyes on task when using lug wrench<br>Keep eyes on task when replacing lug nuts<br>Keep eye on task and highway when lowering car<br><br>**Tools and Equipment**<br>Select the correct tool for replace lug nuts<br>Ensure that jack is positioned on solid surface in required set-point of car before continuing | L |

Specific statements must describe the actions, what actions will be taken, and how the steps/tasks are to be safely performed:

- Are the at-risk events and preventive measure specifically designed around what the person is to actually do?
- Is the objective or the general guidance statement well defined?
- Are action words used describing how to prevent an employee from an injury?
- Are preventive measures practical?
- Is there a common sense approach, given the nature of the tasks performed?
- Do employees have control over at-risk events and preventative measures?
- Can employees control their actions or does it take management commitment and leadership to provide the tools and resources necessary to perform the job safely?

# O.4 ANNOTATED JHA - CHANGING A TIRE EXAMPLE

| Brief Description: Changing a Tire | Department: Highway | Date April 2007 | Last revised: |
|---|---|---|---|
| Performed By: J. Roughton/Nathan Crutchfiled | Employee Signature: Joe Tire | Supervisor Signature: Jim Production | |
| Personal Protective Equipment None required | | Note: Recommended PPE will be listed under each specific task based PPE Assessment | |

*Note: NR = Non-Routine Task – Place a check mark for these types of task. RA = Risk Analysis, RR = Residual Risk (After JHA Development)*

| Job Steps and Task-Specific Description | NR | RA | Existing and Potential Hazards and/or Consequences of Exposure Consideration | Potential At-Risk Events and Preventive Measures | RR |
|---|---|---|---|---|---|
| **Pull car off the road** | | | | **Operating Vehicle Safety** | M |
| Park clear of traffic | | H | Getting hit by an oncoming vehicle | Ensure that brake is set so that vehicle will not move. | |
| Park on level ground | | | | Make sure that vehicle is parked off of the roadway so that it is clear of traffic. | |
| Set parking brake | | | Employee participation and buy-in via signing the JHA to verify agreement of hazard. | Ensure that vehicle is parked on a level surface to prevent it from moving when it is jacked up | |
| Turn off ignition | | | | | |
| Remove the key | | | | | |
| Turn on the hazard warning lights | | | | | |
| **Check for Traffic** | | | | **Exit/Entry from Car** | H |
| Exit car | X | H | Getting hit by an oncoming vehicle | Pay attention to on-coming traffic when existing vehicle | |
| Block wheel | | | | | |
| Set flares/reflectors | | | | | |
| **Open Trunk** | X | M | Strain/sprain from removing tire and tools from trunk | **Eyes on Task** | M |
| Loosen spare tire | | | | Keep eyes on task when removing tire | |
| Remove tire | | | | **Tools and Equipment** | |
| Retrieve Jack and lug wrench | | | | Keep back as straight possible when removing tire and tools from trunk | |
| | | | | **Lifting/Lowering** | |
| | | | | Use proper lifting techniques when lifting tire from trunk | |

At-Risk events are integrated into each task to highlight specific behavior(s) associated with the hazard.

Details the preventative measures that are to be used to ensure that the tasks/steps are performed safely.

The PPE Section has two specific uses:

1. **Facility-wide specific PPE**: PPE listed in this section is required for the entire facility, based on the consequences of exposure and the PPE assessment.
2. **Task-Specific PPE**: Specific PPE will be identified by each task in this section. This section will ensure that the user understands task-specific PPE as identified in the PPE assessment and how it related to the at-risk event and preventative measures associated with the hazard.

*Annotations:*

This section is used to break down the job steps into the individual tasks.

For nonroutine task (not completed frequently) an "X" placed in this section.

This section is where the initial risk assessment is noted. Refer to Chapter 6 for details.

This section highlights existing and potential hazards and consequences of exposure that may result from the specific hazards.

This section is where the residual risk assessment is noted. Refer to Chapter 6 for details.

# O.5 COMPARISON JHA ON CHANGING A TIRE: TRADITIONAL vs. NEW VERSION

**Traditional Version**

| Sequence of Events | Potential Accidents or Hazards | Preventive Measures |
|---|---|---|
| Park vehicle | Vehicle too close to passing traffic<br>Vehicle on uneven, soft ground<br>Vehicle may roll | Drive to area well clear of traffic. Turn on emergency flashers<br>Choose a firm, level area<br>Apply the parking brake; leave transmission in gear or in PARK; place blocks in front and back of the wheel diagonally opposite to the flat |

**New Version**

| Job: Changing a Tire | Department: Highway | Date April 2007 |
|---|---|---|
| Performed By: J. Roughton/Nathan Crutchfield | | Supervisor Signature: Jim Production |
| Personal Protective Equipment: None required | | Note: Recommended PPE will be listed under each specific task based PPE Assessment |

*Note: NR = Non-Routine Task – Place a check mark for these types of task. RA=Risk Analysis, RR=Residual Risk (After JHA Development)*

| Job Steps and Task-Specific Description | NR | RA | Existing and Potential Hazards and/or Consequences of Exposure Consideration | Potential At-Risk Events and Preventive Measures | RR |
|---|---|---|---|---|---|
| Pull car off the road<br><br>Park clear of traffic<br>Park on level ground<br>Set parking brake<br>Turn off ignition<br>Remove the key<br>Turn on the hazard warning lights | | H | Getting hit by an oncoming vehicle | Operating Vehicle Safety<br><br>Ensure that brake is set so that vehicle will not move. Make sure that vehicle is parked off of the roadway so that it is clear of traffic.<br>Ensure that vehicle is parked on a level surface to prevent it from moving when it is jacked up | M |

## O.6 JOB HAZARD ANALYSIS, CANADIAN CENTRE FOR OCCUPATIONAL HEALTH & SAFETY (CCOHS), REPRINTED WITH PERMISSION

One way to increase the knowledge of hazards in the workplace is to conduct a job hazard analysis on individual tasks. A job hazard analysis (JHA) is a procedure that helps integrate accepted safety and health principles and practices into a particular operation. In a JHA, each basic step of the job is examined to identify potential hazards and to determine the safest way to do the job. Other terms used to describe this procedure are job safety analysis (JSA) and job hazard breakdown.

Some individuals prefer to expand the analysis into all aspects of the job, not just safety. This approach, known as total job analysis, job analysis or task analysis, is based on the idea that safety is an integral part of every job and not a separate entity. In this document, only health and safety aspects will be considered.

The terms "job" and "task" are commonly used interchangeably to mean a specific work assignment, such as "operating a grinder," "using a pressurized water extinguisher," or "changing a flat tire." JHAs are not suitable for jobs defined too broadly, for example, "overhauling an engine"; or too narrowly, for example, "positioning car jack."

### O.6.1 What are the benefits of doing a Job Hazard Analysis?

The method used in this example is to observe a worker actually performing the job. The major advantages of this method include that it does not rely on individual memory and that the process prompts recognition of hazards. For infrequently performed or new jobs, observation may not be practical. With these, one approach is to have a group of experienced workers and supervisors complete the analysis through discussion. An advantage of this method is that more people are involved, allowing for a wider base of experience and promoting a more ready acceptance of the resulting work procedure.

Members of the joint occupational safety and health committee should participate in this process.

Initial benefits from developing a JHA will become clear in the preparation stage. The analysis process may identify previously undetected hazards and increase the job knowledge of those participating. Safety and health awareness is raised, communication between workers and supervisors is improved, and acceptance of safe work procedures is promoted.

The completed JHA, or better still, a written work procedure based on it, can form the basis for regular contact between supervisors and workers on health and safety. It can serve as a teaching aid for initial job training and as a briefing guide for infrequent jobs.

It may be used as a standard for health and safety inspections or observations and it will assist in completing comprehensive accident investigations.

### O.6.1.1  What are the four basic steps?

Four basic stages in conducting a JHA are:

- selecting the job to be analyzed
- breaking the job down into a sequence of steps
- identifying potential hazards
- determining preventive measures to overcome these hazards

*What is important to know when "selecting the job"?*

Ideally, all jobs should be subjected to a JHA. In some cases there are practical constraints posed by the amount of time and effort required to do a JHA. Another consideration is that each JHA will require revision whenever equipment, raw materials, processes, or the environment change. For these reasons, it is usually necessary to identify which jobs are to be analyzed. Even if analysis of all jobs is planned, this step ensures that the most critical jobs are examined first.

Factors to be considered in assigning a priority for analysis of jobs include:

- Accident frequency and severity: jobs where accidents occur frequently or where they occur infrequently but result in disabling injuries.
- Potential for severe injuries or illnesses: the consequences of an accident, hazardous condition, or exposure to harmful substance are potentially severe.
- Newly established jobs: due to lack of experience in these jobs, hazards may not be evident or anticipated.
- Modified jobs: new hazards may be associated with changes in job procedures.
- Infrequently performed jobs: workers may be at greater risk when undertaking non-routine jobs, and a JHA provides a means of reviewing hazards.

*How do I break the job into "basic steps"?*

After a job has been chosen for analysis, the next stage is to break the job into steps. A job step is defined as a segment of the operation necessary to advance the work. See examples below.

Care must be taken not to make the steps too general, thereby missing specific steps and their associated hazards. On the other hand, if they are too detailed, there will be too many steps. A rule of thumb is that most jobs can be described in less than ten steps. If more steps are required, you might want to divide the job into two segments, each with its separate JHA, or combine steps where appropriate. As an example, the job of changing a flat tire will be used in this document.

An important point to remember is to keep the steps in their correct sequence. Any step which is out of order may miss potential hazards or introduce hazards which do not actually exist.

Each step is recorded in sequence. Make notes about what is done rather than how it is done. Each item is started with an action verb. Appendix O-A illustrates a format which can be used as a worksheet in preparing a JHA. Job steps are recorded in the left hand column, as shown below:

| Sequence of Events | Potential Accidents or Hazards | Preventive Measures |
|---|---|---|
| Park vehicle | | |
| Remove spare and tool kit | | |
| Pry off hub cap and loosen lug bolts (nuts) | | |
| And so on. . . .. | | |

This part of the analysis is usually prepared by watching the worker do the job. The observer is normally the immediate supervisor but a more thorough analysis often happens by having another person, preferably a member of the joint occupational health and safety committee, participate in the observation. Key points are less likely to be missed in this way.

The worker to be observed should be experienced and capable in all parts of the job. To strengthen full cooperation and participation, the reason for the exercise must be clearly explained. The JHA is neither a time and motion study in disguise, nor an attempt to uncover individual unsafe acts. The job, not the individual, is being studied in an effort to make it safer by identifying hazards and making modifications to eliminate or reduce them. The worker's experience can be important in making improvements.

The job should be observed during normal times and situations. For example, if a job is routinely done only at night, the JHA review should also be done at night. Similarly, only regular tools and equipment should be used. The only difference from normal operations is the fact that the worker is being observed.

When completed, the breakdown of steps should be discussed by all the participants (always including the worker) to make sure that all basic steps have been noted and are in the correct order.

*How do I "identify potential hazards"?*

Once the basic steps have been recorded, potential hazards must be identified at each step. Based on observations of the job, knowledge of accident and injury causes, and personal experience, list the things that could go wrong at each step.

A second observation of the job being performed may be needed. Since the basic steps have already been recorded, more attention can now be focused on potential hazards. At this stage, no attempt is made to solve any problems which may have been detected.

To help identify potential hazards, the job analyst may use questions such as these (this is not a complete list):

- Can any body part get caught in or between objects?
- Do tools, machines, or equipment present any hazards?
- Can the worker make harmful contact with objects?
- Can the worker slip, trip, or fall?
- Can the worker suffer strain from lifting, pushing, or pulling?
- Is the worker exposed to extreme heat or cold?
- Is excessive noise or vibration a problem?
- Is there a danger from falling objects?
- Is lighting a problem?
- Can weather conditions affect safety?
- Is harmful radiation a possibility?
- Can contact be made with hot, toxic, or caustic substances?
- Are there dusts, fumes, mists, or vapours in the air?

Potential hazards are listed in the middle column of the worksheet, numbered to match the corresponding job step. For example:

| Sequence of Events | Potential Accidents or Hazards | Preventive Measures |
| --- | --- | --- |
| Park vehicle | a) Vehicle too close to passing traffic<br><br>b) Vehicle on uneven, soft ground<br><br>c) Vehicle may roll | |
| Remove spare and tool kit | a) Strain from lifting spare | |

| Pry off hub cap and loosen lug bolts (nuts) | a) Hub cap may pop off and hit you | |
|---|---|---|
| | b) Lug wrench may slip | |
| And so on. . . .. | a) . . . | |

Again, all participants should jointly review this part of the analysis.

*How do I "determine preventive measures?"*

The final stage in a JHA is to determine ways to eliminate or control the hazards identified. The generally accepted measures, in order of preference, are:

1. Eliminate the hazard.

   This is the most effective measure. These techniques should be used to eliminate the hazards:

   - Choose a different process
   - Modify an existing process
   - Substitute with less hazardous substance
   - Improve environment (ventilation)
   - Modify or change equipment or tools

2. Contain the hazard.

   If the hazard cannot be eliminated, contact might be prevented by using enclosures, machine guards, worker booths or similar devices.

3. Revise work procedures.

   Consideration might be given to modifying steps which are hazardous, changing the sequence of steps, or adding additional steps (such as locking out energy sources).

4. Reduce the exposure.

   These measures are the least effective and should only be used if no other solutions are possible. One way of minimizing exposure is to reduce the number of times the hazard is encountered. An example would be modifying machinery so that less maintenance is necessary. The use of appropriate personal protective equipment may be required. To reduce the severity of an accident, emergency facilities, such as eyewash stations, may need to be provided.

In listing the preventive measures, use of general statements such as "be careful" or "use caution" should be avoided. Specific statements which describe both what action is to be taken and how it is to be performed are preferable. The recommended measures are listed in the right-hand column of the worksheet, numbered to match the hazard in question. For example:

| Sequence of Events | Potential Accidents or Hazards | Preventive Measures |
|---|---|---|
| Park vehicle | a) Vehicle too close to passing traffic<br>b) Vehicle on uneven, soft ground<br>c) Vehicle may roll | a) Drive to area well clear of traffic. Turn on emergency flashers<br>b) Choose a firm, level area<br><br>c) Apply the parking brake; leave transmission in gear or in PARK; place blocks in front and back of the wheel diagonally opposite to the flat |
| Remove spare and tool kit | a) Strain from lifting spare | a) Turn spare into upright position in the wheel well. Using your legs and standing as close as possible, lift spare out of truck and roll to flat tire. |
| Pry off hub cap and loosen lug bolts (nuts) | a) Hub cap may pop off and hit you<br>b) Lug wrench may slip | a) Pry off hub cap using steady pressure<br>b) Use proper lug wrench; apply steady pressure slowly. |
| And so on. . . .. | a) . . . | a) . . . |

*How should I make the information available to everyone else?*

JHA is a useful technique for identifying hazards so that measures can be taken to eliminate or control them. Once the analysis is completed, the results must be communicated to all workers who are, or will be, performing that job. The side-by-side format used in JHA worksheets is not an ideal one for instructional purposes. Better results can be achieved by using a narrative-style format. For example, the work procedure based on the partial JHA developed as an example in this document might start out like this:

1. Park vehicle.
   a) Drive vehicle off the road to an area well clear of traffic, even if it requires rolling on a flat tire. Turn on the emergency flashers to alert passing drivers so that they will not hit you.
   b) Choose a firm, level area so that you can jack up the vehicle without it rolling.

c) Apply the parking brake, leave the transmission in gear or PARK, place blocks in front and back of the wheel diagonally opposite the flat. These actions will also help prevent the vehicle from rolling.

2. Remove spare and tool kit.

   a) To avoid back strain, turn the spare up into an upright position in its well. Stand as close to the trunk as possible and slide the spare close to your body. Lift out and roll to flat tire.

3. Pry off hub cap, loosen lug bolts (nuts).

   a) Pry off hub cap slowly with steady pressure to prevent it from popping off and striking you.

   b) Using the proper lug wrench, apply steady pressure slowly to loosen the lug bolts (nuts) so that the wrench will not slip and hurt your knuckles.

4. And so on.

## Appendix O-A: Sample form for Job Hazard Analysis Worksheet

| Job Hazard Analysis Worksheet | | |
|---|---|---|
| Job: | | |
| Analysis By: | Reviewed By: | Approved By: |
| Date: | Date: | Date: |
| Sequence of Events | Potential Accidents or Hazards | Preventive Measures |
| | | |

## Appendix O-B: Sample forms for Tasks and Job Inventory

| Tasks with Potential Exposure to Hazardous Materials or Physical Agents | | |
|---|---|---|
| Analysis By: | Reviewed By: | Approved By: |
| Date: | Date: | Date: |
| Sequence of Events | Potential Accidents or Hazards | Preventive Measures |
| | | |

| Job Inventory of Hazardous Chemicals | | |
|---|---|---|
| Analysis By: | Reviewed By: | Approved By: |
| Date: | Date: | Date: |
| Sequence of Events | Potential Accidents or Hazards | Preventive Measures |
|  |  |  |

What is a Job Hazard Analysis? http://www.ccohs.ca/oshanswers/ hsprograms/ job-haz.html, OSH Answers, Canadian Centre for Occupational Health & Safety (CCOHS), 2006, Reproduced as is with the permission of CCOHS, October 2006.

# 11

# Standard or Safe Operating Procedures (SOP)

The JHA pulls together all the elements of a job, defines the steps, tasks, and all combinations of elements that create the desired job outcome. The next step in the process is to ensure that the JHA is translated into the Standard Operating Procedures (SOP) of the organization. If the process is not part of the current standards process, then it remains an "add-on," and safety will continue to be viewed as separate from production. At the end of the chapter you will be able to:

- Identify the key considerations for SOPs
- Develop an effective SOP that aligns with the JHA.

"Forget past mistakes. Forget failures. Forget everything except what you're going to do now and do it."

—William Durant

## 11.1 HOW FAR IS FAR ENOUGH? WHY DEVELOP AN SOP?

The final phase of the JHA process is to convert the JHA into an effective SOP. The JHA process pulls together the specific details of the job and provides an assessment of each step and the tasks required to complete each step. Through the use of the Cause and Effect Diagram, the Risk Assessment, and Hierarchy of Controls, the scope and severity of the risk and hazards are determined. Controls are determined and implemented. Following the JHA process in addition provides a strong foundation for the SOP development.

The rationale for developing an SOP is to detail the job steps and tasks and highlight specifically how to do the job through the use of essential descriptions that include diagrams, pictures and other techniques to transfer the information. The JHA's primary focus is on the control, reduction or removal of potential or existing hazards and identifying the consequences of exposure. The SOP combines with all other elements of the job under review. The SOP details the quality control requirements, criteria for the service or specific product production or delivery, the time requirements, personnel requirements and all other details necessary to do the job.

While the pre-assessment and the JHA tie nicely into the overall operational goals and objectives of the organization, your organization may not use or have an SOP process. If that is the case, the JHA may be the stopping point providing only for the safe completion of the job. However, in the course of completing the JHA, you will have developed an in-depth assessment of how that particular job is currently being completed. The possibility exists that the job is not being completed in an efficient manner and indicators come to light that show training, quality, efficient use of time, materials, or personnel really need an overhaul.

As the job cause and effect diagrams are developed, you may hear team participants debating the sequence of steps or leaving out steps based on observations. The tasks necessary to complete each step may be glossed over or may bring to light ergonomic requirements that impact the hiring process. The interrelationships between the environments, the tools/equipment/materials, and who is doing the job may be vague, confused or misunderstood. Further, the job may be found to have developed potential legal liability issues that have opened the organization up for lawsuit or regulatory problems. These are all indicators that the job is not only being completed in an unsafe, inefficient, or ineffective manner, but is also having a negative impact on costs and profits through poorly designed procedures.

The JHA purposely keeps its text limited and terse as it brings together job details. The SOP is essentially more complete and complex as it must fully develop the scope of the job. Step 1 is to compare the current SOP to the final JHA. A SOP review is part of the original job data research. Using the final JHA, gaps and problems with the SOP can be assessed. SOPs have the same problems that JHAs face. They may be out of date, or the job methods, technology, tools, people, etc. may have changed. "Scope drift" may have occurred, where the job as initially designed has gradually changed for various reasons. Management, supervision and employees may each have a different perspective on how the job is to be completed. Using the JHA process brings clarity to the current job and its needs.

Developers of the current SOPs may not have involved the safety professional in their layout and development. The JHA process reduces the potential to leave out core safety criteria. If an SOP process is used for job basic training, the safety professional must work to ensure that an alignment of the JHA and the SOP is present. If this is not done, safety training may inadvertently be at odds with the "official" job training criteria and delivery. Refer to Appendix P for a sample SOP format. If you want someone to learn the hazards of the job you need to make the training readily understood and clearly coordinated with all other disciplines that provide training.

Areas where the JHA assists the SOP process include:

- Job orientation criteria.
- Enhances training criteria.
- Aligns steps/tasks, tools/equipment/materials, job environment, current policies/procedures, employee performing the task.
- Points out potential gaps in efficient task completion.
- Allows updating of inspection forms, checklists, guidelines and job observation criteria.

## 11.2 ELEMENTS OF AN SOP

What if your organization does not use the SOP process? Step 2 is to determine from management if the implementation of an SOP process would benefit the company. If the decision is made to not develop a formal SOP process, then the JHA becomes the documentation of the safety requirements. A management decision is made and no further action is taken. The JHA process becomes a de facto SOP. In this case you still may need to provide additional detail to assure the JHA criteria are implemented. As Step 3 in this case, you would develop a "Safe Job Procedure," limiting its main focus to the safety criteria and select a standard operating procedure format that best suits your safety program needs.

We are not suggesting that you develop an SOP for each job. What we are suggesting is that you review those job tasks that are deemed a high priority based on your risk assessment and build formal procedures that will tie into and use the JHA as a backup material.

The SOP is a narrative or written summary of the JHA worksheets. The general criterion for an effective SOP includes:

- Write the SOP in a detailed, step-by-step format. Usually this means writing a number of short paragraphs that describe the steps and tasks. A reality check is to have those employees that do the job read and make comment. Can the SOP be used? Is it too complex? Does it accurately reflect how the job is completed and clearly indicate where hazards exist and how they are to be controlled?

- If there are no hazards or possible at-risk events in a specific task, after fully ensuring that is the case, simply state the actions to be taken in that step.
- If a potential and existing hazard does exist in a step, state the action and identify the consequences of exposure to the following elements:
  1. The hazard(s), clearly and accurately describing the hazards(s).
  2. The possible injury or damage it could cause. Identify the severity of any loss-producing event that could develop.
  3. Preventative measures to prevent the injury or damage. State the types of controls, regulatory compliance references and details and actions mandated for this job.
- Try to paint a word picture. The description of the job must be in terms that can be readily understood and at the level of the reader. To the degree possible, make it "concrete" and avoid "abstract" wording.
- Write in the active voice, i.e. "take," not "should be taken."
- Write as clearly as possible using simple words, i.e. *"use,"* not *"utilize."* Use the old KISS principle—"Keep It Simple and Streamlined"—as much as possible.
- Provide detail and attempt to keep sentences short. Use no more than 7–15 words.
- Try to write in a less technical, more conversational style. The objective is that the safety criteria be clearly presented. The document is to be used in training, for disciplinary actions, legal defense and a number of other uses.

The U.S. Environmental Agency suggests the following format for SOPs. Depending on the complexity of the job, elements such as Title page, Table of Contents or a Summary may not be needed. Keep it simple! As we stated above with the JHA development, if the number of steps begins to exceed 15 or so, you should consider breaking the SOP into several parts. If the JHA process has been followed, the following will have already been considered:

- Title Page
- Table of Contents
- Purpose
- Applicability (when the SOP procedure is to be followed)
- Summary
- Definitions (defining words, phrases, or acronyms)
- Personnel Qualifications/Responsibilities
- Procedure
- Criteria, checklists, or other standards
- Records Management (forms to be used and locations of files)
- Quality Control and Quality Assurance Section
- Reference Section

The definition of Insanity: "If you do the same things over and over you will get the same results."

—Deming

The JHA process must be fully incorporated into the SOP of the organization. Stressed throughout this book is the effort to involve all levels of the organization, management, supervision and employees. At this stage, problems and issues have been brought to light and, using the JHA process, controls and job modifications can be assessed and implemented.

## 11.3 SUMMARY

The JHA and SOP provide the data not just for safety improvement but also provide the foundation for overall job improvement. By leveraging the rapport developed between all the employees involved, providing visual aids and structured assessments, the implementation of improved safety criteria has a greater potential for success. The safety profession now has a full process from data gathering to final operating efficiency.

## CHAPTER REVIEW QUESTIONS

1. What is the rationale for developing an SOP?
2. What is the general criterion for an effective SOP?

## REFERENCES

1. Adapted from Oregon OSHA Website, Job Hazard Analysis, http://www.cbs.state. or.us/external/osha/pdf/workshops/103w.pdf, public domain
2. *The Leader's Handbook*, Peter Scholtes, McGraw Hill 1998

## ADDITIONAL REFERENCES

1. *Guidance for Preparing Standard Operating Procedures* (SOPs) EPA QA/G-6; U.S. Environmental Protection Agency Quality System Series, March 2001; http://www.epa.gov/quality/qs-docs/g6-final.pdf, public domain

2. Cornell University Environmental Safety and Health Standard Operating Procedures: http://www.ehs.cornell.edu/lrs/chp/12.sop.form.intro.htm

3. Kenneth Friedman, Ph.D., Department of Journalism and Communication, Lehigh University, Bethlehem, Pa., http://www.lehigh.edu/~kaf3/sops/sop_index.html

4. Guide to Developing Effective standard Operating Procedures for Fire And EMS Departments, Federal Emergency Management Agency United States Fire Administration; http://www.usfa.dhs.gov/downloads/pdf/publications/fa-197.pdf, public domain

# Appendix P

## P.1 STANDARD OPERATING PROCEDURE FOR SPLICING: {INSERT SIMPLE INSTRUCTIONS}

1. Car has flat tire

> Insert picture that fits this instruction

2. Pull car off of roads

> Insert picture that fits the instructions

3. Prepare to remove tire

---

Insert picture that fits the instructions

---

**When all steps and tasks are defined, the procedure starts over at Step #1.**

# Part 4

## Additional Tools That Can Be Used to Develop a Successful JHA

# 12

## Overview of a Safety Management Process

A number of models are available that can be followed to both assess and develop a safety management process. This chapter provides an overview of the OSHA Voluntary Protection Program (VPP). At the end of the chapter you will be able to:

- Identify the basic criteria for VPP
- Identity the warning signs for system issues
- Discuss the PDSA (Plan-Do-Study-Act) method for assessing a problem.

"To provide the difference we may need to consider some alternative ideas about what we know; to what we have always believed about safety; about what works and what does not. I cannot conceive of any organization getting a handle on its safety problems without turning in its old ideas and beliefs first."

—Dan Petersen

"Unless commitment is made, there are only promises and hopes . . . but no plans."

—Peter Drucker

"The first responsibility of a leader is to define reality. The last is to say thank you."

—Max DePree

When you manage any type of business you take many risks. You wager on your business success against other organizations or investors, compete for new employees, and work to keep current employees. With all of the efforts for attention, employee safety can be an afterthought and is not considered as part of the overall business plan.

"Honesty is the cornerstone of all success, without which confidence and ability to perform shall cease to exist."

—Mary Kay Ash

To be successful your JHA process must rest within a more comprehensive safety management system. Several OSHAs management systems can be implemented with a level management commitment. One system is the "Five-Point Workplace Program" based on the Safety Process Management Guidelines originally issued by OSHA. These guidelines are voluntary and represent OSHA's policy on what each worksite should have in place to protect employees from work-related hazards. The guidelines are based on OSHA's experience with the Voluntary Protection Program (VPP) and are designed to help a business to recognize and promote effective safety management as means of ensuring a safe workplace.

OSHA encourages organizations to implement and maintain a system that provides systematic policies, procedures, and practices that are adequate to protect employees from safety hazards. An effective system identifies provisions for the systematic identification, evaluation, and prevention or control of hazards in the workplace, specific job hazards, and existing and potential hazards and/or consequences of exposures that may arise from foreseeable conditions. Compliance with OSHA standards is an important objective. If you develop a successful safety management system, compliance becomes a non-issue because a good safety culture will tend to fix the issues when identified [1].

There must to be a strong management commitment to the safety process, as with any successful management system. Refer to Figure 12-1 for an overview

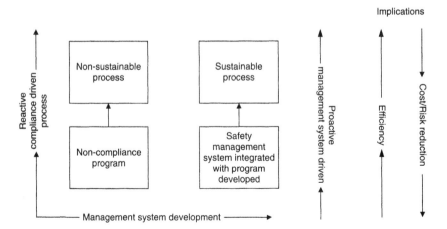

**Figure 12-1** Journey to a Successful Management System With a Sustainable Process

of a successful safety management system. This chapter will help you get started on the right track to developing this successful safety management system.

## 12.1 PROCESS ELEMENTS

If you are not familiar with the VPP program, we encourage you to review that program framework. For more details on VPP refer to OSHA's website http://www.vpppa.org/About/index.cfm.

### 12.1.1 What Are the Voluntary Protection Programs?

Voluntary Protection Programs (VPPs) represent OSHA's efforts to extend worker protection beyond the minimum required by OSHA standards. VPP, along with onsite consultation services, full-service area offices, and OSHA's Strategic Partnership Program (OSPP) represents a cooperative approach which, when coupled with an effective enforcement program, expands employee protection to help meet the intended goals of the OSH Act [2, 4].

### 12.1.2 How Does VPP Work?

There are three levels of VPP recognition: Star, Merit, and Demonstration. All levels are designed to:

- Recognize employers who have successfully developed and implemented effective and comprehensive safety and health management systems;
- Encourage these employers to continuously improve their safety and health management systems;
- Motivate other employers to achieve excellent safety and health results in the same outstanding way; and
- Establish a relationship between employers, employees, and OSHA that is based on cooperation [2, 3, 4].

Refer to Section 12.15, Voluntary Protection Program, for a detail discussion.

### 12.1.3 How Does VPP Help Employers and Employees?

According to OSHA, VPP participation can mean the following to an organization:

- Reduced numbers of worker fatalities, injuries, and illnesses;
- Lost-workday case rates generally 50 percent below industry averages;

- Lower workers' compensation and other injury- and illness-related costs;
- Improved employee motivation to work safely, leading to a better quality of life at work;
- Positive community recognition and interaction;
- Further improvement and revitalization of already-good safety and health programs; and a
- Positive relationship with OSHA [4].

OSHA suggests a five-point workplace program to create a successful management system. OSHA's VPP management and employee participation is complementary and forms the core of an effective safety system. We will discuss management and employee participation as individual elements. There is a clear and distinct difference between management of the operation and employee participation. Refer to Figure 12-2 for key elements to a successful safety culture. If one part of the system is missing, it represents a serious gap that must be addressed. Refer to Figure 12-3 for an overview of the key elements in a successful safety culture. The following are the management process core elements that are common to all VPPs:

- Management leadership.
- Employee participation.
- Hazard identification and assessment (refer to Chapters 1, 2, 3).
- Hazard prevention and control (refer to Chapters 1, 2, 3).
- Information and training (refer to Chapter 7).
- Evaluation of process effectiveness.

Effective management of employee safety is a decisive factor in reducing the extent and severity of work-related injuries. An effective management system addresses work-related hazards, including existing or potential hazards and/or their consequences of exposures that could result from workplace conditions or practices. In addition, it addresses hazards that are not regulatory driven [2].

Whether or not a safety process is in writing, it is important that it is effectively implemented, managed, and practiced. It should be obvious that, as the size of the organization, the number of employees, or the complexity of an operation increases, the need for written guidance will increase. The process should ensure there is clear communication of consistent application of policies and procedures to all employees [1].

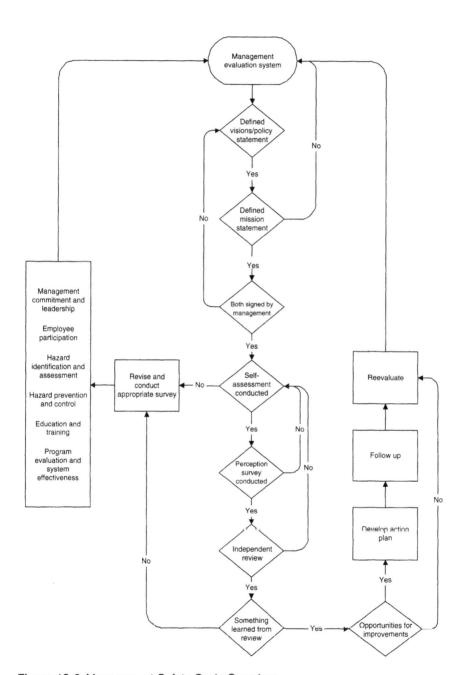

**Figure 12-2** Management Safety Cycle Overview

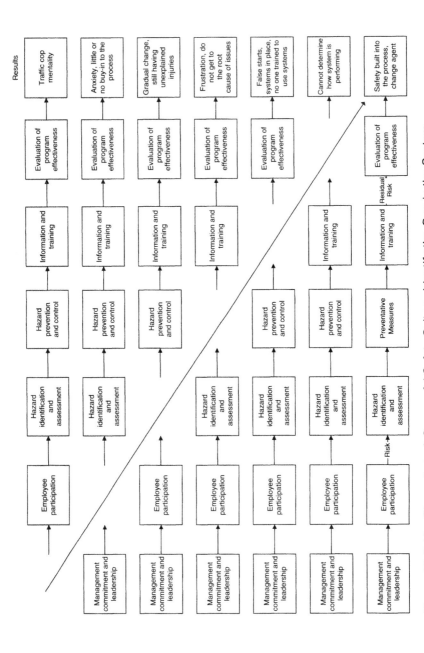

**Figure 12-3** Key Elements to A Successful Safety Culture: Identify the Gap in the System

OSHA Web site, http://www.osha-slc.gov/SLTC/safetyhealth_ecat/mod3.htm, public domain

Roughton, James, James Mercurio, *Developing an Effective Safety Culture: A Leadership Approach*, Butterworth-Heinemann, 2002

## 12.2 MANAGEMENT COMMITMENT AND LEADERSHIP

Management commitment and leadership means from the top leadership down through all ranks. If management demonstrates commitment, provides the motivating force, and the needed resources to manage safety, an effective system can be developed and sustained. This demonstration of management commitment and leadership should include the following elements, which are consistent with an effective safety process:

- Establishing the responsibilities of managers, supervisors, and employees for safety and holding them accountable for carrying out assigned responsibilities.
- Providing managers, supervisors, and employees with the authority, access to relevant information, education and training, and resources needed to carry out their safety responsibilities.
- Identifying at least one manager, supervisor, or qualified employee to receive and respond to reports about safety conditions and, where appropriate, to initiate corrective action [2, 3, 4].

We use the word "demonstration" to ensure that all individuals, not just the production employees, understand operational hazards [3]. This is an important concept no matter what you are trying to accomplish; always "walk-the-walk" and "talk-the-talk."

> "Every sale has five basic obstacles: no need, no money, no hurry, no desire, no trust."
>
> —Zig Ziglar

## 12.3 EMPLOYEE PARTICIPATION

As we discussed in Chapter 5, employees should be provided an opportunity to participate in establishing, implementing, and evaluating the safety process. Employee participation provides the means that allows them to develop and/or express their safety commitment to themselves and/or their fellow employees. To fulfill and enhance employee participation, management should implement the following elements:

- Regularly communicating with all employees concerning safety matters.
- Providing employees with access to information relevant to the safety system.

- Providing ways for employees to become involved in hazard identification and assessment, prioritizing hazards, education and training, and participation in the safety management system.
- Establishing procedures where employees can report work-related injury promptly and ways they can make recommendations about appropriate solutions to control the hazards identified.
- Providing prompt responses to reports and recommendations.

It is important to remember that under an effective safety management system employers do not discourage employees from reporting safety hazards, or from making recommendations about near misses or hazards, or from participating in the safety process.

## 12.4 HAZARD IDENTIFICATION AND ASSESSMENT

As discussed through the JHA process, hazard analysis of the work environment involves a variety of elements to identify hazards changes that might create new hazards. The following measures are recommended to help identify hazards:

- Conducting comprehensive baseline workplace assessments, updating assessments periodically, and allowing employees to participate in the assessments.
- Analyzing planned and/or new facilities, process materials, and equipment.
- Assessing risk factors of all applications related to the employees tasks.
- Conducting regular site safety inspections so that new or previously missed hazards are identified and corrected.
- Providing a reliable system for employees to notify management about conditions that appear hazardous and to receive timely and appropriate responses.
- Encouraging employees to use the notification system without fear of reprisal. This system utilizes employee insight and experience in safety and allows employee concerns to be addressed.
- Investigating injuries and "near misses" so that the root causes and means of prevention can be identified and an action plan developed to eliminate the hazard.
- Analyzing injury trends to identify patterns with common causes so that they can be reviewed and prevented [2].

Hazards should systematically be identified and evaluated. This evaluation can be accomplished by assessing the following activities and reviewing safety information:

- The establishment's injury experience, OSHA 300 logs.
- Workers' compensation claims (Employers First Report of Injury).
- First aid logs.
- Results of any medical screening/surveillance with proper medical privacy regulations maintained.
- Employee safety complaints and reports.
- Environmental and biological exposure data (industrial hygiene).
- Information from workplace safety inspections.
- Materials Safety Data Sheets (MSDSs).
- Results of employee safety perception surveys.
- Safety manuals, guidelines, and materials.
- Safety warnings provided by equipment manufacturers, chemical, and other suppliers.
- Safety information provided by trade associations or professional safety organizations.
- Evaluating new equipment, materials, and processes for hazards before they are introduced into the workplace.
- Assessing the severity of identified hazards and ranking those that cannot be corrected immediately according to their severity [2].

It is important to evaluate regulatory requirements that may impose additional and specific requirements for hazard identification and assessment. The hazard identification and assessment analysis should be conducted as follows:

- As often as necessary to ensure that there is compliance with specific requirements and Best Management Practices (BMP).
- When workplace conditions change that could create a new hazard or when there is increased risk of hazards.

## 12.5 HAZARD PREVENTION AND CONTROL

As we discussed in Section 1, effective planning and design of the workplace or job task can prevent hazards. Where it is not feasible to eliminate hazards, action planning can control unsafe conditions.

Elimination or control should be accomplished in a timely manner once a hazard or potential hazards are identified. The following procedures are examples of measures to be taken:

- Using engineering techniques where feasible and appropriate.
- Establishing safe work practices and procedures that can be understood and followed by all affected employees.
- Providing PPE when engineering controls are not feasible.
- Using administrative controls: for example, reducing the duration of exposure.
- Maintaining the facility and equipment to prevent equipment breakdowns.
- Planning and preparing for emergencies, and conducting training and emergency drills, as needed, to ensure that proper responses to emergencies will be "second nature" for all employees involved.
- Establishing a medical surveillance process that includes handling first aid cases onsite and off-site at a nearby physician and/or emergency medical care to help reduce the risk of any incident that may occur [2].

Once identified, an action plan should be developed to help solve the issues, or to come into compliance with applicable requirements. These plans can include setting priorities and deadlines and tracking progress in controlling hazards.

## 12.6 EDUCATION AND TRAINING

As discussed in Chapter 7, safety training is an essential component of an effective safety process. This training should address the roles and responsibilities of both the management and the employees. It will be most effective when combined with other training on performance requirements and/or job practices. The complexity depends on the size and the nature of the existing and potential hazards and/or consequences of exposures.

The following section provides a brief explanation of specific training. You should review your operation and expand on the brief summary. Refer to Chapter 7 for a detailed treatment of safety training.

### 12.6.1 Employee Training

Employee training programs should be designed to ensure that everyone understands and is aware of the hazards that they may be exposed to, and the proper methods for avoiding such hazards.

## 12.6.2 Management Training

Management must be trained to understand the key role and responsibilities that they have in safety and to enable them to carry out their safety responsibilities effectively. A training program for management should include the following elements:

- Analyzing of the work under their supervision to anticipate and identify existing and potential hazards and/or consequences of exposure.
- Reinforcing employee training on the nature of potential hazards associated with their work and on protective measures. The reinforcement is done through continual positive feedback and, as necessary, through enforcement of safe work practices, such as tool box meeting or sometimes called 2-minute drills, one-on-one contacts, etc.
- Understanding their safety responsibilities.

Anyone who has responsibilities for education and training should be provided the level of training necessary to carry out their safety responsibilities.

## 12.7 EVALUATION OF PROCESS EFFECTIVENESS

The safety management system should be evaluated to ensure that it is effective and appropriate to address specific workplace conditions. As the result of the review, the system should be revised in a timely manner to correct any deficiencies as identified.

## 12.8 THE NATURE OF ALL SAFETY SYSTEMS

The essential key is to fix the system and not place the blame. Many of us lose sight of the system because we tend to operate in a reactionary mode [3]. Refer to Table 12-1 on the nature of all safety systems. This table provides an overview of the inputs, processes, and outputs as the centerpoint of a management system.

## Table 12-1
## Nature of All Safety Systems

| |
|---|
| **Inputs**: Act as antecedents or activators (Scott Geller) – Clarify the process<br><br>• Standards = Vision, mission, objectives, strategies, policies, plans, programs, budgets, processes, procedures, appraisals, job descriptions, and rules.<br>• People = Management and employees.<br>• Resources = Tools, equipment, machinery, materials, facilities, and environment. |
| **Process**: The steps in the transformation of inputs into outputs.<br><br>• Procedures = Steps in a process.<br>• Consequences = Natural (injuries), System (reward/discipline), and exposure (risk). |
| **Outputs**: Results of a process.<br><br>Products and services = The intended system output<br><br>• Conditions = Objects/employees in the workplace. The "states of being" in the physical and psychosocial workplace environment<br>• Behaviors = Action of employees taken at all levels of the organization. What they do or do not do in the workplace affects safety. |
| Adapted from OR-OSHA web site, http://www.cbs.state.or.us/external/osha/educate/training/pages/materials.html, OR-OSHA 116, Safety and Health Program Evaluation, Rev. 1/00 sig, public domain |

Refer to Figure 12-4 for a sample of the inputs, outputs, and consequences of a management system. Refer to Figure 12-5 for an overview of a Risk Assessment Review Process, as used in the JHA process.

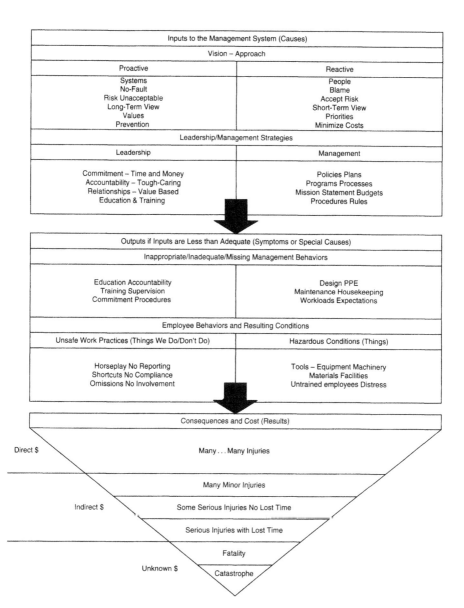

**Figure 12-4** Inputs, Outputs, and Consequences of a Management System

Adapted from Inputs, Outputs, and Consequences of a Management System High Level, The Big Picture: Inputs, Outputs, Consequences, OR OSHA 119, pp. 29, http://www.cbs.state.or.us/ external/osha/educate/training/ pages/materials.html, public domain

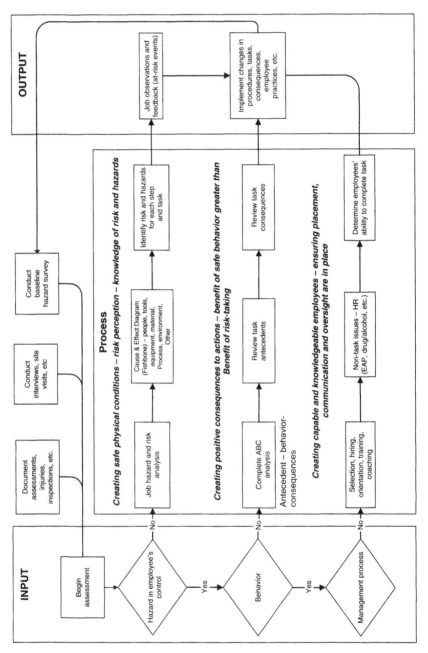

**Figure 12-5** Risk Assessment Review Process, Lower Level Process

## 12.9 INDICATORS AND MEASURES

Indicators are the metrics that relate to the elements and describe how to tell if the sub-element is in place and working.

Managers and employees should conduct an assessment of the management system together and develop a joint rating. It may be important to have a selected number of employees to evaluate the system and use sampling techniques to verify that the sample size or number to be used is large enough to validate findings. If you talk with 20 employees and all 20 say the same thing, you can only make the assumption that most employees support that position if the sample size is statistically viable. Whatever you do, be flexible with how you establish your evaluation.

[OSHA web site, http://www.osha-slc.gov/SLTC/safetyhealth_ecat/asmnt_worksheet_instructions.htm, public domain]

Reviewing and revising goals and objectives is critical to achieving your vision of a safe workplace. To maintain a high-quality process demands continuous improvement; therefore organizations should never be complacent about safety. A periodic review of each program component will help you to achieve and maintain the vision reflected in your policy statement [6].

## 12.10 ASSESSMENT TECHNIQUES

Many methods are available to help assess the management system and safety program effectiveness. The basic methods for assessing the management system effectiveness include:

- Reviewing documentation of specific activities (safety committees, business contracts, activity-based safety system, behavior-based safety, etc.)
- Interviewing employees in the organization for their knowledge, awareness, and perceptions of what has happen to the safety management system. One method will be the employee safety perception survey.
- Reviewing site conditions for hazards and finding opportunities for improvement in the management systems that allowed the hazards to occur or to be uncontrolled [2].
- Arrange for an independent review.

Evaluations can provide a grading system, so that each year's results can be quantitatively compared to previous years. This report should be available to all employees and should be written in a way that is understandable to all employees. Only use terms understood and used on a daily basis.

## 12.11 MULTI-EMPLOYER WORKPLACE

In a multi-employer workplace, the primary responsibility for the host employer is to:

- Provide information about potential hazards, company controls, safety rules, and emergency procedures to all employers.
- Make sure that safety responsibilities are assigned to specific employees as applicable.

The responsibility of the on-site contractor is to:

- Ensure that the host (controlling) employer is aware of the hazards associated with the contractor's work and any identified hazards are being addressed.
- Advise the host or controlling employer of any previously unidentified hazards that the contract employer identifies at the workplace.

## 12.12 EMPLOYEE RIGHTS

Employees have the right to report to their employers, their unions, OSHA, or other governmental agency about workplace hazards. Section 11(c) of the OSHA Act of 1970 makes it illegal for employees to be discriminated against for exercising their rights and for participating in job safety-related activities. These activities include:

- Complaining individually or with others directly to management concerning job safety conditions
- Filing of formal complaints with government agencies, such as OSHA or state safety agencies, fire departments, etc. (An employee's name is kept confidential.)
- Participating in union committees or other workplace committees concerning safety matters
- Testifying before any panel, agency, or court of law concerning job hazards
- Participating in walk-around inspections
- Filing complaints under Section 11(c) and giving evidence in connection with these complaints [2].

Employees also cannot be punished for refusing a work assignment if they have a reasonable belief that it would put them in danger or cause them serious physical harm, provided that they have requested that the employer remove the danger and the employer has refused; and provided that the danger cannot be eliminated quickly enough through normal OSHA enforcement procedures.

If an employee is punished or discriminated against anyway for exercising their rights under the OSH Act, the employee can report it to OSHA within 30 days. OSHA then is to investigate the report. If the employee has been illegally punished, OSHA can seek appropriate relief for the employee and if necessary can go to court to protect the rights of the employee.

## 12.13 HEALING A SICK SYSTEM

A number of reasons may exist that allow a safety management system to suffer or fail to produce an injury-free environment. It is important to implement an effective safety action plan to ensure that the "prognosis" for all systems in an organization remains positive. A safety process evaluation is used to determine the "wellness" of the safety system [3]. Refer to Table 12-2 for symptoms of an ailing system.

### Table 12-2
### Symptoms, Obvious and Underlying Warning Signs to the Cause

| |
|---|
| Primary Symptoms – Specific hazardous unsafe conditions or At-risk events. <br><br> • Conditions are allowed to exist at the site of the potential injury and at-risk events are performed by the potential victims (employee). <br> • The result of poor safety process (no management system) design or implementation. |
| Secondary Symptoms – Underlying hazardous unsafe conditions and At-risk events. <br><br> • Due to inadequate implementation of policies, plans, processes, programs, and procedures, at-risk events are performed by employees; co-workers, supervisors, and middle management contribute to, or produce the primary symptoms. <br> • May exist and/or occur in any department. |

Adapted from OR-OSHA web site, http://www.cbs.state.or.us/external/osha/educate/training/pages/materials.html, OR-OSHA 116, Safety and Health Program Evaluation, Rev. 1/00 sig, public domain

## Table 12-3
## Causes: Poor Corporate "Nutrition" Producing the Symptoms

| |
|---|
| System Root Causes – Management Process defects<br><br>• Are unknowingly formulated by upper management from inadequate design of policies, plans, processes, programs, and procedures.<br>• Produce secondary symptoms.<br>• May exist in programs in any department. |
| Deep Root Causes – Unsupportive vision, mission, strategies, objectives<br><br>• Formulated by top management commitment and discussions with employees. Create reactive leadership, an unattainable vision, mission, strategies, or objectives, and poor budgets allow product and program design defects and allow harmful factors external to the organization - materials, industry, community, society, etc. |
| Adapted from OR-OSHA web site, http://www.cbs.state.or.us/external/osha/educate/training/pages/materials.html, OR-OSHA 116, Safety and Health Program Evaluation, Rev. 1/00 sig, public domain |

Refer to Table 12-3 for an overview of causes, with poor corporate "nutrition" producing the symptoms.

When developing a successful safety management system, you must first determine what gap is present between where you are and where you should be. To ensure that your system changes result in the effects you intend, you must understand and see the opportunities for improvements as outlined in the following sequence:

• Determine where you are now by analyzing your management system

  *Where are you now? What are your program metrics telling you?*
  *What does the safety system look like now? Describe it in detail.*

• Decide where you want to be in the future. Remember, "Begin with the end in mind, Steven Covey."

  *Have a vision where you want to be in the next 3–5 years.*
  *What do you want your management system to look like?*

• Understand the opportunity for improvement; evaluate the safety management system

  *Hold to the vision that there is always a better way*
  *What cultural values are missing? Note: we did not state that safety is a "priority."*
  *What system components are inadequate?*

• Develop an action plan to get from "here" to "there." [3]

## 12.14 THE PDSA CYCLE

Let's take a look at a simple management system approach called the Shewhart cycle, or PDSA cycle (for Plan-Do-Study-Act) for learning and improving. Refer to Figure 12-6 for the Shewhart Process Cycle [8]. This cycle has been adapted within the ANSI Z10-2005 Occupational Safety Process.

**Step 1: Plan**
*Design the change or test*

• Purpose: Take time to thoroughly plan the proposed change before it is implemented.
• Pinpoint specific conditions, behaviors, and the results you expect to see as a result of the change.
• Plan to make sure that there is a successful transition to the changes desired.

**Step 2: Do**
*Carry out the change or test*

• Purpose: Implement the change or test on a small scale to see how it works.
• Educate, train, and communicate the change to all affected employees. This will help all affected employees transition to the new programs, procedures, etc.
• Keep the initial changes small so that variability can be measured.

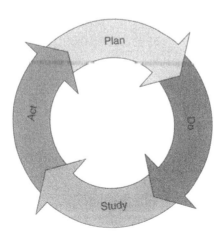

**Figure 12-6** Shewhart, PDSA Cycle for Learning and Improvement

**Step 3: Study**
*Examine the effects or results of the change or test*

- Purpose: To determine what was learned: what went right or wrong.
- Statistical process analysis, surveys, questionnaires, interviews.

**Step 4: Act**
*Adopt, abandon, or repeat the cycle*

- Purpose: Integrate what works into the safety system.
- Ask: are you doing the right things, as well as ask if you are doing things right.
- If the result was not as intended, abandon the change or begin the cycle again with the new knowledge gained [3].

In Chapter 13 we will show you how to go beyond the cycle using Six Sigma.

## 12.15 VOLUNTARY PROTECTION PROGRAM

One method to ensure a safety management system complies with OSHA mandates is to become an OSHA Voluntary Protection Process (VPP) member. Refer to Table 12-4 for a summary of what it takes to certify under the VPP process. Refer to Figure 12-7 for a flow diagram of the VPP system. This

**Table 12-4**
**4 Voluntary Protection Process**

| |
|---|
| Pre-Application Stage |
| 1. VPP Education: Management and Employees |
| 2. Communicate VPP to All Employees <br><br> • Management's roles and responsibilities <br> • Employees roles, responsibilities and rights <br> • Union's roles, responsibilities, and rights |
| 3. VPP Mock Evaluation/Assessment <br> Process deficiencies identified and corrected <br><br> • Major deficiencies – 6 to 12 months <br> • Major deficiencies – up to 6 months <br> • Major deficiencies – up to 3 months |

**Table 12-4**
*Continued*

| |
|---|
| Application Stage |
| 1. Correct Process Deficiencies |
| • 3 months<br>• 6 months<br>• 12 months<br>• Longer as applicable |
| 2. Prepare Application |
| • Applications to read like a "Resume"<br>• Every item in application to have documented proof/verification. |
| 1. Submit Application |
| • Three-ring binder with tabs by subject<br>• Make 5 copies<br>• Send certified mail. |
| **Post-Application Stage** |
| 2. Contact OSHA office |
| • Determine evaluation date<br>• Determine team<br>• Determine who receives application |
| 1. Review Process |
| • Implement revisions<br>• Document revisions |
| 2. Prepare for VPP Evaluation |
| • Provide conference room<br>• Provide computer<br>• Organize documents<br>• Prepare list of all employees for interviews<br>• Provide VPP team members w/escorts for plant walk-around. |

## Table 12-4
### *Continued*

Evaluation Stage

1. Introduction meeting

- Introduce OSHA VPP team
- Let OSHA VPP team discuss the logistics of this evaluation
- Present overview of organization
- Management structure
- Safety and health process
- Safety team members
- Request daily briefing and draft pre-approval report

2. Evaluation

- Provide all documents in conference room
- Provide escorts
- Provide lunch

1. Closing Conference

- CELEBRATE that it is over

2. Post-Evaluation Stage

- Merit (with plan) approval
- Develop plan of action (POA)
- Monitor and review progress
- Document POA and activities
- Submit POA proof
- Provide for VPP evaluation

1. Star Approval

- Monitor safety and health process quarterly
- Conduct annual VPP safety and health process review and revise as needed
- Prepare for VPP evaluation.

Refer to Figure 12-7 for a flow diagram of the VPP system. Note that each number corresponds to each item as listed in Table 12-4 to make it easy to read.

Roughton, James, James Mercurio, *Developing an Effective Safety Culture: A Leadership Approach*, Butterworth-Heinemann, 2002

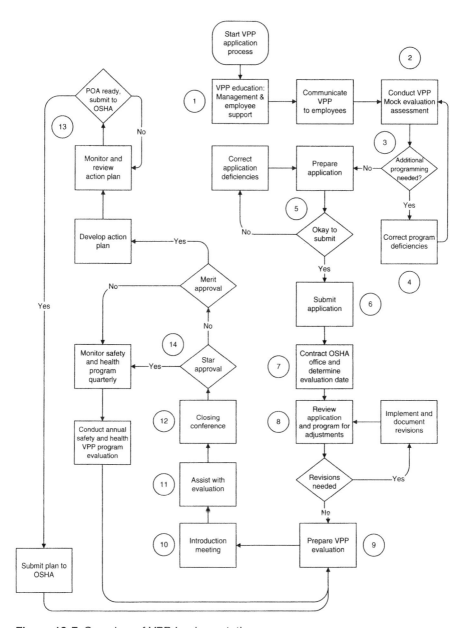

**Figure 12-7** Overview of VPP Implementation

Adapted from Roughton, James, James Mercurio, "Developing an Effective Safety Culture: A Leadership Approach," Butterworth-Heinemann, 2002

diagram summarizes the information as listed in Table 12-4 and helps you to apply for and successfully obtain VPP status.

## 12.16 SUMMARY

This chapter has defined an overview of what a successful safety management system should look like. It has outlined what should be evaluated, who should do the evaluation, what tools can be used, how the evaluation should be conducted, how to use the results, and how to heal a sick system.

The following is a line-by-line summary of what we discussed in this chapter. By following the elements presented, you will be able to establish a successful safety management system:

- Providing visible top management involvement in implementing and sustaining the safety management system so that all employees understand management's commitment is serious.
- Arranging for and encouraging employee participation in the structure and operation of the safety management system and in decisions that affect the employee's safety. This will help to commit their insight and energy to achieving the safety process goals and objectives.
- Clearly stating a policy and/or vision statement on safety expectations so that all employees can understand the value of safety activities and programs.
- Establishing and communicating a clear goal for the safety process and defining objectives for meeting the established goals so that all employees understand the desired results and measures planned for achieving them.
- Assigning accountability and communicating responsibility for all aspects of the safety process so that managers, supervisors, and employees know what performance is expected.
- Holding managers, supervisors, and employees accountable for meeting their safety responsibilities so essential tasks will be performed.
- Providing adequate authority and resources to responsible parties so that assigned duties can be met.
- Reviewing management system elements at least annually to evaluate their effectiveness in meeting the intended goals and objectives so that deficiencies can be identified and the process and/or the objectives can be revised when they do not meet the goal of an effective safety process.

Although compliance with specific OSHA requirements is an important objective, an effective management system looks beyond specific requirements and targets the development of an effective safety culture. For a detailed review of how an effective safety culture can be accomplished refer to my book, *Developing an Effective Safety Culture: A Leadership Approach*, published by Butterworth-Heinemann in 2002.

## CHAPTER REVIEW QUESTIONS

1. What does OSHA encourage organizations to implement?
2. What is VPP?
3. What are the three levels of VPP recognition?
4. What are the core elements of a management process?
5. What are three parts of a safety system?
6. In a multi-employer workplace what is the primary responsibility of the host employer?
7. What are a system's primary and secondary warnings?
8. What are the main parts of the Shewhart cycle?
9. What are the elements of a practical hazard analysis?
10. After the closing for a successful VPP effort, what should you do?

## REFERENCES

1. Oklahoma Department of Labor, Safety and Health Management: Safety Pays, 2000, http://www.state.ok.us/~okdol/, Chapter 2, pp. 12–15, public domain
2. OR-OSHA web site, http://www.cbs.state.or.us/external/osha/educate/training/pages/materials.html, OR-OSHA 116, Safety and Health Program Evaluation, Rev. 1/00 sig, Modified with Permission, public domain
3. OSHA Web site, http://www.osha-slc.gov/SLTC/safetyhealth_ecat/comp1_review_program.htm#, public domain
4. Roughton, James, James Mercurio, *Developing an Effective Safety Culture: A Leadership Approach*, Butterworth-Heinemann, 2002
5. The Joint Commission, http://www.jointcommission.org/PerformanceMeasurement/PerformanceMeasurementSystems/, public domain

# 13

# Six Sigma as a Management System: A Tool for Effectively Managing a JHA Process

Six Sigma is a process that brings together an array of problem-solving and assessment tools and concepts. It can be a formal process, and also can provide a great toolkit that can be used to improve the JHA process. At the end of the chapter you will be able to:

- Discuss the basic framework used by Six Sigma
- Discuss the process improvement criteria
- Identify what DMAIC stands for and how it can be used
- Discuss the XY Matrix and how it can be used with the JHA
- Discuss various tools used in Six Sigma
- Cite the levels of Six Sigma and what they mean.

"If you break it, you own it,"

—Colin Powell

When talking about the Six Sigma Black Belt certification, the response often is: "Just what the heck is Six Sigma?" and "How does it apply to safety?" This chapter provides a brief overview of the Six Sigma process and a snapshot of the tools that can be used in JHA development. It is not within the scope of this book to go into Six Sigma in detail. However, Six Sigma philosophy and methods can become a major contributor to the study of jobs and their analysis.

## 13.1 SIX SIGMA EXPOSED

This chapter is intended for those who want to go beyond traditional compliance and basic programs. We introduced a number of new concepts used in developing a JHA. These new concepts will help to make the JHA a centerpiece of the safety process for everyone reading this book. However, the methods used to get to this point can appear difficult without the proper structure. In this chapter, we will present an overview of Six Sigma and how you can use it to logically collect the data for your JHA [1].

In keeping with our "how to" approach, this chapter will provide a basic foundation for using the Six Sigma tools to solve safety issues. The discussion of Six Sigma methodology in this chapter will help to tie all the chapters together and will set the stage for using Six Sigma tools to develop JHAs. In safety, you have to use your imagination.

### 13.1.1 The Beginning

Six Sigma was invented by Motorola in 1986 and was popularized by General Electric in the 1990s. Six Sigma started as a defect reduction effort in manufacturing and was then applied to other business processes for the same purpose. Now, Six Sigma is applied to almost any type of business process. Organizations such as Honeywell, Citigroup, Motorola, DuPont, Dow Chemical, Kodak, Sony, IBM, and Ford have implemented Six Sigma programs across diverse business operations ranging from highly industrial or high-tech manufacturing facilities to service and financial operations [2].

The Six Sigma methodology's main objective is to help improve a process so that a problem does not recur. Human behavior tends to drive individuals to find short-term solutions by putting a bandage on problems or taking "short cuts" that may lead to an injury or increased risk. We all know what happens when a bandage is pulled off—it hurts, and then the wound has to heal again! Therefore, Six Sigma is a system that allows a team (employee participation) to get to the root cause of a problem, instead of just finding short-term solutions. Only when the cause(s) of "variation" has been identified can the process be improved so that variation is reduced or does not recur in the future [2, 3].

The approach and tools described in this chapter will follow a Basic Improvement Model that is closely aligned with the Six Sigma methodology. This Six Sigma model differs in many respects from the Continuous Process Improvement (CPI) model as developed by some of the best quality masterminds (such as Deming, Juran, and Crosley). We briefly discussed the Plan-Do-Study-Act (PDSA) model in Chapter 12. This chapter goes beyond this concept and discusses how PDSA has changed and evolved into Six Sigma [1].

## 13.1.2 What Does Process Improvement Mean?

Everything that we do in life is a *process*. A process is basically the steps and decisions that everyone makes each day, either at home or at work, to accomplish tasks; it is simply how we do things in life. Further, a process involves constantly changing occurrences, which means we must learn how to adapt to our surroundings. Everything we do in our lives involves a process. It is all about changing behaviors to adapt to changes in the organization. The following are some examples to think about:

- Getting out of bed in the morning. (Sometimes it is tough.)
- Going to work on Monday morning.
- Using your cell phone while driving a car. (This process involves a series of at-risk events.)
- Driving your car too fast. (Another series of at-risk events that many of us engage in.)
- Reading a standard operating procedure (SOP). (To get the most benefit you will read it in a logical format.)
- Taking a certification test.
- Preparing an email message.
- Developing a job hazard analysis [3].

Think about your own processes. The list above can be expanded into a larger list. Some of these processes involve at-risk events, as described earlier in this book. A challenge is to count the number of times that you put yourself at risk. How many at-risk events are you exposed to each day?

Think about it this way: in a traditional role, management views safety and quality as a program, and some managers think that you can run a safety system from a set of standards published in a manual or SOP. This is okay when there are specific compliance issues. Most of our mentors did not understand how to move safety and quality efforts from a traditional (reactive) state (a program) to a proactive state (a process). Some managers are still in the mode of reacting (fighting fires) in every situation.

As an analogy, picture yourself on a journey, driving your car from the East coast to the West coast. On this journey you need to stop for rest breaks, food, gas, etc. In the process of the journey, you stop to eat and then get back on the road in the same direction. As applied to safety or quality, if an injury occurs (or a product exhibits poor quality), what are you going to do? Naturally you will stop and try to fix the issue. Under the traditional method, we try to fix the issue, but seem to always stay in the reactive mode and never come back to the original process. Under the new approach (proactive), you would react to the issue, fix the issue, and then move back to the process. In many cases, you may find that you have to modify your process [3].

Instilling a process improvement mentality requires a different way of thinking about an organization. In other words, organizations must change their behavior in order to keep the process consistent. In order to think about the system as a process and not just a program, the focus must be on improving the process long term (something that will stand the test of time) and not just updating procedures to solve short-term issues which are typically administrative controls. To get started, management must have an open mind and consider the following options:

- What is the best way to learn about the process so that the appropriate process can be identified for improvement?
- What resources are required to complete any improvement efforts?
- Are the right individuals available and trained that can help in any improvement efforts?
- What type of team must be assembled to accomplish improvements? [1]
- How can the improved process be institutionalized in the organization?

Real improvement will take all considerations into account and will require everyone to become a "fire preventer" and not a "fire fighter." While we have learned through the years to become better time managers, sometimes we all get caught up in the old routine and have to react. Many current mandates take time and resources away from process improvement. It is the nature of the culture in which we live. For example, with OSHA's recordkeeping requirements, you are always challenged to determine if a case is recordable based on the rules and letters to interpretations. Using downstream data that comes from many low quality sources, OSHA then measures success by using Total Case Incident Rate (TCIR). It is our position to initially forget about TCIR measurements and implement a management system such as VPP discussed in Chapter 12. The TCIR is a post-loss indicator and should be only one of many pieces of data collected. The same issue occurs in the quality environment, where we tend to measure customer complaints and react to each situation, not applying techniques for total improvement [2].

One quick story. I (James Roughton) worked for General Electric (GE) in the late 1970s as a production line manager and used the statement "I do not have the time" constantly. It became a way of life for me. One day GE hired a consultant who came up with a time management scheme. As you can imagine, at first all of us supervisors reacted with, "Here is yet another program," and "We do not like it." However, we all started the program and were given $3 \times 5$ cards and asked to stop every 15 minutes to write down what we had accomplished in that period. My response was, "I have a lot of things that I can write down, so give me a lot of cards." Unexpectedly, during the first several days, I would have to stop and think about what I had done that was actually productive. Each time I would stop, it would remind me of the non-productive things that I was doing. Unfortunately, everyone had to document something to turn in to the manager. As time went by, some of us began to realize that we were doing many things that did not add value to our work day. From a personal standpoint, I stopped doing those things and found that I had a lot of time to spare. Consequently, I stopped fighting as many fires. So, today "I do not have the time" is not in my vocabulary. This may not work for everyone, but it worked for me. One thing that I learned is that you as an individual have to make what you do your own and it must come from the heart.

Depending on the situation, each process can differ in importance. A process can either be simple or complicated. For example, repairing a piece of equipment may be a simple task involving only a few people, if the instructions are clearly written in a concise manner. As we have discussed in this book, without identifying structure and methods, hazard identification is a very complicated process because you must know the work elements of the job steps and associated tasks to be able to identify the consequences of physical exposure, at-risk events of the employees, and all of the elements that affect the behavior of the employees.

## 13.1.3 What Does Process Improvement Look Like?

"Process improvement" simply means a way of making things work better. It is a means of attempting to change behaviors in the organization away from always placing the blame on employees for failures in the system. It is a way of looking at how a process can be changed to work more efficiently.

If a logical problem-solving approach is not in place, the root causes of problems cannot be defined. If the root cause is not identified, then the problem will re-occur. With time constraints, we try to find a quick fix ("just do it") for what is broken and/or what went wrong, and put a bandage on the problem. Without evidence or supporting facts, the situation could get worse when we make changes without fully understanding how the process components interrelate.

When you engage in a true Six Sigma approach, you will learn what causes things to happen and how to use your new knowledge to reduce variation in the process (remove activities or methods that contribute no value to the system). To begin, a team (not a single individual), as discussed in Chapter 5, develops a list of factors, called "critical X's." In the scheme of JHAs, these critical X's could be a list of jobs, the steps, and the tasks. Once these critical X's are identified, the gaps in the process can be determined. The team can then examine all of the factors affecting the specific process, the materials used in that process, and the methods and/or equipment used to transform the materials into a more effective and efficient process. The changes can then be taught to employees who perform the task.

## 13.1.4 Benefits of Improving a Process

The major benefit of improving a process is that there is a standardized approach used to evaluate how well an organization performs work activities. The organization will in turn learn to perform the task in a more efficient manner.

A team approach is intrinsic to any process. Using Six Sigma tools and methods of initiating change reinforces teams (safety committees) to work together, thereby creating awareness. Using team members' collective (tribal) knowledge, experiences, and data collected is a powerful approach to improving any process.

## 13.1.5 Improving the Process Using the Six Sigma Methodology

The most important and the most essential element of any process is to ensure that senior management and employees buy into what is to be accomplished. Management commitment and leadership must foster an organizational environment where an improvement mentality can thrive and employees are using the Six Sigma tools and techniques on a regular basis.

For an organization to reach this level, the management team must ensure that everyone receives the necessary education and training that will help all employees to carry out process improvement efforts effectively [2].

## 13.2 A BASIC SIX SIGMA PROCESS IMPROVEMENT MODEL

At the heart of Six Sigma is DMAIC. DMAIC is an acronym that describes the five phases of the model: Define, Measure, Analyze, Improve, and Control. Refer to Figure 13-1 for a graphical overview of Six Sigma. DMAIC is a problem-solving method that uses specific tools to turn a persistent problem into a statistical-based problem, generate a statistical solution, and then convert the statistical solution back into a practical solution.

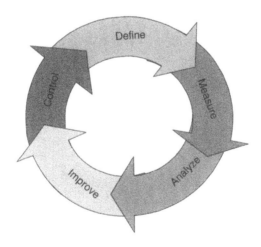

**Figure 13 1** Graphical Overview of Six Sigma Process Elements

Let's begin the use of Six Sigma by describing what is in each step of the model and how it works. The model outlines in detail the steps that must be followed to complete a Six Sigma project. As we walk through the Six Sigma concepts, we will highlight the tools used and how they are tied to the JHA process.

Once the process is understood in detail, it can provide a resource for the average user and a guide for selecting the appropriate tools that can be used in the JHA development. The concepts can then be applied to other safety-related efforts.

## 13.2.1 DMAIC Methodology

We have broken the five-step process—Define, Measure, Analyze, Improve, and Control (DMAIC)—into 15 steps to put a difference highlight on a typical safety process, in particular the JHA development. Refer to Figure 13-2 for a basic Six Sigma Process Model.

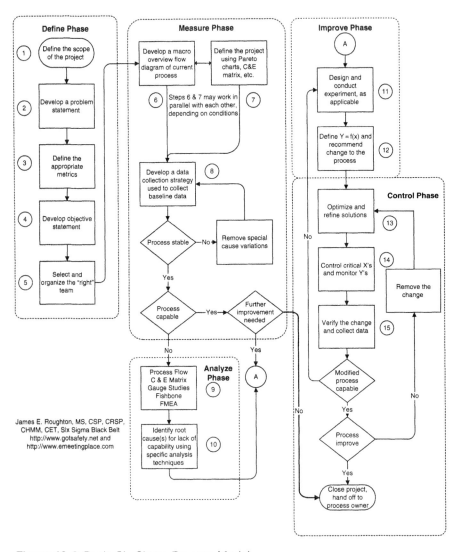

**Figure 13-2** Basic Six Sigma Process Model

Adapted from Breakthrough Management Group, Inc., "Black Belt Training Manual, Define Phase, Black Belt DAMIC Methodology," 1999–2004, pp 17–21, 31. Adapted from "Handbook for Basic Process Improvement," U.S. Navy, http://www.balancedscorecard.org/files/, May 1996, pp. 5, Public Domain

Using the 15 steps of the basic Six Sigma Model will increase your knowledge of the process through an easy-to-use format. The following sections will provide a summary of the 15 steps in conducting a Six Sigma Project [1].

### 13.2.1.1 Defining the Project

Step 1: Define the scope of the project.
Step 2: Develop a problem statement. What is to be measured and in what format?
Step 3: Define the appropriate metric.
Step 4: Develop objective statement.
Step 5: Select and organize the "right" team.

### 13.2.1.2 Measuring the Project

Step 6: Develop a macro overview flow diagram of current process.
Step 7: Define the project with Pareto charts, XY matrix, etc.
Step 8: Develop a data collection strategy used to collect specific baseline data.

### 13.2.1.3 Analyzing the Project

Steps 9: Further define the process using process flow, XY matrix, gauge studies, Fishbone, FMEA, etc.
Step 10: Identify root cause(s) for lack of capability using specific analysis techniques.

### 13.2.1.4 Improving the Project

Step 11: Design and conduct an "experiment," as applicable.
Step 12: Define the $Y = f(x)$ and recommend changes to the process. ($Y$ is a function and outcome of $X$.)

### 13.2.1.5 Controlling the Project

Step 13: Optimize and redefine solutions to be implemented.
Step 14: Control critical X's (Inputs) and monitor Y's (Outputs for effectively solving a problem).
Step 15: Verify the change and collect ongoing data on solution effectiveness.

## 13.2.2 Define Phase

The purpose of the Define phase is to develop a problem statement containing the requirements and the objectives of a project. This phase should focus on critical issues, with the problem statement establishing a defined project strategy and a list of customer requirements. The Define phase includes obtaining senior management support, buy-in, and approval to define and continue with the project.

### 13.2.2.1 Step 1: Define the Scope of the Project

When an organization undertakes Six Sigma improvement projects, the areas designated for improvement must be clearly identified and decisions made on which issues are most critical to the operation.

Step 1 begins with selecting the elements that must be improved (injury reductions, product improvement, inventory management, etc.) To accomplish this, a well-defined scope must be established that will validate the business issues at hand. In many cases the scope and objectives can be established by a designated team, the process owner, and/or top management. We have used the risk and hazard control improvement process as the target for improvement using the JHA.

**Selecting the Process**

The important considerations that must be taken into account when selecting a process include:

- Process improvement is predicated on understanding what is important to the customer. Every work environment, large or small, has two types of customers, internal customers (employees) and external customers (product, services, buyers, contractors, vendors, etc.). The starting point in selecting an improvement process is to obtain information from the customer about their likes or dislikes concerning the current process, in our case a safe operation, by asking the question: What do you want to see as an end result? [1]
- Each project is clearly defined. As the project moves forward, other issues that become visible can be identified and put on a "parking lot" list and then addressed at a later time. The key is to ensure that the project is not too broad. This is one of the most common mistakes made by individuals or managers, where they try to solve the world's problems all at one time. It is critical to ensure that the steps involved in meeting the objective are

located inside the boundaries of the established requirements. If you have ever watched the reality TV show, Donald Trump's "The Apprentice," you will see a common trend: individuals are divided into teams and these teams try to do too many things at one time and fail because they cannot control everything. The smart thing to do is define the appropriate boundaries, which must include a start and end point [1].

Once a project is identified, there must be a thorough investigation into the current scope of the issues to determine the feasibility of the project and the needs of the organization. Knowing the importance of the problem will aid in establishing a well-defined process improvement objective. The definition of the objective should answer this question: "What improvement do we want to accomplish?"

The objective is frequently determined by listening (most importantly) to customers during the discovery phase. The team can design a questionnaire and/or use interviews to identify primary metrics (what you want to measure as well as what is being measured) to use to set goals and benchmarks for improving the defined issues [1].

A Pareto diagram or a series of Pareto diagrams are used to allow the team to identify one or more factors which may occur frequently and need to be investigated further. This analysis would be based on preliminary data collected by the team. This data could be injury data from various plants or areas within the plant, and the types of injuries that have occurred. The point to all of this is to make sure that you have enough data to justify what area(s) need to be improved.

> A Pareto chart is used to graphically summarize and display the relative importance of the differences between groups of data. (Kerri Simon, http://www.isixsigma.com/library/content/c010527a.asp.) The Pareto chart ranks issues from the largest to the smallest using a bar chart that shows largest on the left in descending order from left to right.

Trend Chart: The trend chart helps to visualize the trend of defects over a specific period of time. It uses a line graph that shows time period selected on the x-axis and items measured on the y-axis (number of injuries, rates, etc.)

> "A trend chart allows a company to engage in visual management. They typically display the value of a quantifier through time, together with a goal line." http://www.isixsigma.com/dictionary/Trend_Charts-420.htm

Process Flow Chart: A process flow chart is designed to help define how the current process functions and show elements in sequence. It shows how the flow of steps in the current process combines to affect the outcome [1]. We have used flow charts throughout this book.

A Process Flow Chart is a graphical representation of a process, depicting inputs, outputs, and units of activity. It represents the entire process at a high or detailed level (depending on the use) or level of observation, allowing analysis for optimization of the workflow.

In addition, it can represent an entire process from start to finish, showing inputs, pathways, action or decision points, and data collection points. It can serve as an instruction manual or a tool for facilitating detailed analysis and optimization of workflow project. http://www.isixsigma.com/dictionary/Flowchart-431.htm

### 13.2.2.2 Step 2: Developing a Problem Statement

The problem statement consists of the following: what, when, where, how much and how do I know [1]. This is the same type of root cause investigations that you conduct each time an injury occurs. The problem statement is not usually set in stone in the initial stage of the project, as it can be redefined as new project specification limits are identified. Once the baseline data is collected, the problem statement can be further defined. At this point the problem statement will shift from "I do not know what I want to do" to "I know exactly what I want to do." [3] At this stage you start to have a data-driven system that will help you to understand how the selected process works.

Recently while listening to talk radio I heard the comment, "You can collect the dots but not connect the dots." There is a tendency to select the root cause of a problem without a full understanding of how all of the elements fit together. This results in difficulty in developing the correct action plan to resolve the right issue.

For any improvement effort to be successful, the project team must start with a clear definition of "What is the problem?" and "What can be expected from the existing process?" Let's take a look at an example:

- Repairing a seal on a high-pressure piece of equipment currently takes six hours. Management would like to reduce the time required, but is concerned that the quality of the product will suffer if the process is changed. The team believes that the repair time can be reduced to four hours by improving the process of how the seal is replaced. The objective can be stated as: "High-pressure air compressor fourth stage seals are repaired in four hours or less, with no increase in the mean time between failures for the repaired parts." [2]

A well-defined problem statement guides the team to proceed in a logical manner. The key is to write a description of the process; for example, "The process by which we change the high-pressure air compressor seal can be modified to change the seal in 4 hours with no consequential performance to the equipment." Specify the end result objective of the process improvement effort.

### 13.2.2.3 Step 3: Define the Appropriate Metric

Defining the appropriate measurement (metrics) to track the progress of the project is as important as the process, as it helps the team to understand how the process is improving. In this step you must determine if the metrics are business-related (costs, defects, etc.) or another type of metric, for example, hazard related injury data. In addition, the team needs to understand the primary and secondary ways of monitoring improvements using the data. A decision must be made to determine if the project needs other measurable objectives such as consequential metrics (a business or process measurement that allows tracking of any negative impacts on the implementation of solutions on a given project). These consequential metrics are used to measure unintended consequences of potential changes [1]. The key is that you do not want to try to improve a process and in the end implement a change that will increase risk or spend more time or money, or require more resources, than the initial problem [2, 3].

### 13.2.2.4 Step 4: Develop Objective Statement

Several rules on developing the objective statement should be followed. One such format is from the Breakthrough Management Group (BMG): Improve {process metric(s)} from {baseline} to {target} by {date}. [2]

It is important to understand the end result. What are you trying to achieve? In the case of safety, it is a specific type of injury reduction. The end result or goal will keep the team focused on what you need to accomplish by gathering the customers' needs in a quantifiable way, associated with a clearly defined problem statement and understanding what potential solution(s) will help improve the issue at hand [3].

"Begin with the end in mind."

—Steven Covey

## 13.2.2.5 Step 5: Select and Organize the "Right" Team

At this point in the project it is important that you identify the "right" people to serve on the team. This is the time to ensure that you are willing to commit the resources for the improvement effort, including time, money, and materials; set reporting requirements; and determine the team's level of authority. These elements should be formalized in a written charter [2]. Refer to Chapter 3, Team Selection.

The applicable tools that can be used in the safety arena during the Define phase include the following:

- Project and team charter development: Creating a team and developing a team charter and identifying an employee who will champion (mentor) the project. The project and team charters are a very important element of any project to ensure that the problem statement is clearly defined, defect definitions are developed (in our case, specific injuries or high risk/hazard), and that the team has a written charter and deliverables for the project [1].
- Team charters include such deliverables as a business case analysis: such as injury reduction, problem and goal statements, scope, milestones, and roles and responsibilities.
- One key element that must be added in the team charter is a plan for communicating information to management that is related to the project. (Patrick Waddick, http://www.isixsigma.com/library/content/c010304b.asp)

As discussed, one method that has been used to collect baseline data in safety is a well-defined questionnaire. This questionnaire can be distributed to the employees with description of an injury type that has actually happened. The employee can think about all of the issues that related to the injury and then provide feedback on what almost happened (near misses or injuries).

Questionnaire (incident recall technique) example: "To assist the organization with identifying at-risk events, please describe at least five near misses you have had and/or scenarios in which you could see someone getting hurt while working [1, 2]." Refer to Table 13-1 for several case examples.

Identifying problems associated with the process helps define the objectives. The employees working in the identified process can help to identify activities that take too long to accomplish, involve too many work-hours, involve redundant or unnecessary steps, at-risk events, and/or are subject to frequent breakdowns or other delays. Why have employees involved? The simple answer is that they should know the process better than anyone else. But this is not just a problem-finding exercise; the problem-generating conditions must be

**Table 13-1**
**Example Questionnaire**

Case 1: When working with a paper cutter, I was making a detailed cut. While trying to stop the paper from moving, I was holding my hand too close to the path of the blade and nicked my thumbnail with the blade. Note: The guard was missing.

Case 2: While reading a sign on the outside of the boiler room in the warehouse I was standing too close to the door. Someone leaving the room did not see me, pushed the door open quickly, and almost hit me with the door.

corrected and improved. Problems (injuries, damage, defects) are symptoms of process failure, and it is the deficiencies in the process that must be identified for improvement. Refer to Chapter 5 on employee participation.

## 13.2.3 Measure Phase

The purpose of the Measure phase is to better understand the current conditions of the issue at hand by identifying numerical data that provides the best measurement of the current performance. This data is used to establish baseline measurement for tracking and measuring the project progress. The measurements developed must be useful and relevant to identify and measure the source of "variation." The Measure phase includes the following elements:

- Identifying the specific performance requirements.
- Mapping the process to define Inputs (X's) and Outputs (Y's), usually documented as $Y = f(x)$. This mapping is conducted at each process step, the relevant outputs and all the potential inputs, critical X's, that might impact each output (Y's) and how they are connected to each other. Refer to Chapter 12, Figures 12-4 and 12-5 for an example of this process.
- Generating a list of existing and potential measurements needed (gathering the right data).
- Validating that a problem exists based on the measurements, for example, at-risk events. Analyzing the measurement system, the capability of the system, and establishing a process capability baseline is necessary for any process improvement. Variation is a measurement of the fluctuation over time of the data. It is essential in determining process improvement.
- Identifying where errors in the measurement system can occur. For example, the job task may have been built on assumptions about what is being done.
- Measuring the inputs, processes and outputs, and collecting the data to identify the process parameters.

- Refining the problem statement or objective (that was initially developed in the Define phase). Sometimes the problem statement and/or objective will need to be modified and/or updated as the project unfolds. What we thought to be the problem may now be part of greater issues or problems. [1]
- Applicable tools that can be used during the Define phase include the following:
- Fishbone diagram: The fishbone is a powerful tool and is used to demonstrate the relationships between inputs and outputs [1]. Refer to Chapter 10 for the example that we used on changing a tire. It can be used to quickly pull together an initial picture (graphics) on all elements involved [2].

> The fishbone is a tool that can be used to solve specific problems by brainstorming causes and effects and logically organizing them by branches. A fishbone is also called the Cause and Effect diagram and Ishikawa diagram, and has been introduced and used earlier. http://www.isixsigma.com/dictionary/Fishbone-64.htm.

- Process Mapping: A process map is used to allow everyone to understand the current process. This will enable the team to pictorially define the hidden causes of variation [1]. The flow diagrams presented in this book are a form of Process Maps. Refer to Appendix Q for a Process Map of changing a tire.

> The process map is also a hierarchical method for displaying a visual illustration of how a transaction is processed. Process Mapping comprises a stream of activities that transforms a defined input or set of inputs into a pre-defined set of outputs. http://www.isixsigma.com/dictionary/Process_Map-101.htm

- XY Matrix: The XY Matrix, not to be confused with the fishbone, is used to quantify how significantly each input may affect the output [1]. One of the keys to a successful risk management system is the ability to establish priorities of related hazards so that everyone has the same opportunities to identify and assess risk. Refer to Chapter 6, Figure 6-4 for an overview of establishing task priorities. To help in categorizing and comparing the probability and the severity of the risk, we have developed an overview of the XY matrix to help provide an overview of the process. We will discuss severity and probability of identifying hazards in a different format than we did in Chapter 6, where we provided several simple methods to determine risk of a task [1].

For a successful JHA process, you must determine where to start. Both risk and loss history must be used. The first objective is to list all of the jobs in the operation and conduct a risk assessment on each job, using the appropriate risk assessment concepts listed in this section. This risk assessment by job is used to also target jobs that have high severity potential, yet may not have created a loss. Once the assessment has been conducted and the risk data has been documented, you will have a better chance of developing priorities for jobs in a logical way using validated data and not just an opinion. The next part is to select a job and break down the job into individual components of steps and then related tasks. Using the fishbone and job process mapping, refer to Appendix Q for a detailed discussion of how to apply the XY matrix [1].

- Gauge Repeatability and Reproducibility (R&R): Gauge Repeatability and Reproducibility is another powerful tool that can be used to analyze the variation of the process and the measurement systems so as to minimize any unreliability in the measurement systems [1]. This can be used to calibrate individuals to ensure that everyone sees the same thing consistently.

> Characterizing measurement error is one of the most important but overlooked and misunderstood concepts of the Six Sigma process. Gauge R&R is used in many forms to assess the measurement precision (spread). Kim Niles, http://healthcare. isixsigma.com/library/content/c020527a.asp

When developing a project, you may need to use steps 6 and 7 interchangeably; i.e., they may work in parallel as they may interreact with each other, depending on conditions.

### 13.2.3.1 Step 6: Develop a Macro Map of Current Process

In step 6, we discussed developing a macro map of the job process (a very high level map) to help everyone understand the actual and element relationships. This is where the team begins to see a picture of the process and is able to ask questions about how the system actually works. Once the team understands the overall process, then they can simplify the flow diagram by removing redundant or unnecessary steps or tasks that create hazards, as appropriate. This can be a real eye-opener for management and the team, as they prepare to take the first steps in improving the process.

As discussed, the flow diagram provides a picture of how the process really works and can help the team spot risk and hazard problems in the process flow. The team can now ask the following questions:

- What steps and tasks or decision points are redundant? The team may find that these steps and tasks contain unnecessary inspections, procedures that were implemented in the past in an attempt to fool-proof the process after a failure in the management system. All of these can eat up vast amounts of resources.
- Where are resources not utilized efficiently? The team may find a weak link in the process that they can bolster by adding or removing one or more steps or tasks. Where hazards are present, are the controls, PPE, equipment, etc. right for the job?

A word of caution is that, before making changes to the process based on this preliminary review of the "as-is" flowchart, the team must answer the following questions for each step of the process. Before improving a process, determine if the entire process is valid:

- Does any event work in parallel rather than in sequence?
- Do these events have to be completed before another event can be started, or can two or more events be performed at the same time?
- What would happen if a series of events are eliminated? Would the output of the process remain the same? What would be the difference? Would the revised output be acceptable or would it be unacceptable due to being incomplete or does the process have too many defects?
- Would eliminating an event achieve the process risk improvement objective?
- Are the events being performed by the appropriate individual? Who else is involved?
- Is the event a work-around (quick fix, "just do it's") because of poor training efforts or administrative programs inserted into the process to prevent recurrence of a failure or other related issues?
- Is the event a single repeated action, or is it a part of the whole system which can be eliminated?
- Does the event add value to the process?

If the answers to these questions indicate opportunities for improvement through elimination, the team should consider suggesting doing away with the event. If an event or decision block can be removed without degrading the process or increasing risk, the team could be recovering resources which can be used elsewhere in the organization.

Eliminating redundant or unnecessary events provides an added benefit: i.e., a decrease in at-risk events (such as sticking your hand into a piece of equipment to unjam a piece of material). We want to remove the at-risk event which caused the injury, not just find a "safe way" to stick the hand into the equipment.

After making preliminary changes to the process, the team should create a flow diagram of the new simplified process and compare it to the current process. A sanity check:

- Is the simplified process acceptable to the business operation? Is it an improvement while removing risk?
- Does it continue to be in compliance with applicable existing company procedures, policies, regulatory requirements, etc.?

If the answer to these questions is "yes," and the customer agrees with the change, and if the team has the authority to make changes, then the simplified change should be implemented as soon as possible. A new flow diagram must be developed that will show the process as the new standard.

Before the proposed change is put in place, everyone (employees, management, engineering, etc.) working in the process must be trained on how to use the new process. It is vital to ensure that everyone understands and adheres to the new way. Otherwise, the process will at some point revert to the old way. In other words, the habitual behavior of the organization must be changed to ensure that the new process does not revert back to the traditional way of doing business. Change is hard and a drift back will occur if all elements identified in the XY matrix and the fishbone are not taken into account.

Steps 1 through 7 have taken the team through a process simplification phase of process improvement. In these steps, decisions are based on experience, qualitative knowledge of the process, and perceptions of the best way to operate.

### 13.2.3.2 Step 7: Define the Project with Pareto Charts, XY Matrix, etc.

In this step a number of basic tools are used to help determine if the project needs to be changed since the project was defined. The project can be redefined using tools such as a flow diagram to generate a step-by-step map of the activities which occur as part of performing the process [2]. These new details will provide more insight into the inputs and outputs detailing the entire process.

These tools will guide team members and others involved in a project to further define and understand how the process is currently working. Using these tools, the team starts to compare proposed changes against older methods. As the project moves forward, the team develops new flow diagrams to document how the team understands how the system actually functions. Flow diagrams have been used throughout this book to graphically show various ideas and concepts.

To develop an accurate flow diagram, the team must assign individuals to the defined work areas and observe the correct work flow detailing each step and task as it is actually performed. When employees involved in a project are new to the area that they are going to evaluate, they will need to observe the flow of activities through the process several times before they fully understand the process. An essential action is to ensure that employees outside of the team are involved in the reviews of the process and that they provide realistic feedback. At that point a flow diagram of what actually occurs can be developed. When the team reviews the detailed flow diagram, a decision needs to be made as to what type of observations have to be performed to ensure that the proper data is collected.

The goal of this step is for the team to fully understand the process before they make any attempt to make any changes. The key is not to change any task(s) before it is fully understood what impact a change could have on the system. At this point a detailed fishbone, process map, trend line, XY matrix, or risk assessments are used by the team to further define the current situation by answering the following questions:

- Does the flowchart show exactly how things are currently being done?
- If not, what needs to be added or modified to make it as close as possible to an "as-is" picture of the process?
- Have all employees in the process been involved and provided their input concerning the process steps and the sequence of events?
- Are other individuals (nondepartmental) involved in the process or a similar process? What did they have to say about how the process really works?
- After gathering this information, once again it may be necessary to rewrite the project objective as discussed in Step 1.

### 13.2.3.3 Step 8: Develop a Data Collection Strategy

Developing a strategy for collecting baseline data is essential to understanding how the process is really working, as the baseline measurement is used for comparison later in the project as changes to the process are made. This step begins the evaluation of the process against the objectives of the project established in Steps 1 and 2. The tools used, detailed in Step 6, guide the team to determine who should collect data and where, as well as how the process data should be collected.

In one author's particular case, the target was to use Six Sigma in a safety-related project. A lot of research was conducted reviewing past history data to determine which type of injuries had occurred. It was determined that 80% of the major injuries were hand related.

The objective was to look at how many times employees were using their hand in a hazardous manner, termed at-risk hand events. We wanted to see if there was a method to reduce the number of at-risk events. A process flow diagram was used to identify individual steps and tasks where measurements should be taken.

The use of behavioral observation data found that the data was not always consistent because of variations between observers. Observers had to be calibrated consistently and this was a long-term process.

The key to the success of the project was to use process knowledge and employee experience to determine where to take measurements.

Once the team determined what type of data had to be collected (why, how, where, and when to collect the data), a data collection plan was developed. To implement the data collection plan, the team developed a data collection sheet. This data collection sheet included explicit directions on when and how to use it. The team tried to make the data collection as user-friendly as possible.

It was decided that the team could only collect baseline data when, and only when, the data collection plan was in place, the data collection sheet was developed, and the data collectors were trained in the procedures. Once all of this was completed, a pilot run was conducted to test the elements of the data worksheet. The data collection worksheet was then modified based on the pilot run and implemented.

## 13.2.4 Analyze Phase

In the Analyze phase, the data is collected so that hypotheses about the root causes of variations in the measurements can be generated and subsequently validated. It is at this stage that issues are analyzed as statistical problems. This analysis can include the following events:

- Generating an hypothesis about possible root causes of variation and potential critical inputs (Critical X's)

- Identifying the vital few root causes and critical inputs that have the most significant impact
- Validating specific hypotheses by performing the appropriate analysis.

In the Analyze phase, specific statistical methods and tools are used to isolate key factors that are critical in understanding the causes of defects (injuries). The most applicable tools at this phase include the following:

- Five Why's: Asking five "Why's" is a powerful tool that can be used to help understand the root causes of defects in a process, and to highlight incorrect assumptions about causes.

> The Five Why's typically refers to the practice of asking "Why" five times—that is, why the failure occurred—in order to get to the root cause/causes of the failure. There can be more than one cause to a problem as well. (Marc R., http://www.isixsigma.com/dictionary/ 5_Whys-377.htm)

- Tests for normality: Tests for normality are statistical methods (descriptive statistics, histograms) used to determine if the data collected is normal or not, so it can be properly analyzed by other tools.
- Correlation/regression analysis: These tools help to identify the relationship between inputs and outputs or the correlation between two different sets of variables.
- Analysis of Variances (ANOVA): ANOVA is an inferential statistical technique designed to test for significance of the differences among two or more sample means.
- Failure Mode and Effect Analysis (FMEA): An FMEA allows the user to identify improvement actions to prevent defects from occurring.

> The FMEA is used to rank and prioritize the possible causes of failures as well as develop and implement preventative actions, with responsible persons assigned to carry out these actions. http://www.isixsigma.com/dictionary/Failure_Modes_and_Effects_ Analysis_(FMEA)-86.htm

- Hypothesis testing: This refers to a series of tests to help identify sources of variability using historical or current data and to provide objective solutions to questions which are traditionally answered subjectively. The bottom line is that we need to understand if our assumptions are proven or disproven, based on the data and proper testing for validity.

From a JHA perspective, the only analysis that is needed is the use of the Five Why's.

"When two planes nearly collide, they call it a 'near miss.' It's actually a NEAR HIT. A collision is a 'near miss.' BOOM! 'Look, they nearly missed!'"
—George Carlin, *The Absurd Way We Use Language*

### 13.2.4.1 Step 9: Tool Use (Process Flow, XY Matrix, Gauge Studies, Fishbone, FMEA)

When special cause variation reduces the effectiveness and efficiency of the process, you should investigate the situation to better understand the root causes and take action to remove it.

If it is determined that the special cause was clearly temporary and one time in nature, no action may be required beyond understanding the reason for this temporary event. In one of our examples, the early phone call when trying to leave for work caused a variation in the data which was easily explained and required no further action.

Special cause variation can signal an improvement in the process, bringing it closer to the process improvement objective. When that happens, the team may want to incorporate the change permanently.

If the team fails to investigate potential special cause variation and continues on with improvement activities, the process may be neither stable nor predictable in the future. This lack of stability and predictability may cause additional problems to occur, preventing the team from achieving the process improvement objective.

Once the process has been stabilized, the data collected in Step 8 is used again. This time the team can plot individual data points to produce a bar graph called a histogram. The collected data, the original baseline, and the specification limits (if applicable) are plotted, and the team determines if the process is capable of being improved. The following questions are used to guide the team in improving the process:

- Are unusual patterns noted in the plotted data?
- Does the bar graph show multiple tall peaks and steep valleys? This may be an indication that other processes or sources may be influencing the studied process.

- Do all of the data points fall inside the upper and lower specification limits (if applicable)? If not, the process may not be capable.
- If all of the data points fall within the specification limits, is the data grouped closely enough to the target value? This is a judgment call by the team. While the process is capable, the team may not be satisfied with the results it produces.
- If there are no specification limits for the process, does the shape of the histogram, a form of yield, conform to a bell curve (normal curve)? After examining the shape, the team has to decide if the shape is satisfactory and if the data points are close enough to the target value.

The team uses the Basic Six Sigma Process Model to establish a scientific methodology for conducting process improvements. They will plan the change, conduct a test run and collect data, evaluate the test results to find out if the process improved, and decide whether to standardize or continue to improve the process. The process model has no limitations on how many times the team can attempt to improve the process incrementally.

### 13.2.4.2 Step 10: Identify Root Cause(s) for Lack of Capability Using Specific Analysis Techniques

Steps 1 through 7 of the model were concerned with gaining an understanding of the process and documenting the results. After the initial investigation and process review has been conducted, it is important to start identifying the root causes of potential gaps in the system which could prevent the process from meeting the objective. The team begins the analysis cycle at this point, using the cause-and-effect diagram, XY matrix, brainstorming, FMEA, Gauge studies, and other tools to generate possible reasons why the process fails to meet the desired objective.

Once the team identifies possible root causes, it collects additional data to determine how much these causes actually affect the results. People are often surprised to find that the data does not substantiate their predictions, or their gut feelings, as to root causes. The key is to follow the data and then make a decision based on the data as to what direction you need to go. Using this data the team can use a Pareto chart to show the relative importance of the causes they have identified.

At various times, do a reality check and review the data. Does the data make sense? Has the data been collected and organized properly? Have the right analysis tools been selected and used? Can you simply make a change based on the issues at hand?

## 13.2.5 Improve Phase

The Improve phase focuses on developing ideas on how to remove sources of variation in the process. This phase deals with testing and standardizing potential solutions. The idea at this point is to understand what is really occurring in the process and not what is perceived to be the root cause(s) of any variation. Once you have identified specific inputs that affect the outputs, you can start to develop a strategy on how to control the process. This phase involves the following events:

- Identifying ways to remove causes (common and special) of variation
- Verifying critical inputs
- Establishing operating parameters (upper and lower specification limits)
- Optimizing critical inputs and/or reconfiguring the relevant process to help in reducing defects.

The most applicable tools to be used at this phase include:

- Process Mapping: Process mapping helps to detail the new process identified for improvements. http://www.isixsigma.com/dictionary/Process_Map-101.htm
- Process Capability Analysis: Process capability analysis helps the user to understand if the process is capable of maintaining the control(s) that have been identified. Cpk is a statistic used to test the capability of a process after improvement actions have been implemented to ensure that a real improvement has been obtained in preventing defects. Capability analysis is a graphical or statistical tool that visually or mathematically compares actual process performance to the performance standards established by the customer.

> "Process Capability index ('equivalent') taking account of off-centeredness: effectively the Cp for a centered process producing a similar level of defects – the ratio between permissible deviation, measured from the mean value to the nearest specific limit of acceptability, and the actual one-sided $3 \times$ sigma spread of the process. As a formula, Cpk = either (USL-Mean)/($3 \times$ sigma) or (Mean-LSL)/($3 \times$ sigma) whichever is the smaller (i.e. depending on whether the shift is up or down). Note this ignores the vanishingly small probability of defects at the opposite end of the tolerance range. Cpk of at least 1.33 is desired."
> http://www.isixsigma.com/dictionary/Cpk-68.htm

### 13.2.5.1 Step 11: Design and Conduct Experiment, as applicable

Design of Experiment (DOE) is a structured, organized method for determining the relationship between factors (Xs) affecting a process and the output of that process (Y). It is a planned set of tests used to define the optimum settings to obtain the desired output and validate improvements.

> DOE refers to experimental methods used to quantify indeterminate measurements of factors and interactions between factors statistically through observance of forced changes made methodically as directed by mathematically systematic tables. http://www.isixsigma.com/dictionary/Design_of_Experiments_-_DOE-41.htm

In this step, you design an experiment to check for understanding of the capability of the project to assess which change will be most effective. Once you have conducted a DOE, the team develops a plan for implementing any changes based on the possible reasons for the process's inability to meet the objective set for it. The improvement plan may be as simple as revising the steps in the simplified flowchart created after changes were completed.

### 13.2.5.2 Step 12: Defining the $Y = f(x)$ of the Process

Processes require inputs (X) and produce outputs (Y). If you control the inputs, you will control the outputs, generally expressed as $Y = f(x)$. In this step, you develop a plan for implementing a change based on the possible reasons for the process's inability to meet the objective set for it.

## 13.2.6 Control Phase

The Control phase is the last phase of the Six Sigma methodology and is used to establish a standard measuring system and control plan for you to ensure the process is capable of maintaining performance and that you can correct issues as needed. The Control phase includes the following elements:

- Validating the measurement systems
- Verifying the process's long-term capability
- Developing and implementing a control plan to ensure that the same issues do not reoccur.

The following are the most applicable tools for the Control phase:

- Control Plans: A control plan is a single document or set of documents that detail the actions identified during the project. This could include schedules and responsibilities needed to control the key process input variables at the optimal settings.
- Process Flow Chart(s) with documented Control Points: This is a single chart or series of charts that visually display the new operating processes and detail the appropriate control points.
- Statistical Process Control (SPC) charts: SPC charts help to track a process by plotting data over time taking into account the lower and upper specification limits of the capability of the process.
- Check Sheets: Check sheets enable systematic recording and compilation of data from historical sources or observations as they happen, so that patterns and trends can be clearly detected and shown [1].

### 13.2.6.1 Step 13: Optimizing and Redefining Solutions

In this step you attempt to optimize and refine solutions that have been collected [1]. The data collection as developed in Step 8 may have to be modified to make sure that the correct data is being collected. The data collection plan was originally developed in Step 5. Since the process is going to change when the planned improvement is instituted, the team must review the original plan to ensure that it is still capable of providing the data the team needs to assess process performance. If the determination is made that the data collection plan should be modified, the team considers the same things and applies the same methodologies as in Step 5.

### 13.2.6.2 Step 14: Control Critical X's and Monitor Y's

Test the changed process and collect data to verify if the X's (inputs) identified are controlled.

### 13.2.6.3 Step 15: Verify the Change and Collect Data

Assess whether the process change is stable. As in Step 7, the team uses a series of analytical methods to determine process stability. If the process is stable, the team moves on to establishing a control plan; if not, the team must return the process to its former state and plan another change.

In some situations, a small-scale test may not be feasible. If that is the case, the team will have to inform everyone involved about the nature and expected effects of the change and conduct training adequate to support a full-scale test.

The change should be implemented on a limited basis before it is applied to the entire organization. If the organization is working on a shift basis, the process change could be tried on one shift while the other shifts continue as before. Whatever method the team uses, the goal is to prove the effectiveness of the change, avoid failure, and maintain the organizational support.

The information developed in Step 9 provides the outline for the test plan. During the test, it is important to collect appropriate data so that the results of the change can be evaluated. The team will have to take the following actions in conducting the test to determine whether the change actually resulted in process improvement:

- Finalize the test plan.
- Prepare the data collection sheets.
- Train everyone involved in the test.
- Distribute the data collection sheets.
- Change the process to test the improvement.
- Collect and collate the data.

If the data collected in Step 11 show that process performance is worse, the team must return to Step 8 and try to improve the process again. The process must be stable before the team goes on to the next step.

## 13.3 KEY AREAS OF SIX SIGMA

Now that we have discussed an overview of the Six Sigma process, let's turn our attention to key areas in the Six Sigma process. The Six Sigma process:

- Provides continuous focus on specific requirements
- Uses real-time measurements, data collection, and statistics to help identify and measure types of variation in a process
- Helps to identify the root causes of issues in a clearly defined and orderly manner
- Places emphasis on process improvement to help remove variation from the process, thereby helping to reduce defects
- Helps management understand how to use proactive tools and techniques to focus on problem solving, continuous improvement, and constant striving for perfection

- Provides cross-functional collaboration within an organization
- Sets high expectations.

"Sigma" means standard deviation, or the spread or variability of data within a sample. Six sigma means six standard deviations of variability. Statistically, this translates into having no more than 3.4 Defects per Million Opportunities (DPMO).

## 13.4 SIX SIGMA LEVELS

As discussed, the objective of Six Sigma is to achieve a level of no more than 3.4 Defects per Million Opportunities (DPMO). The level of 3.4 DPMO translates into 99.99966% defect free, which is considered to be perfection. To understand the effects of other sigma levels, refer to Table 13-2, which lists the various sigma levels so that you can understand the consequences of exposure of a process.

An important clarification is needed at this time. Six Sigma measures defects per million opportunities (i.e., the number of times that something can go wrong in one million opportunities). The more complex the risk, the more "defect" opportunities for exposure in the system [1]; i.e., there is a higher potential for injuries in a larger facility than in a smaller facility.

After a collective 60 plus years in the safety field and working in various industries and in various capacities, we have come to the conclusion that there

### Table 13-2
### Sigma Levels

| Sigma Level | Defects per Million Opportunities |
|---|---|
| Six Sigma | 3.4 |
| Five Sigma | 230 |
| Four Sigma | 6,210 |
| Three Sigma | 66,800 |
| Two Sigma | 308,000 |
| One Sigma | 690,000 |

Mekong Capital Ltd, Section 1.3, Six Sigma Levels, www.mekongcapital.com and Sigma Performance Levels – One To Six Sigma, http://www.isixsigma.com/library/content/c020813a.asp

is no such thing as "luck" in safety. It is not like playing the lottery. You either have a good management system in place or you have not been able to develop a good management system. The message for safety managers: "Hope is not a strategy, Rick Page." A structured, methodical, data-based process is essential to increase your opportunities for success.

"Responsibility, Not Luck"

—Roughton

## 13.5 INVESTING IN PREVENTION PAYS OFF!!

According to the OSHA, "establishing a Safety and Health program to prevent occupational injuries and illnesses is not only the right thing to do; it's the profitable thing to do. Studies have shown a $4 to $6 return for every dollar invested in safety and health." [2, 3]

In addition to preventing pain and suffering, OSHA estimates that each fatality avoided saves approximately $910,000. Each injury prevented that would have involved recovery time away from work saves $28,000. For each other serious injury and illness avoided, $7,000 is saved [3].

### Table 13-3
### Examples of Losses

| |
|---|
| If you have 1500 employees in your organization, you would expect injury at a certain rate based on your industry risk and hazard and the quality of your work environment. Based on average industry profits, for each $500 in direct loss: |
| A soft drink bottler would have to bottle and sell over 61,000 cans of soda |
| A food packer would have to can and sell over 235,000 cans of corn or a bakery would have to bake and sell over 235,000 donuts |
| A contractor would have to pour and finish 3,000 square feet of concrete |
| A ready-mix company would have to deliver 20 truckloads of concrete |
| A paving contractor must lay 900 feet of two-lane asphalt road |
| http://www.osha.gov/SLTC/safetyhealth_ecat/images/safpay1.gif [4], public domain |

## Table 13-4
## Examples of Zero Defects

Is it truly necessary to go for zero defects? Why isn't 99.9% defect-free good enough? Here are examples of what life would be like if 99.9% were "good enough:"

- 1 Hour Of Unsafe Drinking Water Every Month
- 2 Long or Short Landings At Every American Airport Each Day
- 400 Letters Per Hour Which Never Arrive At Their Destination
- 500 Incorrect Surgical Operations Each Week
- 3,000 Newborns Accidentally Falling From The Hands Of Nurses Or Doctors Each Year
- 4,000 Incorrect Drug Prescriptions Per Year
- 22,000 Checks Deducted From The Wrong Bank Account Each Hour
- 32,000 Missed Heartbeats Per Person Per Year

Examples Of What Life Would Be Like At Six Sigma

- 13 Wrong Drug Prescriptions Per Year
- 10 Newborns Accidentally Falling From The Hands Of Nurses Or Doctors Each Year
- 1 Lost Article Of Mail Per Hour
- The Quest For **Six** Sigma becomes essential!

http://www.sixsigmaspc.com/six-sigma/sixsigma.html

Costs of injuries include not only direct medical expenses and workers' compensation cost, but also the indirect costs (commonly known as hidden cost), which can increase total costs by as much as four or more times. Indirect costs include but are not limited to: training and paying replacement workers, investigating the incident, and interrupted production, schedule delays, managing the claim, legal fees, and increased insurance costs [3].

The issues faced by safety professionals parallel those of the quality control process. Injuries and damages when viewed as defects created by the process can be addressed using Six Sigma.

When considering direct and indirect costs, the following illustrations show how much employers must increase profit and sales just to recover from a direct loss of $500 [2, 3].

Refer to Table 13-3 for some examples of losses. Refer to Table 13-4 for examples of zero defects.

## 13.6 POSITIVE CHANGES TO CORPORATE CULTURE

Last but not least, whether using a Six Sigma or other types of methodologies, working closely with employees will allow a better safety culture to emerge. Both the organizational and individual behaviors can start to change. Six Sigma is all about full participation and communication. Employees facing difficult work problems can find solutions more effectively when they are provided the necessary training techniques, and shown how to ask the right questions, measure the right things, and correlate an issue with a control plan.

With a Six Sigma focus, an organization's culture can start to shift to one with a systematic approach to problem-solving and a proactive attitude among all employees. Successful Six Sigma implementation success can contribute to the overall sense of pride of the employees.

Six Sigma transforms the way an organization thinks and works on major issues. This is accomplished through the following areas:

- Designing processes to have the best and most consistent outcomes from the beginning.
- Investigating and conducting studies to identify the cause of variation and how the variations can interact with each other.
- Using "just-the-facts" with data (analysis and reasoning) to find and support the root causes of variations, instead of educated guesses or intuition. Experience is a great source of information and is very useful in many respects. However, when this knowledge is repeatedly used to fix problems that continue, it represents a "bandage approach." The problem is briefly fixed but no permanent controls are put in place to prevent recurrence.
- Focusing on process improvement as key to excellence in safety, quality, customer satisfaction, and/or services.
- Encouraging employees to be proactive about preventing potential problems instead of waiting for problems to occur and then reacting.
- Providing a broad participation in problem solving by getting more employees involved in finding causes and solutions for problems.
- Learning and sharing new knowledge in terms of best practices.
- Making decisions based on data collected and analysis, not "gut" reaction. This does not mean it will negatively impact a company's ability to make quick decisions. In contrast, by applying the DMAIC principles, the decision makers are more likely to have the data they need to make a well-informed decision.
- "Just-do-it's" are identified during the Six Sigma process and implemented. These are issues that do not require a full analysis, just implementation. For example, a new tool can be substituted and the problem

will be corrected. Instead of going through the entire process, the issue is resolved quickly and you move on to other important issues. This could also apply to ergonomics issues.


"Just the facts, ma'am."
—*Dragnet* TV episode, Detective Sergeant Joe Friday


## 13.7 SUMMARY


In Six Sigma, a process is represented in terms of $Y = f(x)$, in which the outputs $(Y)$ are determined by input variables $(X)$, known as critical X's. If we suspect that there is a relationship between outcome and input—an example could be a hand injury (Y) with the potential cause being sticking your hand into a running machine (X)—we must collect and analyze data by using Six Sigma tools and techniques to prove the hypothesis

If we want to change the outcome, we need to focus on identifying and controlling the causes of the at-risk events (hand into machine), rather than checking the outcomes (hand injury). When we have collected enough data and have a good understanding of the critical X's (where, where, who, what, how) then we can more accurately predict Y. Otherwise, we continue to focus our efforts on activities like inspection, auditing, testing, and reworking that may or may not reveal the underling problem of why hands are being stuck into equipment.

Many organizations have recurring problems associated with injuries, shipping products to customers which do not meet specifications, etc., causing customers to be unhappy. By reducing the various defect rates, an organization will be able to consistently reduce employee injuries and damage, ship products to customers within specifications and increase customer satisfaction.

By lowering defect rates or injuries, an organization can reduce medical costs as well as the inefficient use of employee knowledge that could assist in defect reduction. This savings reduces the cost of goods manufactured and/or sold for each unit of output, injuries (lost workdays), and adds significantly to bottom-line.

Implementation of Six Sigma represents a long-term commitment and culture shift. The success of Six Sigma projects depends on the level of commitment by the senior management. General Electric's success with Six Sigma is due in large part to the role that Jack Welch (former CEO) played in advocating and integrating it into the core of the company's strategy. Six Sigma requires a serious commitment by top management as well as other key employees.

This chapter provided a basic overview of Six Sigma for the interested safety professional. Coupled with the "toolkit" discussions in this chapter, the safety professional is provided with a practical step-by-step procedure to initiate and carry out improvement activities in an organized and more effective manner.

## CHAPTER REVIEW QUESTIONS

1. Why was Six Sigma started?
2. What is Six Sigma's main objective?
3. What is the major benefit of improving a process?
4. What does DMAIC stand for?
5. What is the purpose of the Define phase?
6. What is the purpose of the Measure phase?
7. What is the purpose of the Analysis phase?
8. Where is the focus of the Improve phase?
9. Why is the Control phase important?
10. How is the XY matrix used in developing the risk assessment?

## REFERENCES

1. *Handbook for Basic Process Improvement*, U.S. Navy, *Handbook for Basic Process Improvement*, http://www.au.af.mil/au/awc/awcgate/navy/bpi_manual/handbook.htm, May 1996, public domain
2. Breakthrough Management Group, Inc., *Black Belt Training Manual*, Define Phase, Black Belt DMAIC Methodology, 1999–2004
3. Mekong Capital Ltd, www.mekongcapital.com
4. OSHA Website, http://www.osha.gov/dts/osta/oshasoft/safetwb.html, public domain

# Appendix Q

## Q.1 XY MATRIX

The XY matrix correlates a process of Y's or critical output factors associated with X's, called the critical input factors. Through a series of rankings, you can generate a score for each X which can be sorted to establish a priority. This matrix can be used in many situations, especially where there is limited data from the process. As with any type of prioritization, developing a matrix is subjective. Using a well-defined XY matrix allows a team to intelligently prioritize hazards in a logical and more succinct manner. The XY matrix can be used to establish a metric to measure a safety process. The XY matrix can be a critical forward-thinking process to help quantify hazards in a more defined manner.

## Q.2 DEVELOPING THE MATRIX

Developing the XY matrix is as simple as having the JHA team (Chapter 5) develop a detailed process map (flowchart) for the job steps and tasks being reviewed. This process map will provide a logical flow as the team starts to quantify the risk of hazards found in the process. Once this is done, the team must agree on operational definitions as to the different way ranking will be recorded. Refer to Chapter 6, Table 6-7 and Table 6-8, for several examples. By using a set of definitions developed by the team, you will have a consistent method that can be used as a guide.

## Q.2.1 Brainstorming

Once the map of the process has been developed, the team can start a brainstorming session to identify the critical X's and Y's of the hazard identified through observation, data collections, etc. Refer to Figure Q.1 for a flow diagram on changing a tire.

As a starting point for brainstorming, the 5M Model introduced in Chapter 6 can be followed. Refer to Chapter 6, Figure 6-2. The model is divided into five areas that can be used to categorize information to help with the brainstorming of the X's. You can mix and match these areas or change them to fit your particular situation.

In our example, we will use the job steps and tasks as shown in the tire-changing example. The brainstorming is used to ensure that all of the tasks have been captured and that risk levels have been assigned. Try to be as specific as possible when listing the tasks (critical X's). In other words, do not just list "tire," but instead use a specific simple definition such as "change a flat tire." This will be consistent with the JHA. The team can use a flipchart for the process mapping and brainstorming.

To complete the assessment, an electronic spreadsheet can be used to calculate the scores. The following is the sequence of events in developing a successful XY matrix:

- Determining the outputs or "Y's" created by the process is the first step. This could be based on past experience, probability, severity, frequency, etc. List the Y's in no particular order in the cells of the spreadsheet along the top. Include all outputs that are necessary to stop an injury. In our example of changing a flat tire, our output is no injuries. Refer to Figure Q.2.
- Next use a scale of 1–10 with 10 being the most critical to the process and 1 being the least critical to the process. This is the list of parameters that closely fit your specific conditions. These rankings are independent of each other and they are not required to be in a sequential order. In other words, it is not necessary to use a ranking such as: 10, 9, 8, 7, etc. You can have Y's with the same ranking. It is up to the team and how you have defined the process. Ensure that whatever criteria is used is consistently applied throughout the project. In our example, we will use High (10), Medium (7), Low (5), and Minor (1). These rankings are used as a weighting factor only for when we later calculate the scores for X. Refer to Figure Q.3.

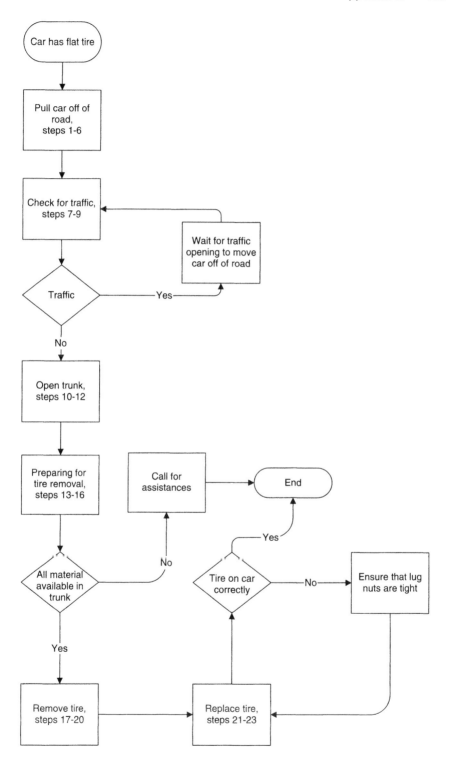

**Figure Q.1** Process Map of Changing a Tire based on XY Matrix (Simplified)

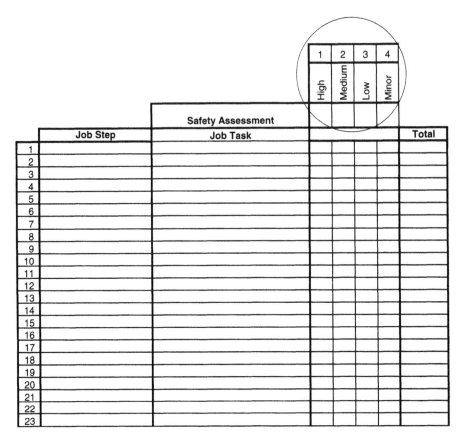

**Figure Q.2** List the Inputs (Y's) to the Process XY Matrix for Assessing Job Steps and Task

- Once you have determined the outputs (Y's,) the next step is to determine a list of inputs, critical X's. List the X's in no particular order down the left column of the spreadsheet.
- Once all of the X's are listed, the team is ready to start the brainstorming. It is important that the team adhere to brainstorming techniques. "All

| | | Safety Assessment | 1 High | 2 Medium | 3 Low | 4 Minor | |
|---|---|---|---|---|---|---|---|
| | Job Step | Job Task | 10 | 7 | 5 | 1 | Total |
| 1 | | | | | | | |
| 2 | | | | | | | |
| 3 | | | | | | | |
| 4 | | | | | | | |
| 5 | | | | | | | |
| 6 | | | | | | | |
| 7 | | | | | | | |
| 8 | | | | | | | |
| 9 | | | | | | | |
| 10 | | | | | | | |
| 11 | | | | | | | |
| 12 | | | | | | | |
| 13 | | | | | | | |
| 14 | | | | | | | |
| 15 | | | | | | | |
| 16 | | | | | | | |
| 17 | | | | | | | |
| 18 | | | | | | | |
| 19 | | | | | | | |
| 20 | | | | | | | |
| 21 | | | | | | | |
| 22 | | | | | | | |
| 23 | | | | | | | |

**Figure Q.3** List the Inputs (Y's) to the Process XY Matrix and Assigning Ranking

ideas are considered," "No debating of ideas," etc. If a team member has no opinion, ideas, or lacks experience, they should pass, since giving an opinion may create a bias to others. Just as we have established the Y's prioritization, the X's will also be ranked in the matrix. In our case, again, we will use the job steps and job tasks which are consistent with the Tire Changing JHA developed in Chapter 10. Refer to Figure Q.4.

| | | | Safety Assessment | 1 | 2 | 3 | 4 | |
|---|---|---|---|---|---|---|---|---|
| | | | | High | Medium | Low | Minor | |
| | | | | 10 | 7 | 5 | 1 | |
| | Job Step | | Job Task | | | | | Total |
| 1 | Pull car off the road | | Park clear of traffic | 9 | 3 | 2 | 2 | 123 |
| 2 | | | Park on level ground | 2 | 3 | 4 | 6 | 67 |
| 3 | | | Set parking brake | 1 | 0 | 0 | 0 | 10 |
| 4 | | | Turn off ignition | 0 | 0 | 0 | 1 | 1 |
| 5 | | | Remove the key | 0 | 0 | 0 | 0 | 0 |
| 6 | | | Turn on the hazard warning lights | 5 | 3 | 3 | 1 | 87 |
| 7 | Check for Traffic | | Exit the car | 8 | 3 | 2 | 5 | 116 |
| 8 | | | Block the wheel | 7 | 5 | 4 | 3 | 128 |
| 9 | | | Set flares/reflectors on the road | 2 | 3 | 4 | 4 | 65 |
| 10 | Open Trunk | | Loosen spare Tire | 0 | 0 | 1 | 1 | 6 |
| 11 | | | Remove tire | 3 | 2 | 1 | 1 | 50 |
| 12 | | | Retrieve Jack and lug wrench | 2 | 2 | 2 | 2 | 46 |
| 13 | Preparing for Tire Removal | | Remove wheel covers | 2 | 2 | 2 | 2 | 46 |
| 14 | | | Place jack under car | 3 | 2 | 4 | 5 | 69 |
| 15 | | | Jack car up | 5 | 6 | 7 | 1 | 128 |
| 16 | | | Take weight off tire | 2 | 2 | 2 | 2 | 46 |
| 17 | Removing Tire | | Use lug wrench | 2 | 2 | 2 | 2 | 46 |
| 18 | | | Loosen lug nuts slightly | 3 | 3 | 3 | 3 | 69 |
| 19 | | | Jack up the car | 7 | 7 | 8 | 9 | 168 |
| 20 | | | Remove lug nuts | 2 | 2 | 2 | 2 | 46 |
| 21 | Replacing Tire | | Replace flat tire | 4 | 3 | 2 | 2 | 73 |
| 22 | | | Replace lug nuts | 2 | 2 | 2 | 2 | 46 |
| 23 | | | Lower the car | 2 | 2 | 2 | 2 | 46 |

**Figure Q.4** List the process Job steps and inputs (Job Task) XY Matrix for Assessing Job Steps and Task

- Rank the X's based on the Y's. The spreadsheet ranking will be calculated in the following manner. Look at the row "Park clear of the traffic." Refer to Figure Q.5. In the example, we used the scoring as High (9), Medium (3), Low (2), and Minor (2). When the assessment has been completed the scores will be calculated as follows: $10*9 + 7*3 + 5*2 + 1*2$ which sums to 123. This is the unbiased weighted score that we have developed as a team.

| | Job Step | Job Task | 1 High 10 | 2 Medium 7 | 3 Low 5 | 4 Minor 1 | Total |
|---|---|---|---|---|---|---|---|
| | | Safety Assessment | | | | | |
| 1 | Pull car off the road | Park clear of traffic | 9 | 3 | 2 | 2 | 123 |
| 2 | | Park on level ground | 2 | 3 | 4 | 6 | 67 |
| 3 | | Set parking brake | 1 | 0 | 0 | 0 | 10 |
| 4 | | Turn off ignition | 0 | 0 | 0 | 1 | 1 |
| 5 | | Remove the key | 0 | 0 | 0 | 0 | 0 |
| 6 | | Turn on the hazard warning lights | 5 | 3 | 3 | 1 | 87 |
| 7 | Check for Traffic | Exit the car | 8 | 3 | 2 | 5 | 116 |
| 8 | | Block the wheel | 7 | 5 | 4 | 3 | 128 |
| 9 | | Set flares/reflectors on the road | 2 | 3 | 4 | 4 | 65 |
| 10 | Open Trunk | Loosen spare Tire | 0 | 0 | 1 | 1 | 6 |
| 11 | | Remove tire | 3 | 2 | 1 | 1 | 50 |
| 12 | | Retrieve Jack and lug wrench | 2 | 2 | 2 | 2 | 46 |
| 13 | Preparing for Tire Removal | Remove wheel covers | 2 | 2 | 2 | 2 | 46 |
| 14 | | Place jack under car | 3 | 2 | 4 | 5 | 69 |
| 15 | | Jack car up | 5 | 6 | 7 | 1 | 128 |
| 16 | | Take weight off tire | 2 | 2 | 2 | 2 | 46 |
| 17 | Removing Tire | Use lug wrench | 2 | 2 | 2 | 2 | 46 |
| 18 | | Loosen lug nuts slightly | 3 | 3 | 3 | 3 | 69 |
| 19 | | Jack up the car | 7 | 7 | 8 | 9 | 168 |
| 20 | | Remove lug nuts | 2 | 2 | 2 | 2 | 46 |
| 21 | Replacing Tire | Replace flat tire | 4 | 3 | 2 | 2 | 73 |
| 22 | | Replace lug nuts | 2 | 2 | 2 | 2 | 46 |
| 23 | | Lower the car | 2 | 2 | 2 | 2 | 46 |

**Figure Q.5** Rank the inputs according to their effect on each output XY Matrix

- Now that you have completed the assessment, you can sort the list in descending order so that all of the X's with the highest scores are on the top of the spreadsheet. Refer to Figure Q.6.

This provides a weighted risk assessment of the tasks. You can quickly look at the list and know which tasks have the highest hazards.

| | | Safety Assessment | 1 | 2 | 3 | 4 | |
|---|---|---|---|---|---|---|---|
| | | | High | Medium | Low | Minor | |
| | | | 10 | 7 | 5 | 1 | |
| | Job Step | Job Task | | | | | Total |
| 1 | Pull car off the road | Park clear of traffic | 9 | 3 | 2 | 2 | 123 |
| 2 | | Park on level ground | 2 | 3 | 4 | 6 | 67 |
| 3 | | Set parking brake | 1 | 0 | 0 | 0 | 10 |
| 4 | | Turn off ignition | 0 | 0 | 0 | 1 | 1 |
| 5 | | Remove the key | 0 | 0 | 0 | 0 | 0 |
| 6 | | Turn on the hazard warning lights | 5 | 3 | 3 | 1 | 87 |
| 7 | Check for Traffic | Exit the car | 8 | 3 | 2 | 5 | 116 |
| 8 | | Block the wheel | 7 | 5 | 4 | 3 | 128 |
| 9 | | Set flares/reflectors on the road | 2 | 3 | 4 | 4 | 65 |
| 10 | Open Trunk | Loosen spare Tire | 0 | 0 | 1 | 1 | 6 |
| 11 | | Remove tire | 3 | 2 | 1 | 1 | 50 |
| 12 | | Retrieve Jack and lug wrench | 2 | 2 | 2 | 2 | 46 |
| 13 | Preparing for Tire Removal | Remove wheel covers | 2 | 2 | 2 | 2 | 46 |
| 14 | | Place jack under car | 3 | 2 | 4 | 5 | 69 |
| 15 | | Jack car up | 5 | 6 | 7 | 1 | 128 |
| 16 | | Take weight off tire | 2 | 2 | 2 | 2 | 46 |
| 17 | Removing Tire | Use lug wrench | 2 | 2 | 2 | 2 | 46 |
| 18 | | Loosen lug nuts slightly | 3 | 3 | 3 | 3 | 69 |
| 19 | | Jack up the car | 7 | 7 | 8 | 9 | 168 |
| 20 | | Remove lug nuts | 2 | 2 | 2 | 2 | 46 |
| 21 | Replacing Tire | Replace flat tire | 4 | 3 | 2 | 2 | 73 |
| 22 | | Replace lug nuts | 2 | 2 | 2 | 2 | 46 |
| 23 | | Lower the car | 2 | 2 | 2 | 2 | 46 |

**Figure Q.6** XY Matrix for Assessing Job Steps and Task, Identify the critical inputs from the totals column

Summary of Creating an XY Matrix

Step 1: Review the process map. If one has not been developed, the team should develop the map.

Step 2: List the Y's (outputs) across the top of the form.

Step 3: Rank each output numerically using the established scale. Rank the outputs according to the priorities, on a 1–10 scale, with the greater value of 10 being the higher priority. When ranking X's, use the data, and accept the findings for any ranking. "What it is, is what it is." Rankings are not to be debated no matter how close they may be. Remember that this is a subjective tool. The key is to try to provide a reasonable score based on the data and team discussions.

Never split the middle of significantly different scores. If several team members think that the ranking should be a 9 and several members think that it is a 2, it should not be stated as a 5. If there is a significant difference in scores, the difference should be reviewed in depth. Team members who gave the very high score should state their case based on their knowledge, research, and experience. Those who provide a low score should do the same. This discussion should not be arbitrary but based on the team's further research into the possible risks and hazards.

If a team spends too much time on any particular X, you should consider skipping it and coming back to it later. The other discussions may trigger different opinions or create new ideas.

Step 4: Identify and rate the effect of potential X's that can impact the various Y's. List the steps and inputs (task) on the left side of the form.

Step 5: Use the ranking to analyze and prioritize any future activities. Identify the critical inputs from the value of the totals column. Rank the contribution of each input on each output.

Tarpley, J. Scott, Light Pharma, Inc., Adapted from "Utilising a "XY Matrix" to Prioritise PAT Deployments," Innovations in Pharmaceutical Technology, pp 123

## Q.2.2 Reality Check

Once the matrix is completed, the team should do a reality check to see if the scores make sense, based on the nature of the job, steps, and tasks.

Since this is a subjective measurement, you need to look at the scores in an objective manner. Trust your data. This can be verified once further data is collected and adjustments can then be made.

While the matrix is a simple process, it will take time to assemble the right team, collect the data (steps and task), brainstorm the data, and rank each task.

This exercise is worth the effort as you now have a JHA that incorporates a risk and hazard assessment.

# Final Words: Can You Develop a Culture That Will Sustain Itself?

"Ignorance isn't what you don't know, it's what you know wrong."

—Yogi Berra

At this point I (the author) would like to share a short story with you that I hope will tie everything together that will be presented in this chapter. All that I ask is for you to think about what has been presented and ask how you can apply it to your training.

A couple of years ago I had a real eye-opener in regards to training when working on my Six Sigma Black Belt Certification Safety Project. A team had spent a lot of time deciding on how to conduct a safety project using Six Sigma tools. There were many naysayers, to say the least; many felt that it just could not be done.

While conducting this project, the team asked the question, "Why are the majority of injuries hand-related?" The team looked for ways to reduce at-risk hand usage. After many months of defining the project and establishing the baseline, we looked at the data collected and it all started to make sense. The data told us that 80% of the employees in the study were putting themselves at-risk while using their hands. Armed with this data, we asked questions about the current training. The typical response was: "We 'train' our employees on a specific safety message all the time. We just do not understand why they still get hurt!"

The real question is: "Why is the safety message not getting through to employees?" If you were to ask any supervisor about a quality issue they have had, or how many times a specific machine has been down, they could probably give you many years of history. But if you were to ask how many injuries they had in the last six months, a typical supervisor would have a hard time discussing any facts concerning injuries. On the other hand, if you were to ask them to state their biggest safety issue, what do you think would be the response? You probably have already guessed the answer. "PPE is my biggest issue." So the point to all of this is that we have to do a better job of training employees.

There may be ways to improve the learning retention rates of employees. In the Six Sigma tool box, the "Gage R& R" (Repeatability and Reproducibility) study is used to determine if excessive variability exists in the measurement system. In our case we used it to assess the instructor-to-employee variability.

The R&R study was conducted using three supervisors and three employees who had knowledge of the process. We found a 60% gap between what the supervisors thought they had trained employees on vs. what the employees retained. The safety message was not getting through to employees. Again, the questions were asked, "Why?" and "Can we make a difference?" A Design of Experiment (DOE) was used to determine the effectiveness of four training methods. This DOE incorporated a series of pictures within four training methods. The objective was to determine the effectiveness of each specific training technique/method.

Based on the results of the DOE, a new R& R study was conducted (using the same employees and same method). The results of the second R& R concluded that 71 percent (71.4%) of the employees could identify at-risk events. This was a 35% improvement from the first gage study.

The surprise was that our current supervisor training did not address the fundamentals of one-on-one communication. Our training assumed a high-level overview was enough: "I read and you listen."

There was no real surprise as to the training method found most effective. If I talk you and show you directly what is expected, you will remember: Just through "Show and Tell."

Several things that were learned from the project included the following:

- Do not overlook the fundamental need for direct experience.
- Everyone must change their behaviors if you want a process to improve.
- Data must drive your decision-making process. To verify findings it is not just training but the method of training that is critical.

Proper training will increase the knowledge and skills and improve the retention needed to fully understand workplace hazards.

## TAKING A CLOSER LOOK AT REALITY

We want to leave you with final thoughts on hazard recognition and how individuals perceive things in life. In a group of people, do you think that everyone in that group will think alike or see things in the same way? As an example: What is the color of a yield sign? You are probably saying to yourself, this is child's play. Everyone knows the answer. I use this example in presentations and/or training sessions that I regularly conduct to illustrate a point on perception in hazard recognition. The color of a yield is yellow and black, right? Stop and think about this for a second. Is this the correct color? Yellow and black is most people's perception. However, the color is red and white. Ask any 10 people and list their response. I think that you will find in a short period of time that approximately 90 percent of the people will not know the correct answer and others will have to stop and think about it. This illustration will grab people's attention and this discussion of the yield sign can lead you into other conversations of hazard identification, such as the fact that the yield sign was changed in 1977 because people do not associate the color yellow with STOP.

In one training session the subject of the yield sign was used as an example. One individual in the session wanted the slide presentation to present at his facility. He went home and gave the same presentation to a group of employees. The same survey revealed that 90 percent of the attendees did not know the correct color. During the break, this gentleman and several individuals (who did not believe the correct color) called the police department and asked the question, "What is the color of a yield sign?" The police department did not know the correct answer. The next call was to the Department of Transportation (DOT), and the person answering the phone there did not know the answer. The last phone call was to the fire department, which was able to answer the question correctly. The perception of safety and at-risk events does not necessarily reflect reality!

Let's take a look at another example: How many times have you driven from point A to point B and missed a landmark that you use to judge your trip based on the length of time to get to your destination? Have you suddenly noticed new buildings or objects that you have never seen before? Maybe this is the case when you drive from point A to point B every day, going to work, school, or other situations. We do not all see the obvious, and we miss many hazards associated with the trip! We become accustomed to our surroundings.

Have you ever investigated an injury and when interviewing the employee you hear: "I have almost been injured on this piece of equipment before." Your question might be, "Why didn't you tell someone about the hazard?" The answer: "We did not think that it was important." Almost getting hurt is not important because they have accepted the risk, because the perception of risk is low.

Hazards exist in our environment and are all around us, every day, and most of the time we do not recognize them. Why? We have a tendency to overlook these hazards because we become accustomed to our surroundings. We get into our comfort zone. We feel comfortable, just like seeing the yield sign every day. We see it every day, and it becomes a way of life and the way we view things. We do not pay close attention to details! When we are accustomed to our surroundings, we screen out hazards that have not cause an injury. The objective is, through discussions, repetition, coaching, and ongoing analysis, to bring the intangible into view, changing the perception of risk and hazards.

Management sets performance measurements based on the level of TCIR/OIR that an organization wants to achieve. This system is sometimes tied to performance, bonus, discipline, etc. If there is no "blood on the floor," the perception becomes NO LOSS = NO RISK and we lose our objectivity and fail to focus on hazard recognition.

We believe that a shift must occur in safety management systems from a focus on *program* to a focus on *job analysis* that goes directly to the foundation of how the job is being done. It was not the scope of this book to develop a management system. However, there are many books on the market that can be used to develop management systems. One such book is *How to Develop an Effective Safety Culture: A Leadership Approach*, which will walk you through the entire process [1].

New circumstances and a recombination of existing circumstances may cause old hazards to reappear and new hazards to appear. You must be ready and able to implement hazard elimination or control measures in a timely and effective manner.

Your JHA process should be the focal point of your safety program. Through hazard identification, employee participation, understating the behavioral aspects of humans and training, core knowledge develops. Implementing the structured JHA that incorporates Cause and Effect diagrams, risk assessment, establishing priorities using the XY Matrix, and data analysis provides the fine details of how the job really gets done. Knowledge, focus, understanding, and implementation now drive your efforts.

We would hope that by this time you are not looking for a "magic bullet." We hate to admit it but as safety professionals we have found that there is no magic solution, just hard work by planting those seeds and watching them grow.

For further reference, Appendix R has been summarized outlining OSHA regional and area offices, approved Safety and Health plans, consultation projects and websites

## REFERENCE

1. Roughton, James, James Mercurio, *Developing an Effective Safety Culture: A Leadership Approach*, Butterworth-Heinemann, 2002

# Appendix R

## R.1 OSHA LISTINGS

**OSHA Regional Offices**

| Region I | Region VI |
|---|---|
| (CT*, MA, ME, NH, RI, VT*)<br>JFK Federal Building, Room E-340<br>Boston, MA 02203<br>Telephone: (617) 565-9860 | (AR, LA, NM*, OK, TX)<br>525 Griffin Street, Room 602<br>Dallas, TX 75202<br>Telephone: (214) 767-4731 or 4736<br>x224 |
| Region II | Region VII |
| (NJ*, NY*, PR*, VI*)<br>201 Varick Street, Room 670<br>New York, NY 10014<br>Telephone: (212) 337-2378 | (IA*, KS, MO, NE)<br>City Center Square<br>1100 Main Street, Suite 800<br>Kansas City, MO 64105<br>Telephone: (816) 426-5861 |
| Region III | Region VIII |
| (DC, DE, MD*, PA*, VA*, WV)<br>The Curtis Center<br>170 S. Independence Mall West<br>Suite 740 West<br>Philadelphia, PA 19106-3309<br>Telephone: (215) 861-4900 | (CO, MT, ND, SD, UT*, WY*)<br>1999 Broadway, Suite 1690<br>Denver, CO 80202-5716<br>Telephone: (303) 844-1600 |

| Region IV | Region IX |
|---|---|
| (AL, FL, GA, KY*, MS, NC*, SC*, TN*)<br>Atlanta Federal Center<br>61 Forsyth Street, SW, Room 6T50<br>Atlanta, GA 30303<br>Telephone: (404) 562-2300 | (American Samoa, AZ*, CA*, HI*, NV*, Northern Mariana Islands)<br>71 Stevenson Street, Room 420<br>San Francisco, CA 94105<br>Telephone: (415) 975-4310 |
| Region V | Region X |
| (IL, IN*, MI*, MN*, OH, WI)<br>230 South Dearborn Street, Room 3244<br>Chicago, IL 60604<br>Telephone: (312) 353-2220 | (AK*, ID, OR*, WA*)<br>1111 Third Avenue, Suite 715<br>Seattle, WA 98101-3212<br>Telephone: (206) 553-5930 |

\* These states and territories operate their own OSHA-approved job safety and health programs (Connecticut, New Jersey, and New York plans cover public employees only). States with approved programs must have a standard that is identical to, or at least as effective, as the federal standard.

## OSHA Area Offices

| | | | |
|---|---|---|---|
| Albany, NY<br>(518) 464–4338 | Allentown, PA<br>(610) 776–0592 | Anchorage, AK<br>(907) 271–5152 | Appleton, WI<br>(920) 734–4521 |
| August, ME<br>(207) 622–8417 | Austin, TX<br>(512) 916–5783/5788 | Avenel, NJ<br>(732) 750–3270 | Bangor, ME<br>(207) 941–8177 |
| Baton Rouge, LA<br>(225) 389–0474 (0431) | Bayside, NY<br>(718) 279–9060 | Bellevue, WA<br>(206) 553–7520 | Billings, MT<br>(406) 247–7494 |
| Birmingham, AL<br>(205) 731–1534 | Bismark, ND<br>(701) 250–4521 | Boise, ID<br>(208) 321–2960 | Bowmansville, NY<br>(716) 684–3891 |
| Braintree, MA<br>(617) 565–6924 | Bridgeport, CT<br>(203) 579–5581 | Calumet City, IL<br>(708) 891–3800 | Carson City, NV<br>(775) 885–6963 |

| Charleston, WV (304) 347–5937 | Cincinnati, OH (513) 841–4132 | Cleveland, OH (216) 522–3818 | Columbia, SC (803) 765–5904 |
|---|---|---|---|
| Columbus, OH (614) 469–5582 | Concord, NH (603) 225–1629 | Corpus Christi, TX (361) 888–3420 | Dallas, TX (214) 320–2400 (2558) |
| Denver, CO (303) 844–5285 | Des Moines, IA (515) 284–4794 | Des Plaines, IL (847) 803–4800 | Eau Claire, WI (715) 832–9019 |
| El Paso, TX (915) 534–6251 | Erie, PA (814) 833–5758 | Fairview Heights, II (618) 8960–7290 | Fort Lauderdale, FL (954) 424–0242 |
| Fort Worth, TX (817) 428–2470 (485–7647) | Frankfort, KY (502) 227–7024 | Greenwood Village, CO (303) 843–4500 | Guaynabo, PR (787) 277–1560 |
| Harrisburg, PA (717) 782–3902 | Hartford, CT (860) 240–3152 | Hasbrouck Heights, NJ (201) 288–1700 | Houston, TX (281) 286–0583/0584 (5922) |
| Houston, TX (281) 591–2438 (2787) | Indianapolis, IN (317) 226–7290 | Jackson, MS (601) 965–4606 | Jacksonville, FL (904) 232–2895 |
| Kansas City, MO (816) 483–9531 | Lansing, MI (517) 327–0904 | Linthicum, MD (410) 865–2055/2056 | Little Rock, AR (501) 324–6291(5818) |
| Lubbock, TX (806) 472–7681 (7685) | Madison, WI (608) 264–5388 | Marlton, NJ (856) 757–5181 | Methuen, MA (617) 565–8110 |
| Milwaukee, WI (414) 297–3315 | Minneapolis, MN (612) 664–5460 | Mobile, AL (251) 441–6131 | Nashville, TN (615) 781–5423 |
| New York, NY (212) 337–2636 | Norfolk, VA (757) 441–3820 | North Aurora, IL (630) 896–8700 | North Syracuse, NY (315) 451–0808 |

| Oklahoma City, OK (405) 278–9560 | Omaha, NE (402) 221–3182 | Parsippany, NJ (973) 263–1003 | Peoria, IL (309) 671–7033 |
|---|---|---|---|
| Philadelphia, PA (215) 597–4955 | Phoenix, AZ (602) 640–2348 | Pittsburgh, PA (412) 395–4903 | Portland, ME (207) 780–3178 |
| Portland, OR (503) 326–2251 | Providence, RI (401) 528–4669 | Raleigh, NC (919) 856–4770 | Sacramento, CA (916) 566–7471 |
| Salt Lake City, UT (801) 530–6901 | San Diego, CA (619) 557–5909 | Savannah, GA (912) 652–4393 | Smyrna, GA (770) 984–8700 |
| Springfield, MA (413) 785–0123 | St. Louis, MO (314) 425–4249 | Tampa, FL (813) 626–1177 | Tarrytown, NY (914) 524–7510 |
| Toledo, OH (419) 259–7542 | Tucker, Ga (770) 493–6644/6742/8419 | Westbury, NY (516) 334–3344 | Wichita, KS (316) 269–6644 |
| Wilkes–Barre, PA (570) 826–6538 | Wilmington, DE (302) 573–6518 | | |

OSHA Websites, http://www.osha.gov/dcsp/smallbusiness/consult_directory.html#IDAHO

**OSHA-Approved Safety and Health Plans**

| State Plan |
|---|
| Alaska Alaska Department of Labor and Workforce Development Commissioner (907) 465–2700 FAX: (907) 465–2784 Program Director (907) 269–4904 FAX: (907) 269–4915 Website: http://labor.state.ak.us/lss/home.htm |

| State Plan |
| --- |
| Arizona<br>Industrial Commission of Arizona<br>Director, ICA<br>(602) 542–4411<br>FAX: (602) 542–1614<br>Program Director<br>(602) 542–5795<br>FAX: (602) 542–1614<br>Website: http://www.ica.state.az.us/ |
| California<br>California Department of Industrial Relations<br>Director<br>(415) 703–5050<br>FAX: (415) 703–5114<br>Chief<br>(415) 703–5100<br>FAX: (415) 703–5114<br>Manager, Cal/OSHA Program Office<br>(415) 703–5177<br>FAX: (415) 703–5114<br>Website: http://www.dir.ca.gov/occupational_safety.html |
| Connecticut<br>Connecticut Department of Labor<br>Commissioner<br>(860) 566–5123<br>FAX: (860) 566–1520<br>Conn-OSHA Director<br>(860) 566–4550<br>FAX: (860) 566–6916<br>Website: http://www.ctdol.state.ct.us/osha/osha.htm |
| Hawaii<br>Hawaii Department of Labor and Industrial Relations<br>Director<br>(808) 586–8844<br>FAX: (808) 586–9099<br>Administrator<br>(808) 586–9116<br>FAX: (808) 586–9104<br>Website: http://hawaii.gov/labor/hiosh/index.shtml |

| State Plan |
| --- |
| Indiana<br>Indiana Department of Labor<br>Commissioner<br>(317) 232–2378<br>FAX: (317) 233–3790<br>Deputy Commissioner (317) 232–3325<br>FAX: (317) 233–3790<br>Website: http://www.in.gov/labor/iosha/index.html,<br>http://www.in.gov/labor/insafe/ |
| Iowa<br>Iowa Division of Labor<br>Commissioner<br>(515) 281–6432<br>FAX: (515) 281–4698<br>Administrator<br>(515) 281–3469<br>FAX: (515) 281–7995<br>Website: http://www.iowaworkforce.org/,<br>http://www.iowaworkforce.org/labor/iosh/consultation/index.htm |
| Kentucky<br>Kentucky Labor Cabinet Secretary<br>(502) 564–3070<br>FAX: (502) 564–5387<br>Federal\State Coordinator<br>(502) 564–3070 ext.240<br>FAX: (502) 564–1682<br>Website: http://www.labor.ky.gov/osh/ |
| Maryland<br>Maryland Division of Labor and Industry<br>Commissioner<br>(410) 767–2999<br>FAX: (410) 767–2300<br>Deputy Commissioner<br>(410) 767–2992<br>FAX: (410) 767–2003<br>Assistant Commissioner, MOSH<br>(410) 767–2215<br>FAX: (410) 767–2003<br>Website: http://www.dllr.state.md.us/labor/mosh.html |

| State Plan |
|---|
| Michigan<br>Michigan Department of Consumer and Industry Services<br>Director (517) 322–1814<br>FAX: (517) 322–1775<br>Website: http://www.michigan.gov/cis/0,1607,7-154-11407-,00.html |
| Minnesota<br>Minnesota Department of Labor and Industry<br>Commissioner<br>(651) 296–2342<br>FAX: (651) 282–5405<br>Assistant Commissioner<br>(651) 296–6529<br>FAX: (651) 282–5293<br>Administrative Director,<br>OSHA Management Team<br>(651) 282–5772<br>FAX: (651) 297–2527<br>Website: http://www.doli.state.mn.us/mnosha.html |
| Nevada<br>Nevada Division of Industrial Relations<br>Administrator<br>(775) 687–3032<br>FAX: (775) 687–6305<br>Chief Administrative Officer<br>(702) 486–9044<br>FAX: (702) 990–0358<br>[Las Vegas (702) 687–5240]<br>Website: http://dirweb.state.nv.us/ |
| New Jersey<br>New Jersey Department of Labor<br>Commissioner<br>(609) 292–2975<br>FAX: (609) 633–9271<br>Assistant Commissioner<br>(609) 292–2313<br>FAX: (609) 292–1314<br>Program Director, PEOSH<br>(609) 292–3923<br>FAX: (609) 292–4409<br>Website: http://www.state.nj.us/labor/lsse/lspeosh.html |

| State Plan |
| --- |
| New Mexico<br>New Mexico Environment Department<br>Secretary<br>(505) 827–2850<br>FAX: (505) 827–2836<br>Chief<br>(505) 827–4230<br>FAX: (505) 827–4422<br>Website: http://www.nmenv.state.nm.us/OHSB_website/ohsb_home.htm |
| New York<br>New York Department of Labor<br>Acting Commissioner<br>(518) 457–2741<br>FAX: (518) 457–6908<br>Division Director<br>(518) 457–3518<br>FAX: (518) 457–6908<br>Website: http://www.labor.state.ny.us/workerprotection/safetyhealth/<br>DOSH_PESH.shtm |
| North Carolina<br>North Carolina Department of Labor<br>Commissioner<br>(919) 807–2900<br>FAX: (919) 807–2855<br>Deputy Commissioner, OSH Director (919) 807–2861<br>FAX: (919) 807–2855<br>OSH Assistant Director<br>(919) 807–2863<br>FAX: (919) 807–2856<br>Website: http://www.nclabor.com/osha/osh.htm |
| Oregon<br>Oregon Occupational Safety and Health Division Administrator<br>(503) 378–3272<br>FAX: (503) 947–7461<br>Deputy Administrator for Policy<br>(503) 378–3272<br>FAX: (503) 947–7461<br>Deputy Administrator for Operations<br>(503) 378–3272<br>FAX: (503) 947–7461<br>Website: http://www.orosha.org/ |

| State Plan |
| --- |
| Puerto Rico<br>Puerto Rico Department of Labor and Human Resources<br>Secretary<br>(787) 754–2119<br>FAX: (787) 753–9550<br>Assistant Secretary for Occupational Safety and Health<br>(787) 756–1100, 1106/754–2171<br>FAX: (787) 767–6051<br>Deputy Director for Occupational Safety and Health<br>(787) 756–1100/1106, 754–2188<br>FAX: (787) 767–6051<br>Website: http://www.dtrh.gobierno.pr/osho.asp |
| South Carolina<br>South Carolina Department of Labor, Licensing, and Regulation<br>Director<br>(803) 896–4300<br>FAX: (803) 896–4393<br>Program Director<br>(803) 734–9644<br>FAX: (803) 734–9772<br>Website: http://www.llr.state.sc.us/ |
| Tennessee<br>Tennessee Department of Labor<br>Commissioner<br>(615) 741–2582<br>FAX: (615) 741–5078<br>Acting Program Director<br>(615) 741–2793<br>FAX: (615) 741–3325<br>Website: http://state.tn.us/labor-wfd/ |

| State Plan |
| --- |
| Utah<br>Utah Labor Commission<br>Commissioner<br>(801) 530–6901<br>FAX: (801) 530–7906<br>Administrator<br>(801) 530–6898<br>FAX: (801) 530–6390<br>Website: http://www.uosh.utah.gov/ |
| Vermont<br>Vermont Department of Labor and Industry<br>Commissioner<br>(802) 828–2288<br>FAX: (802) 828–2748<br>Project Manager<br>(802) 828–2765<br>FAX: (802) 828–2195<br>Website: http://www.labor.vermont.gov/ |
| Virgin Islands<br>Virgin Islands Department of Labor<br>Acting Commissioner<br>(340) 773–1990<br>FAX: (340) 773–1858<br>Program Director<br>(340) 772–1315<br>FAX: (340) 772–4323 |
| Virginia<br>Virginia Department of Labor and Industry<br>Commissioner<br>(804) 786–2377<br>FAX: (804) 371–6524<br>Director, Office of Legal Support<br>(804) 786–9873<br>FAX: (804) 786–8418<br>Website: http://www.doli.state.va.us/index.html |

| State Plan |
|---|
| Washington<br>Washington Department of Labor and Industries<br>Director<br>(360) 902–4200<br>FAX: (360) 902–4202<br>Assistant Director<br>(360) 902–5495<br>FAX: (360) 902–5529<br>Program Manager, Federal–State Operations<br>(360) 902–5430<br>FAX: (360) 902–5529<br>Website: http://www.lni.wa.gov/Safety/default.asp |
| Wyoming<br>Wyoming Department of Employment<br>Safety Administrator<br>(307) 777–7786<br>FAX: (307) 777–3646<br>Website: http://wydoe.state.wy.us/doe.asp?ID=7 |

## OSHA Consultation Projects

| | | | |
|---|---|---|---|
| Albany, NY<br>(518) 457–2238 | Anchorage, AK<br>(907) 269–4957 | Atlanta, GA<br>(404) 894–2643 | Augusta, ME<br>(207) 624–6400 |
| Austin, TX<br>(512) 804 4640 | Baton Rouge, LA<br>(225) 342 9601 | Bismarck, ND<br>(701) 328 5188 | Boise, ID<br>(208) 426 3283 |
| Brookings, SD<br>(605) 688–4101 | Charleston, WV<br>(304) 558–7890 | Cheyenne, WY<br>(307) 777–7786 | Chicago, IL<br>(312) 814–2337 |
| Christiansted St.<br>Croix, VI<br>(809) 772–1315 | Columbia, SC<br>(803) 734–9614 | Columbus, OH<br>(614) 644–2631 | Concord, NH<br>(603) 271–2024 |
| Des Moines, IA<br>(515) 281–7629 | Fort Collins, CO<br>(970) 491–6151 | Frankfort, KY<br>(502) 564–6895 | Hato Rey, PR<br>(787) 754–2171 |
| Helena, MT<br>(406) 444–6418 | Henderson, NV<br>(702) 486–9140 | Honolulu, HI<br>(808) 586–9100 | Indiana, PA<br>(724) 357–2396 |

| Indianapolis, IN (317) 232–2688 | Jefferson City, MO (573) 751–3403 | Lansing, MI (517) 322–1809 | Laurel, MD (410) 880–4970 |
|---|---|---|---|
| Lincoln, NE (402) 471–4717 | Little Rock, AR (501) 682–4522 | Madison, WI (608) 266–9383 | Montpelier, VT (802) 828–2765 |
| Nashville, TN (615) 741–7036 | Oklahoma City, OK (405) 528–1500 | Olympia, WA (360) 902–5638 | Pearl, MS (601) 939–2047 |
| Phoenix, AZ (602) 542–1695 | Providence, RI (401) 222–2438 | Raleigh, NC (919) 807–2905 | Richmond, VA (804) 786–6359 |
| Sacramento, CA (916) 263–2856 | Saint Paul, MN (651) 284–5060 | Salem, OR (503) 378–3272 | Salt Lake City, UT (801) 530–6901 |
| Santa Fe, NM (505) 827–4230 | Tampa, FL (813) 974–9962 | Tiyam, GU 9–1–(671) 475–1101 | Topeka, KS (785) 296–2251 |
| Trenton, NJ (609) 292–3923 | Tuscaloosa, AL (205) 348–3033 | Washington, DC (202) 541–3727 | Waukesha, WI (262) 523–3044 |
| West Newton, MA (617) 727–3982 | Wethersfield, CT (860) 566–4550 | Wilmington, DE (302) 761–8219 | |

| States | Other OSHA Websites |
|---|---|
| Alabama | http://alabamasafestate.ua.edu/safe_state_osha.htm |
| Arkansas | http://www.arkansas.gov/labor/divisions/osha_p1.html |
| Colorado | http://www.bernardino.colostate.edu/public/ |
| Delaware | http://www.delawareworks.com/industrialaffairs/services/ OSHAConsultation.shtml |
| Florida | http://www.usfsafetyflorida.com/Index.aspx |
| Georgia | http://www.oshainfo.gatech.edu/ |
| Idaho | http://www2.boisestate.edu/OSHConsult/ |
| Illinois | http://www2.illinoisbiz.biz/osha/index.htm |

| States | Other OSHA Websites |
|---|---|
| Kansas | http://www.dol.ks.gov/wc/html/wcish_ALL.html |
| Maine | http://www.safetyworksmaine.com/consultations/ |
| Massachusetts | http://www.mass.gov/dos/ |
| Mississippi | http://www.msstate.edu/dept/csh/ |
| Missouri | http://www.dolir.mo.gov/ls/safetyconsultation/ |
| Montana | http://erd.dli.mt.gov/safetyhealth/sbhome.asp |
| Nebraska | http://www.dol.state.ne.us/nwd/center.cfm?PRICAT=4& SUBCAT=4F& ACTION=osha |
| New Hampshire | http://www.des.state.nh.us/ |
| North Dakota | http://www.bismarckstate.edu/ndsafety/ |
| Ohio | http://www.ohiobwc.com/employer/programs/safety/ SandHOSHAandPERRP.asp |
| Oklahoma | http://www.labor.ok.gov/ |
| Pennsylvania | http://www.hhs.iup.edu/sa/OSHA/index.htm |
| Rhode Island | http://www.health.ri.gov/environment/occupational/ consultation.php |
| Texas | http://www.tdi.state.tx.us/wc/services/oshcon.html |
| West Virginia | http://www.labor.state.wv.us/ |
| Wisconsin | http://www.commerce.state.wi.us/MT/MT-FAX-0928- Working.html |

# Glossary

| | |
|---|---|
| Acceleration | When we sometimes speed up or slow down too quickly, for example, with equipment. The changing of speed. |
| Acceptable Risk | The risk that remains after risk controls are applied. Risk can be determined "acceptable" when further efforts to reduce a hazard reach a point of diminishing returns. Extreme care must be used in defining a risk as acceptable. |
| Analysis | The breaking down of a job into its component steps/tasks and then evaluating each step/task, looking for specific hazards. Each hazard is controlled (avoid, substitution, engineering) or a method of employee protection (administrative practices, or PPE) is identified and documented as part of the standard operation. |
| Audit | An evaluation tool involving assigning a numerical rating to items that are being reviewed. While inspections involve identifying hazards, audits are more generally involved in locating ineffective or missing safety programs, procedures, protocols. |
| Behavior | The manner in which one behaves; the observable actions or reactions of persons or things in response to internal or external stimuli. |
| Behavioral Modification | The use of techniques, such as conditioning, biofeedback, reinforcement, or aversion therapy, to alter human action or response. |

| Biological | Living things, human vegetation, animals, insects, virus, bacteria, etc. |
|---|---|
| Bodily reaction | Caused solely from stress imposed by free movement of the body or assumption of a strained or unnatural body position. A leading source of injury. |
| Caught-between | A person is crushed, pinched or otherwise trapped between a moving and a stationary object, or between two moving objects. |
| Caught-in | A person or part of him/her is trapped, or otherwise caught in an opening or enclosure. |
| Caught-on | A person or part of his/her clothing or equipment is caught on an object that is either moving or stationary. This may cause the person to lose his/her balance and fall, be pulled into a machine, or suffer other harm. |
| Change analysis | Review/assessment of the impact of change. |
| Checklists | Specific guidelines that state conditions that are acceptable and records their status. |
| Chemical Reactions | Results of chemicals in combination or on exposure to incomplete material. Chemical reactions can be violent and can cause explosions, dispersion of materials and emission of heat. |
| Contact-by | Contact by a substance or material that, by its very nature, is harmful and causes injury. |
| Contact-with | A person comes in contact with a harmful substance or material. The person initiates the contact. |
| Continuous Process | Sustaining constant vigilance. Identifying and managing risks routinely through all phases of the project's life cycle. |
| Decision Matrix | A method to objectively evaluate two criteria at the same time resulting in a combined rating for that criterion. |
| Electrical Contact | Inadequate insulation, broken electrical lines or equipment, lightning strike, static discharge, etc. |

| | |
|---|---|
| Employee Participation | Provides the means by which employees can develop and express their own commitment to safety for both themselves and their co-workers. |
| Explosives | Explosions result in large amounts of gas, heat, noise, light and over-pressure. |
| Exposure | When an employee, contractor, or visitor enters a "danger zone" by virtue of their proximity to the hazard. |
| Fall-to-below | A person slips or trips and falls to a level below the one he/she was on. |
| Fall-to-surface | A person slips or trips and falls to the surface he/she is standing or walking on. |
| Flammability/Fire | Complex interaction of fuel, oxidizers, ignition source under specific conditions. |
| Forward-Looking View | Thinking toward tomorrow, identifying uncertainties, anticipating potential outcomes. Managing project resources and activities while anticipating uncertainties. |
| Frequent | Steps/tasks that occur often, i.e., 8, 10, 12 hours per day. Likely to occur many times in a short period of time. |
| Habit | A recurrent, often unconscious pattern of behavior acquired through frequent repetition; an established disposition of the mind or character. |
| Hazard | Any thing or condition that increases the probability of injury, illness, or loss. An unsafe condition or practice (personal behavior) that could cause an injury, property damage, near miss, etc. A condition, event, or circumstance that could lead to or contribute to an unplanned or undesired event. A potential energy source that could cause harm to employees. It is usually associated with a condition or activity that, if left uncontrolled, can result in an injury. |
| High Catastrophic | Death, permanent disability, more serious consequence, lost workday cases, multiple injuries, and/or major property damage. If the factors considered are severe this is the worst case in the event of an incident. |

| Identified Risk | Combination of potential or probability and severity. Risk that has been determined to exist using analytical tools such as the JHA and supporting matrix. Refer to Figure 6-2 for one example of a Risk matrix. The time and costs of analysis efforts, the quality of the risk management system, and the state of the technology involved affect the amount of risk that can be identified. |
|---|---|
| Inspections | Most frequently used to assess workplace hazards by means of an examination of every part of the workplace to identify conditions that do not comply with safety requirements. Range from general to specific. |
| Integrated Management Support | Making risk management an integral and vital part of project management. Adapting risk management methods and tools to a project's infrastructure and culture. |
| Job Hazard Analysis | The focal point of understanding the complexity and interactions of a job. A technique that is used to focus on job steps and/or tasks as a way to help identify hazards before they occur. |
| Job | A job is defined as any specific activity, mental or physical or both, that has been assigned to an employee as a responsibility and defines the activities, tools, equipment, materials, procedures, and structure for a service or product... |
| Low, Marginal | Other than critical, minor injuries, potential restricted workday cases, minor property damage. If the factors considered are not severe it would be unlikely that serious injury or damage would occur. |
| Mechanical | Pinch points, sharp points and edges, weight, rotating parts, stability, ejected parts and materials, impact. |
| Medium, Critical | Permanent partial and temporary disability cases, more serious physical harm and/or major property damage. If the factors considered are moderate in severity, injury or damage is not likely but not catastrophic or severe. |

| More Serious Consequences | Define based on specific operation, for example, potential fatality. |
|---|---|
| Negligible, Minor | No harmful injury, potential first aid and medical treatment, near miss, minor property damages. |
| Non-Routine or Occasional | Steps/Tasks occur infrequently. This task may occur each week, month, yearly. Likely to not routinely occur in the life of the task. |
| Other More Serious Situations | Define based on specific operation, for example, potential fatality. |
| Other Than Serious | Conditions that could cause injury or illness to employees but would not include serious physical harm; for example, first aid cases and/or near misses. |
| Over-exertion | A person over-extends or strains himself/herself while performing activity. |
| Over-exposure | Over a period of time, a person is exposed to harmful stress, noise, heat, cold, or toxic chemicals/atmospheres, etc. |
| Pressure | Increased pressure in hydraulic and pneumatic systems. Physical force against a fluid. |
| Priority | Precedence, established by order of importance or urgency. |
| Probability Rating Factors | Low: If the factors considered indicate it would be unlikely that an accident could occur; Medium: If the factors considered indicate it would be likely that an accident could occur; or High: If the factors considered indicate it would be very likely that an accident could occur. |
| Probability | The possible chance that a given event will occur. |
| Probable, Likely | Steps/Tasks occur often over course of the workday, i.e., less than 8, 10, 12 hours per day. Will occur often in the life of the task. |
| Radiation | Non-ionizing (laser), Ionizing (nuclear). |

| Residual Risk | The portion of total risk that remains after control efforts have been employed. Residual risk comprises acceptable risk and unidentified risk [4]. |
| --- | --- |
| Risk | An expression of the impact of an undesired event in terms of event severity and event likelihood or probability [3]. Risk is a measure of the probability and severity of adverse effects. |
| Seldom, Remote | Steps/Tasks occurs sometime, sporadically or several times. Unlikely to occur, but possible. |
| Severity | The scope and degree of injury or damage which is reasonably predictable. |
| Steps | Steps are specific ways that the job is performed. They allow a reviewer to understand each element of the job and how it relates to specific hazards. Steps define the main action categories. |
| Struck-against | A person forcefully strikes an object. The person provides the force or energy. |
| Struck-by | A person is forcefully struck by an object. The force of contact is provided by the object. |
| Tasks | Tasks define actions required to complete each step. |
| Teamwork | Working cooperatively to achieve common goal. Pooling talents, skills, and knowledge. |
| Total Risk | The sum of identified and unidentified risk. Ideally, identified risk will comprise the larger proportion of the two. |
| Tribal Knowledge | Any unwritten information that is not commonly known by other individuals. This is past experience, based on reacting to a problem. This term is used most when referencing information that may need to be known by others in order to produce a quality product or service. The information may be the key to quality performance but it may also be incorrect. Data does not support this knowledge. |
| Unacceptable Risk | That portion of identified risk that cannot be tolerated, but must be avoided–either eliminated or controlled. |

| | |
|---|---|
| Unidentified Risk | That risk that has not yet been identified. Some risk is not identifiable or measurable, but is no less important. Near miss (mishap) investigations may reveal some previously unidentified risks. |
| Unlikely, Improbable | Assume that the steps/tasks might occur but improbable, very rarely. So unlikely, that it can be assumed that it will not occur. |
| Value | A principle, standard quality considered worthwhile or desirable: permanent and based on fundamental principle for good. |
| On-the-job training | A well-trained and excellent supervisor periodically spends time with an employee to discuss safe work practices, and provide additional instruction to counteract any observed unsafe practices. One-on-one training is most effective when applied to all employees under supervision and not just those with whom there appears to be a problem. Properly used, it provides the positive feedback given for safe and good work practices. It helps workers establish new safe behavior patterns. It immediately recognizes and thereby reinforces the desired behavior. |
| | It requires knowledgeable, mutative supervisors with good communication skills. Poorly used, it can mask serious workplace deficiencies. |

# Index

Printed and bound by CPI Group (UK) Ltd, Croydon, CR0 4YY

03/10/2024

01040434-0010